FOOD ADDITIVE

Edited by **Yehia El-Samragy**

Food Additive
Edited by Yehia El-Samragy

Published by InTech

Janeza Trdine 9, 51000 Rijeka, Croatia

Publishing Process Manager Romana Vukelic

Technical Editor Teodora Smiljanic

Cover Designer InTech Design Team

Additional hard copies can be obtained from orders@intechopen.com

Food Additive, Edited by Yehia El-Samragy
p. cm.
ISBN 978-953-51-0067-6

For EU product safety concerns:
IN TECH d.o.o., Prolaz Marije Krucifikse Kozulić 3, 51000 Rijeka, Croatia,
info@intechopen.com or visit our website at intechopen.com.

Meet the editor

Dr. El-Samragy has over 35 years of a professional career bridged between academia and industry. He is a Professor of Food Science at Ain Sham University, Cairo, Egypt, and visiting professor at Cornell University, Ithaca, NY and Utah State University, Logan, UT, USA. He worked as an expert at some national and international organizations including JECFA, FAO, UNIDO and ISO, in different projects of technology transfer, food standards, food product development, waste utilization, cleaner production, implementation of integrated quality, environmental and food safety management systems. He is a Certified Lead Tutor and Lead Auditor of QMS and FSMS (IRCA Certificate # 01182132). He has extensive experience in delivering training courses on QMS, EMS, OHSAS, HACCP, FSMS and FSSC in Egypt, Sudan, Libya, Saudi Arabia, Yemen, Jordan, Dubai, Sharjah, Syria, Zambia, Tanzania, South Africa, Kazakhstan, Russia and USA.

Contents

Preface

Food additives are natural or manufactured substances, which are added to foods to restore colors lost during processing, provide sweetness, prevent deterioration during storage and guard against food poisoning (preservatives). A food additive is defined as "any substance not normally consumed as a food in itself and not normally used as a characteristic ingredient of food, whether or not it has nutritive value, the intentional addition of which to food for a technological purpose in the manufacture, processing, preparation, treatment, packaging, transport or storage of such food results, or may be reasonably expected to result, in it or its by-products becoming directly or indirectly a component of such foods." (Food Safety Authority of Ireland, 2011).

This book provides a review of traditional and non-traditional food preservation approaches, including physical methods such as non-thermal pasteurization, chemical methods using natural food preservatives and their combinations for extension of shelf life of fruit juices and beverages. Evaluation of agro-industrial wastes, considering their great potential for the production of industrially relevant food additives, has been presented. The isolation of an albumin fraction with high foaming ability and foam stabilizing ability from guar meal, and designated guar foaming albumin as food additive was studied. Detailed knowledge about the composition, structure, classification, biosynthesis, genetic determinants, mechanisms of production, secretion, enzymes for production, methods of production, purification, analysis and characterization of exopolysaccharides, as well as the potential applications for production of functional food, e.g. prebiotics and immunomodulators, have been covered. Effects of some food colorings such as tartrazine, as food additives, on hepatic function, lipid profile, and biomarkers of oxidative stress in blood and different tissues, have been evaluated. The problems and control of contamination of other enzymes during phytase preparation have been inspected for the possibility of using fungal phytase preparation as food additive. Exposure of humans to aluminum from different sources, i.e. food, water, airborne dust, antiperspirants etc., and related immunizations, allergy injections, and elevated antacids, were investigated. It was concluded that there is no good evidence of adverse health effects attributed to aluminum from food, that this issue has not been adequately assessed and that those who lack good renal function are at greatest risk of aluminum-induced toxicity, perhaps from food. The assessment of potential reproductive and developmental toxicity perspectives of some newly synthesized food additives on the market, i.e. lutein, caramel colors, erythrosine, green S, amaranth, brilliant blue FCF, curcumin, canthaxanthin, aspartame, paraben, coriander essential oil, allyl isothiocyanate, and tricalcium phosphate were overviewed in relevance to the six main categories of traditional food additives.

Furthermore, potential utilization of lactophorin and lactadherin from sweet whey, a byproduct of cheese manufacturing against human rotavirus gastroenteritis in infants and young children, was discussed.

In the attempt to identify future research areas of interest, it should be mentioned that this book pointed out the fact that more information is needed to explore the possibility of using some other materials as food additives. One of these issues is the full potential of the cell wall degrading enzymes in cereal processing such as bio-polishing of cereal grain and pre-degradation for animal feeds using enzymatic polishing technique. More and more information on Spirulina is becoming available, which is essentially an exceptional simple extract of blue-green algae used worldwide as a food product and as a dietary supplement. It also has biological, biotechnological and nutrition applications, as well as some remote applications like bio-fuel production and functioning as a life support system in space studies. This interest will continue without any doubt. Research along these lines will be rewarding both socially and professionally. Another area of microorganism use proves that Lactobacillus strains play an important role in food fermentation processes and L. rhamnosus is a potential candidate for probiotic product preparation or as food additives as proven by its ability to acid-bile tolerance, salt, antimicrobial activity against human pathogens and food spoilage organisms. The characteristics of the antimicrobial protein(s) that are produced at different growth phases of LAB, require more work in the future.

Prof. Dr. Yehia El-Samragy
Ain Shams UniversityFaculty of Agriculture
Food Science Department
Cairo,
Egypt

1

Food Additive

R. M. Pandey and S. K. Upadhyay
*Division of Genetics, Plant breeding & Agrotechnology,
National Botanical Research Institute, Lucknow,
India*

1. Introduction

Substances which are of little or no nutritive value, but are used in the processing or storage of foods or animal feed, especially in the developed countries; includes antioxidants; food preservatives; food coloring agents; flavoring agents; anti-infective agents; vehicles; excipients and other similarly used substances. Many of the same substances are pharmaceutics aids when added to pharmaceuticals rather than to foods. Food additives are substances added to food to preserve flavor or enhance its taste and appearance. Some additives have been used for centuries; for example, preserving food by pickling with vinegar, salting, as with bacon, preserving sweets or using sulfur dioxide as in some wines. With the advent of processed foods in the second half of the 20th century, many more additives have been introduced, of both natural and artificial origin. It is sometimes wrongly thought that food additives are a recent development, but there has certainly been an increase in public interest in the topic. Not all of this has been well-informed, and there are signs that commercial interests have been influenced by consumer pressure, as well as food producers manipulating the situation by marketing techniques. Various labeling regulations have been put into effect to ensure that contents of processed foods are known to consumers, and to ensure that food is fresh-important in unprocessed foods and probably important even if preservatives are used. In addition, we also need to add some preservatives in order to prevent the food from spoiling. Direct additives are intentionally added to foods for a particular purpose. Indirect additives are added to the food during its processing, packaging and storage. Food Preservatives are the additives that are used to inhibit the growth of bacteria, molds and yeasts in the food. Some of the additives are manufactured from the natural sources such as corn, beet and soybean, while some are artificial, man-made additives. Most people tend to eat the ready-made food available in the market, rather than preparing it at home. Such foods contain some kind of additives and preservatives, so that their quality and flavor is maintained and they are not spoiled by bacteria and yeasts. More than 3000 additives and preservatives are available in the market, which are used as antioxidants and anti-microbial agents. Salt and sugar the most commonly used additives. Some of the commonly used food additives and preservatives are aluminum silicate, amino acid compounds, ammonium carbonates, sodium nitrate, propyl gallate, butylated hydrozyttoluene (BHT), butylated hydroxyanisole (BHA), monosodium glutamate, white sugar, potassium bromate, potassium sorbate, sodium benzoate, etc. Some artificial colors are also added to the foods to give them an appealing look. These coloring substances are erythrosine (red), cantaxanthin (orange), amaranth (Azoic red), tartrazine (Azoic yellow) and annatto bixine (yellow orange). When the food is to be stored for a prolonged period, use of additives and preservatives is essential in order to maintain its quality and flavor. The excess water in the foods can cause the growth of bacteria, fungi and yeasts. Use of additives and preservatives prevents spoiling of the foods due to the growth of bacteria and fungi. Additives and preservatives maintain the quality and consistency of the foods. They also maintain palatability and wholesomeness of the food, improve or maintain its nutritional value, control appropriate pH, provide leavening and color, and enhance its flavor. There are even foods products that are made entirely from chemicals. Coffee creamers, sugar substitutes, and candies consist almost completely of artificial ingredients. Such manipulation of our food can have a profound effect on our body's unique biochemical balance. When we need to store any food for a longer time, it should be properly processed. During this processing, some substances and chemicals, known as additives, are added to the food. Additives consistently maintain the high quality of foods.

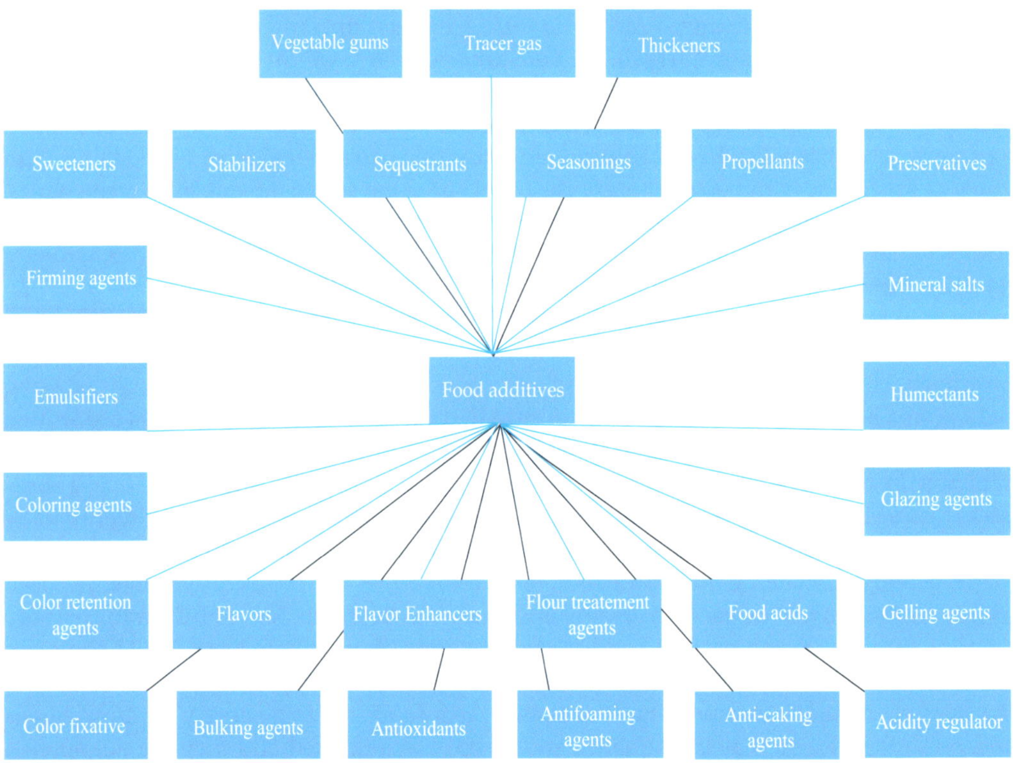

Fig. 1. Showing classification of food additives

2. Classification of food additives

Additives are classified as antimicrobial agents, antioxidants, artificial colors, artificial flavors and flavor enhancers, chelating agents and thickening and stabilizing agents (Fig. 1). Antimicrobial agents such as salt, vinegar, sorbic acid and calcium propionate are used in the products such as salad dressings, baked goods, margarine, cheese and pickled foods. Antioxidants including vitamin C, E, BHT and BHA are used in the foods containing high fats. Chelating agents such as malic acid, citric acid and tartaric acid are used to prevent the flavor changes, discoloration and rancidity of the foods. These are very important in food manufacturing companies. The food Additives is used to retard spoilage, enhance food flavors, replace nutrient lost in processing and makes the food more visually appealing.

2.1 Acidity regulator

Acidity regulators are used to change or otherwise control the acidity and alkalinity of foods.

Types of acidity regulator

i. acid
ii. acidifier
iii. acidity regulator
iv. alkali
v. base
vi. buffer
vii. buffering agent
viii. pH adjusting agent

2.2 Anti-caking agents

Anti caking agents, prevents the formation of lumps making these products manageable for packaging, transport, and for use by end consumer. Anticaking Agent is the food additive that prevents agglomeration in certain solids, permitting a free-flowing condition. It reduces the tendency of particles of food to adhere to one another.

Types of anti-caking agent

i. Anti-caking agent
ii. Anti-stick agent
iii. Drying agent
iv. Dusting agent

Anti-caking agents consist of such substances as starch, magnesium carbonate, and silica and are added to fine-particle solids, such as food products like table salt, flours, coffee, and sugar. Some of the common examples of foods that contain anti-caking agents include:

i. Vending machine powders (coffee, cocoa, soup)
ii. Milk and cream powders
iii. Grated cheese
iv. Icing sugar
v. Baking powder
vi. Cake mixes
vii. Instant soup powders
viii. Drinking chocolate
ix. Table salt

2.3 Antifoaming agents

Antifoaming agents reduce or prevent foaming in foods.

Types of anti- foaming agent

i. Antifoaming agent
ii. Defoaming agent

2.4 Antioxidants

A food additive, which prolongs the shelf-life of foods by protecting against deterioration caused by oxidation. Antioxidants are used to preserve food for a longer period of time. Antioxidants act as oxygen scavengers as the presence of oxygen in the food helps the bacteria to grow that ultimately harm the food. In the absence of antioxidant food additive oxidation of unsaturated fats takes place rendering to foul smell and discoloration of food. Different kinds of antioxidants foods act in a different ways but the end result is to delay or minimize the process of oxidation in food. Some antioxidants foods additives combine with oxygen to prevent oxidation and other prevent the oxygen from reacting with the food leading to its spoilage.

Types of anti-oxidant agent

i. Anti-browning agent
ii. Antioxidant
iii. Antioxidant synergist

Some popular antioxidant foods

Antioxidant vitamins

a) Ascorbic acid- E300

Antioxidant vitamins include Ascorbic acid (vitamin C) this antioxidant vitamin is used in beers, cut fruits, dried potatoes and jams. The antioxidant vitamins in these foods helps in preventing the discoloration of food by preventing the oxidation. It can also act as a substitute of vitamin C in potatoes that is lost during processing.

b) Citric acid - E330

It is used in biscuits, jams, tinned fruits, alcoholic drinks, cheese and dried soup. It has many uses like it prevents the discoloration of food, increases the anti-oxidant effect of other substances and regulates pH in jams and jellies.

c) Tocopherols – E307

(307a, d-alpha-Tocopherol; 307b, Tocopherol concentrate and 307c, dl-alpha-Tocopherol)

This antioxidant food additive is used in the meat pies and oils to reduce the oxidation of fatty acids and vitamins.

d) Butylated hydroxyanisole (BHA) - E320

It is used in margarine, oils, crisps and cheese. This antioxidant helps in preventing the reactions leading to the breakdown of fats.

Antioxidants benefits

There are many benefits of using antioxidant food additives. Antioxidants prevent the blockage of arteries with fatty deposits that prevents the heart-attacks. Also these are associated with the prevention of certain types of cancers, arthritis and more conditions of these kinds.

2.5 Bulking agents

A food additive, which contributes to the bulk of a food without contributing significantly to its available energy value. Bulking agents such as starch are additives that increase the bulk of a food without affecting its nutritional value.

Types of bulking agents

i. Bulking agent
ii. Filler

2.6 Color retention agents

A food additive, which stabilizes, retains or intensifies the colour of a food. In contrast to colorings, color retention agents are used to preserve a food's existing color.

Types of color retention agents

i. Color adjunct
ii. Colour fixative
iii. Colour retention agent
iv. Colour stabilizer

2.7 Coloring

A food additive, which adds or restores colour in a food. Colorings are added to food to replace colors lost during preparation, or to make food look more attractive, more visually appealing.

Types of coloring agents

i. Colour
ii. Decorative pigment
iii. Surface colorant

Beta carotene, Caramel, Carrot oil, Citrus red # 1, Dehydrated beets, FD&C colors: Blue # 1, 2; Red # 3, 40; Yellow # 5, 6 - used in processed foods, especially sweets and products marketed for children, soft drinks, baked goods, frosting, jams, and margarine.

Though there is a growing realization that the color additives should be used to the minimum, the fact is that the food doesn't even look presentable at times without it and appears inedible.

2.8 Emulsifiers

A food additive, which forms or maintains a uniform emulsion of two or more phases in a food. Emulsifiers allow water and oils to remain mixed together in an emulsion, as in mayonnaise, ice cream, and homogenized milk. It stops fats from clotting together.

Types of emulsifiers

i. Clouding agent
ii. Crystallization inhibitor
iii. Density adjustment agent (flavouring oils in beverages)
iv. Dispersing agent
v. Emulsifier
vi. Plasticizer
vii. Surface active agent
viii. Suspension agent

2.9 Emulsifying salt

A food additive, which, in the manufacture of processed food, rearranges proteins in order to prevent fat separation.

Types of emulsifying salt

i. Emulsifying salt
ii. Melding salt

2.10 Firming agents

A food additive, which makes or keeps tissues of fruit or vegetables firm and crisp, or interacts with gelling agents to produce or strengthen a gel.

2.11 Flavors

Flavors are additives that give food a particular taste or smell, and may be derived from natural ingredients or created artificially.

2.12 Flavor enhancers

Flavor enhancers enhance a food's existing flavors. They may be extracted from natural sources (through distillation, solvent extraction, maceration, among other methods) or created artificially.

Types of flavor enhancing agents

i. Flavour enhancer
ii. Flavour synergist

Some flavor enhancers are as follows:

a) Dioctyl sodium-sulfosuccinate - used in processed foods.
b) Disodium guanylate - used in canned meats, meat based foods.

c) Hydrolyzed vegetable - used in mixes, stock, processed meats.
d) Monosodium glutamate (MSG) - used in Chinese food, dry mixes, stock cubes, and canned, processed, and frozen meats.

2.13 Flour treatment agents

A food additive, which is added to flour or dough to improve its baking quality or colour.

Types of flour treatment agent

i. Dough conditioner
ii. Dough strengthening agent
iii. Flour bleaching agent
iv. Flour improver
v. Flour treatment agent

2.14 Food acids

Food acids are added to make flavors "sharper", and also act as preservatives and antioxidants. Common food acids include vinegar, citric acid, tartaric acid, malic acid, fumaric acid, and lactic acid.

2.15 Gelling agents

Gelling agents are food additives used to thicken and stabilize various foods, like jellies, desserts and candies. The agents provide the foods with texture through formation of a gel. Some stabilizers and thickening agents are gelling agents.

2.16 Glazing agents

A food additive, which when applied to the external surface of a food, imparts a shiny appearance or provides a protective coating. Glazing agents provide a shiny appearance or protective coating to foods.

Types of glazing agent

i. Coating agent
ii. Film forming agent
iii. Glazing agent
iv. Polishing agent
v. Sealing agent
vi. Surface-finishing agent

2.17 Humectants

A food additive, which prevents food from drying out by counteracting the effect of a dry atmosphere.

Types of humectants

i. Humectant
ii. Moisture/water retention agent
iii. Wetting agent

2.18 Mineral salts

Mineral salts are added as nutritional additives though they may have other properties like an anti-oxidant or a preservative. Many of them are essentials that need to be included in our daily diets, as they are the source of important nutrients required for the body. The important natural mineral salts that should be consumed are sodium, phosphorus, potassium, chlorine, sulphur and calcium. While the above mentioned happen to be the macro elements of the natural mineral salts, the micro elements are the ones that are essential nutrients for the human body. The micro elements in the minerals salts consist of iodine, iron, fluoride and zinc.

2.19 Preservatives

A food additive, which prolongs the shelf-life of a food by protecting against deterioration caused by microorganisms. It prevents or inhibits spoilage of food due to fungi, bacteria and other microorganisms. It stops microbes from multiplying and spoiling the food.

Types of preservatives

i. Antimicrobial preservative
ii. Antimicrobial synergist
iii. Antimould and antirope agent
iv. Antimycotic agent
v. Bacteriophage control agent
vi. Fungistatic agent
vii. Preservative

Some preservatives are:

a) Benzoic acid and benzoates - are found in soft-drinks, beer, margarine and acidic foods. They are use to extend shelf life and protect food from fungi and bacteria.
b) Nitrites and nitrates - are found in processed meats, such as sausages, hot dogs, bacon, ham, and luncheon meats, smoked fish. They are used to extend shelf life and protect food from fungi and bacteria; preserve color in meats and dried fruits.
c) Sulfites - are found in dried fruits, shredded coconut, fruit based pie fillings. They are used to extend shelf life and protect food from fungi and bacteria.

2.20 Propellants

It helps propel food from a container.

2.21 Seasonings

Seasoning is the process of imparting flavor to, or improving the flavor of food.

2.22 Sequestrants

A sequestrant is a food additive whose role is to improve the quality and stability of the food products. Sequestrants form chelate complexes with polyvalent metal ions, especially copper, iron and nickel, which serve as catalysts in the oxidation of the fats in the food. Sequestrants are a kind of preservative.

2.23 Stabilizers

A food additive, which makes it possible to maintain a uniform dispersion of two or more components. Stabilizers, like agar or pectin (used in jam for example) give foods a firmer texture. While they are not true emulsifiers, they help to stabilize emulsions.

i. Colloidal stabilizer
ii. Emulsion stabilizer
iii. Foam stabilizer
iv. Stabilizer

2.24 Sweeteners

Sweeteners are added to foods for flavoring. Sweeteners other than sugar are added to keep the food energy (calories) low, or because they have beneficial effects for diabetes mellitus and tooth decay and diarrhea. These are the substances that sweeten food, beverages, medications, etc., such as sugar, saccharine or other low-calorie synthetic products. They in general can be termed as sweetening agents. They all are called artificial sweeteners as they are usually not a component of the product they are added to. As per the source, these substances can be classified as natural and artificial sweeteners. Natural sweeteners are obtained from the natural sources like sugarcane and sugar beet and from fruits (fructose) and the artificial ones have a chemical origin. Artificial sweeteners are further of two type namely non-caloric sweeteners and sugar alcohols. Noncaloric sweeteners do not add calories to foods. They are used in snack foods and drinks. Sweeteners like saccharine and aspartame fall under this category. Sugar alcohols are used in chewing gums and hard candies and have almost same calories as sugar. Examples of sugar alcohols are sorbitol and mannitol.

Commonly used sweeteners

a) Acesulfame K - It is a 0 calorie sweetener, 130- 200 times sweeter than sucrose. It is not metabolized by the body. The only limitation it has is that if used in large quantities, it has an after taste. It is used in fruit preserves, dairy products and all types of beverages. It is used to reduce the calories of the products. It is heat resistant and enhances flavors.
b) Aspartame - It is a low calorie sweetener about 200% more sweet than the sugar. It is disintegrated into aspartic acid, fenylalanine and methanol in the body on digestion. It's taste is similar to sugar only more sweet. It is used in all types of foods and beverages and medicines. It is found naturally in protein rich foods.
c) Cyclamate - This is a calorie free sweetener 30-50 times sweeter than sugar. It is metabolized in the gut by few individuals and generally expelled as such. It is generally used in combination with other sweeteners. It has a pleasant taste, and is stable at high temperatures and is economical.
d) Saccharin - It is one of the earliest low calorie sweeteners that is 300-500 times more sweet than sugar. It doesn't metabolize and absorption is slow. Owing to this it is expelled as such from the body. Saccharin is the most widely used sweetener. It was earlier banned in certain countries but now is used quite commonly. There are other sweeteners like Stevioside, Alitame, Thaumatin, Sucralose, Neohesperidine DC and Aspartame-Acesulfame Salt. All artificial sweeteners have been approved by the U.S. Food and Drug Administration (FDA). They are considered harmless if taken in limited quantities.

2.25 Thickeners

Thickeners are substances which, when added to the mixture, increase its viscosity without substantially modifying its other properties.

Types of thickners

i. Binder
ii. Bodying agent
iii. Texturizing agent
iv. Thickener

2.26 Tracer gas

Tracer gas allows for package integrity testing preventing foods from being exposed to atmosphere, thus guaranteeing shelf life.

2.27 Vegetable gums

Vegetable gums come from the varied sources that can be on land or in sea. Some of the seaweeds are the excellent sources of food gums in which comes the carrageenan and alginates. Whereas guar, locust bean gum, pectin are obtained from the plants. Xanthan gum is obtained by the process of microbial fermentation. The source of gelatin is animal tissue. Vegetable gums are the polysaccharides that have the natural origin and used to increase the viscosity of the solution or food even if used in a very small concentration. Major Vegetable Gums are:

a. Cellulose Gum
b. Xanthan Gum
c. Locust Bean Gum
d. Pectin

3. 'E' numbering

To regulate these additives, and inform consumers, each additive is assigned a unique number, termed as "E numbers", which is used in Europe for all approved additives. This numbering scheme has now been adopted and extended by the Codex Alimentarius Commission to internationally identify all additives, regardless of whether they are approved for use.

E numbers are all prefixed by "E", but countries outside Europe use only the number, whether the additive is approved in Europe or not. For example, acetic acid is written as E260 on products sold in Europe, but is simply known as additive 260 in some countries. Additive 103, alkanet, is not approved for use in Europe so does not have an E number, although it is approved for use in Australia and New Zealand. Since 1987, Australia has had an approved system of labelling for additives in packaged foods. Each food additive has to be named or numbered. The numbers are the same as in Europe, but without the prefix 'E'.

The United States Food and Drug Administration listed these items as "Generally recognized as safe" or GRAS; they are listed under both their Chemical Abstract Services number and Fukda regulation under the US Code of Federal Regulations.

4. Dangers of food additives and preservatives

Although additives and preservatives are essential for food storage, they can give rise to certain health problems. They can cause different allergies and conditions such as hyperactivity and Attention Deficit Disorder in the some people who are sensitive to specific chemicals. The foods containing additives can cause asthma, hay fever and certain reactions such as rashes, vomiting, headache, tight chest, hives and worsening of eczema. Some of the known dangers of food additives and preservatives are as follows:

- Benzoates can trigger the allergies such as skin rashes and asthma as well as believed to be causing brain damage.
- Bromates destroy the nutrients in the foods. It can give rise to nausea and diarrhea.
- Butylates are responsible for high blood cholesterol levels as well as impaired liver and kidney function.
- Caffeine is a colorant and flavorant that has diuretic, stimulant properties. It can cause nervousness, heart palpitations and occasionally heart defects.
- Saccharin causes toxic reactions and allergic response, affecting skin, gastrointestinal tract and heart. It may also cause tumors and bladder cancer.
- Red Dye 40 is suspected to cause certain birth defects and possibly cancer.
- Mono and di-glycerides can cause birth defects, genetic changes and cancer.
- Caramel is a famous flavoring and coloring agent that can cause vitamin B6 deficiencies. It can cause certain genetic defects and even cancer.
- Sodium chloride can lead to high blood pressure, kidney failure, stroke and heart attack.

To minimize the risk of developing health problems due to food additives and preservatives, you should avoid the foods containing additives and preservatives. Before purchasing the canned food, you must check its ingredients. You should buy organic foods, which are free from artificial additives. Try to eat the freshly prepared foods as much as possible rather than processed or canned foods.

5. Effects of food additives

Avoiding or minimizing toxins in your diet is an important step toward enhancing your health and lowering your risk of disease. Foods, amongst other things (cosmetics & medications), represent a source of these toxins. Effects of food additives may be immediate or may be harmful in the long run if you have constant exposure. Immediate effects may include headaches, change in energy level, and alterations in mental concentration, behavior, or immune response. Long-term effects may increase your risk of cancer, cardiovascular disease and other degenerative conditions. Although it may seem difficult to change habits and find substitutes for foods you enjoy, remind yourself that you will be adding to your diet some new wholesome foods that you will come to enjoy even more. Look for foods that are not packaged and processed, but enjoy nature's own bounty of fresh fruits, vegetables, grains, beans, nuts and seeds. Find foods that resemble what they looked like when they were originally grown.

6. Cytotoxic effects of food additives

Exposure to non-nutritional food additives during the critical development window has been implicated in the induction and severity of behavioral disorders such as attention deficit hyperactivity disorder (ADHD). Although the use of single food additives at their regulated concentrations is believed to be relatively safe in terms of neuronal development, their combined effects remain unclear. The neurotoxic effects of four common food additives in combinations of two (Brilliant Blue and L-glutamic acid, Quinoline Yellow and aspartame) has been assess forpotential interactions. Mouse NB2a neuroblastoma cells were induced to differentiate and grow neurites in the presence of additives. After 24 h, cells were fixed and stained and neurite length measured by light microscopy with computerized image analysis. Neurotoxicity was measured as an inhibition of neurite outgrowth. Two independent models were used to analyze combination effects: effect additivity and dose additivity. Significant synergy was observed between combinations of Brilliant Blue with L-glutamic acid, and Quinoline Yellow with aspartame, in both models. Involvement of N-methyl-D-aspartate (NMDA) receptors in food additive-induced neurite inhibition was assessed with a NMDA antagonist, CNS-1102. L-glutamic acid- and aspartame induced neurotoxicity was reduced in the presence of CNS-1102; however, the antagonist did not prevent food color-induced neurotoxicity. Theoretical exposure to additives was calculated based on analysis of content in foodstuff, and estimated percentage absorption from the gut. Inhibition of neurite outgrowth was found at concentrations of additives theoretically achievable in plasma by ingestion of a typical snack and drink. In addition, Trypan Blue dye exclusion was used to evaluate the cellular toxicity of food additives on cell viability of NB2a cells; both combinations had a straightforward additive effect on cytotoxicity. These data have implications for the cellular effects of common chemical entities ingested individually and in combination. (Lau *et al.*, 2006)

Gallic acid is added to foods to prevent oxygen-induced lipid peroxidation and can be obtained by the hydrolysis of tannic acid which can be found in tea, coffee, red wine, and immature fruits. Tannic acid has also been used as a food additive. The effect of gallic acid on mouse spermatogonia, mouse spermatocytes, and mouse Sertoli cells *in vitro* was investigated. First, each cell line was cultured with predetermined concentrations of gallic acid for 3 h to access the effects of gallic acid on *in vitro* growth of testicular cells and MTT cytotoxicity assay was used to measure cell viability. Secondly, intracellular levels of hydrogen peroxide in mouse spermatogonia, mouse spermatocytes, and mouse Sertoli cells treated with gallic acid were analyzed using dihydrorhodamine 123 as a probe to evaluate the pro-oxidative property of gallic acid. The results obtained indicate that gallic acid inhibits the growth and proliferation of testicular cells in a dose-dependent manner and increases the intracellular level of hydrogen peroxide in mouse spermatogonia significantly ($p < 0.05$). It can be suggested that gallic acid exerts cytotoxic effects on testicular cells by its pro-oxidative activity. In conclusion, gallic acid-induced cytotoxicity in mouse spermatogonia, mouse spermatocytes, and mouse Sertoli cells *in vitro* may be of toxicological research interest considering the testicular toxic potential of gallic acid. (Park *et al.*, 2008)

In recent years, the use of carthami flos (the flowers of *Carthamus tinctorius*) as a colouring and flavouring agent in foods has increased in Iran. In order to evaluate its safety, the teratogenic effects of carthami flos on central nervous system development in mice were investigated. Furthermore, its cytotoxic effect on a rat nerve cell culture was studied to complete safety evaluations. For teratogenic studies, after natural mating, pregnant mice were divided into test and control groups. The groups were treated with different dosage regimens of aqueous carthami flos extract during 0-8 days of gestation. Embryos were then isolated at the 13th day of gestation and evaluated for macroscopic, microscopic and morphometric characteristics. The results showed that at higher doses (1.6 and 2 mg/kg/day) the embryos were absorbed, whereas at a lower dose (1.2 mg/kg/day) changes in external, internal and longitudinal diameters, open neuropore, changes in cellular orientation and cellular degeneration were observed. The results obtained from the cytotoxic assay also demonstrated a concentration-dependent cytotoxic effect of carthami flos extract. It is concluded that the use of carthami flos as a food additive should be reconsidered. (Nobakht *et al.*, 2000)

The cytotoxicity of 11 dyes, used as food dyes in Japan, on cultured fetal rat hepatocytes was studied. Xanthene dyes containing halogen atoms in their molecules such as phloxin, rose bengal, and erythrosine were more toxic than other groups of food dyes. The effect of food dyes on the cell growth of hepatocytes was also examined. Phloxin was especially toxic to the cell growth and a dose-response relation was observed between the concentration of phloxin and the cell growth of hepatocytes when the dye was added 3 days after plating. (Sako et al., 1980)

The synergistic effect of food additives or food colors on the toxicity of 3-amino-1,4-dimethyl-5H-pyrido[4,3-b]indole (Trp-P-1) was investigated using primary cultured rat hepatocytes. When hepatocytes from rats fed a standard diet were treated with a mixture of four major food additives (sorbitol, sodium 1.(+)-glutamate, benzoic acid, and propylene glycol) or a mixture of six typical artificial food colors (erythrosine, allura red, new coccine, brilliant blue, tertrazine, and fast green), the in vitro treated food-color mixture itself showed cytotoxicity: the reduction of cell viability and decreases in the activities of gluconeogenesis and ureogenesis. The food-color mixture enhanced cytotoxicity of Trp-P-1 obviously. The effects of in vivo-dosed food additives or food colors on Trp-P-1-caused toxicity. Hepatocytes were isolated and cultured from rats fed a diet containing a mixture of food additives or a mixture of food colors with half the amount of their respective acceptable daily intake for 4 wk. Trp-P-1 was administered to the hepatocytes at various concentrations for 12 h. Synergistic effects of in vivo-dosed food additives and food colors were not observed on Trp-P-1-caused cytotoxicity as estimated by a loss of cell viability and the reductions of DNA and protein syntheses. On the contrary, it was observed that in vivo administered food colors synergistically facilitated to reduce the activities of gluconeogenesis and ureogenesis in Trp-P-1-treated hepatocytes. These results suggest that the daily intake of artificial food colors may impair hepatic functions such as gluconeogenesis and ureogenesis, when dietary carcinogens are exposed to the liver cells. (Ashida *et al.*, 2000)

Study was conducted to investigate the impact of food additives like boric acid, citric acid and sodium metabisulphite individually, in different concentrations, on root tips of *Vicia faba* L. Cytological studies revealed significant decrease in mitotic index with an increase in concentration of the food additives. Most frequent cytological abnormalities observed are fragments, disturbed metaphase, C-mitosis, laggards, bridges, stickiness, precocious movement of chromosomes, unequal and late separation of chromosomes. Bridges and fragments were more frequent at anaphase. The percentage of chromosomal aberrations at mitosis increased with an increase in concentration of the food additives. (Pandey and Upadhyay, 2007)

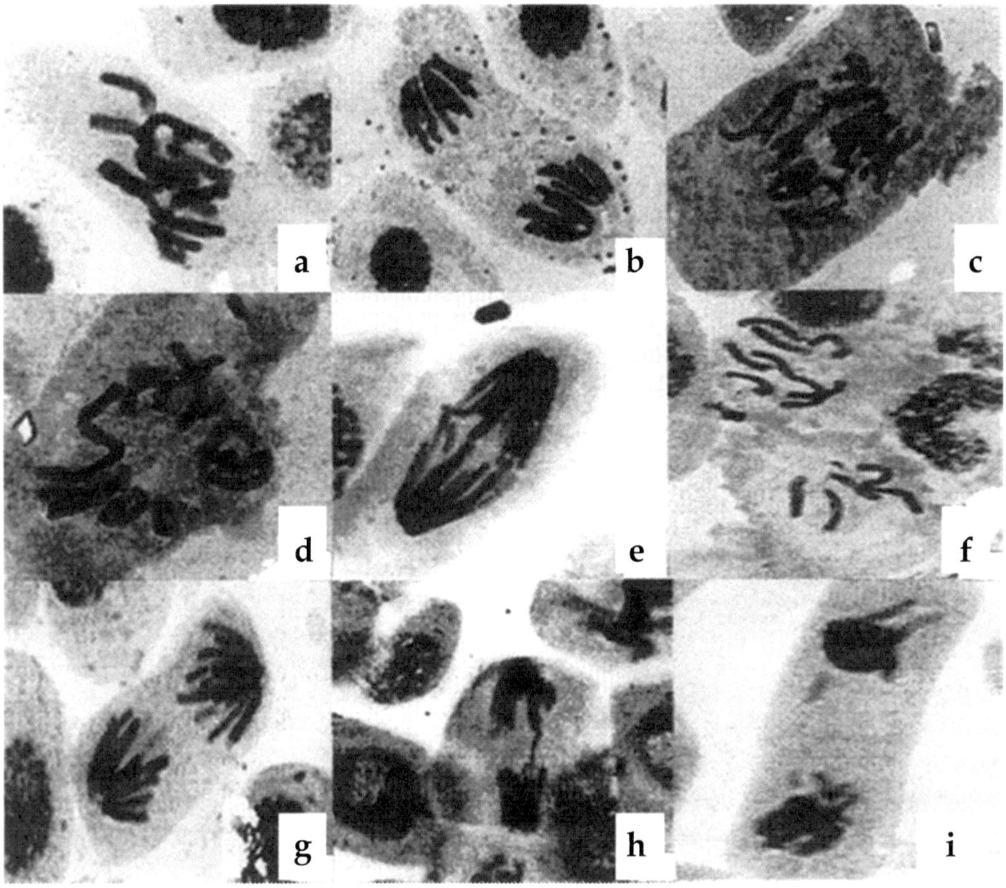

Fig. 2. **a-i** – Showing cytological effects of food additives in Vicia faba L., **a**. Normal metaphase, **b**. Normal anaphase, **c**. Disturbed metaphase, **d**. C-mitosis, **e**. Late separation at anaphase, **f**. Unequal separation, **g**. Fragmentation at anaphase, **h**. Chromatid bridge, **i**. Stickiness and Laggard at anaphase

Clastogenic properties of the food additive citric acid, commonly used as an antioxidant, were analysed in human peripheral blood lymphocytes. Citric acid induced a significant increase of chromosomal aberrations (CAs) at all the concentrations and treatment periods tested. Citric acid significantly decreased mitotic index (MI) at 100 and 200 lg ml-1 concentrations at 24 h, and in all concentrations at 48 h. However, it did not decrease the replication index (RI) significantly. Citric acid also significantly increased sister chromatid exchanges (SCEs) at 100 and 200 lg ml-1 concentrations at 24 h, and in all concentrations at 48 h. This chemical significantly increased the micronuclei frequency (MN) compared to the negative control. It also decreased the cytokinesis-block proliferation index (CBPI), but this result was not statistically significant. (Serkan et al., 2008)

Ester gum (EG) is used in citrus oil-based beverage flavourings as a weighting or colouring agent. In this study, concentrations of 50, 100 and 150 mg/kg body weight were administered orally to male Swiss albino mice, and sister chromatid exchange and chromosomal aberration were used as the cytogenetic endpoints to determine the genotoxic and clastogenic potential of the food additive. Although EG was weakly clastogenic and could induce a marginal increase in sister chromatid exchange frequencies, it was not a potential health hazard at the doses tested. (Mukherjee *et al.*, 1992)

Acesulfame-K, a sweetening agent, was evaluated *in vivo* for its genotoxic and clastogenic potentials. Swiss albino male mice were exposed to the compound by gavage. Bone marrow cells isolated from femora were analysed for chromosome aberrations. Doses of 15, 30, 60, 450, 1500 and 2250 mg of acesulfame-K/kg body weight induced a positive dose-dependent significant clastogenicity (trend test $\alpha < 0.05$). These doses were within the no-toxic-effect levels (1.5-3 g/kg body weight in rats) reported by the Joint Expert Committee for Food Additives of the World Health Organization and the Food and Agriculture Organization of the United Nations. In view of the present significant *in vivo* mammalian genotoxicity data, acesulfame-K should be used with caution. (Mukherjee and Chakrabarti, 1997)

Aspartame and acesulfame-K, non-nutritive sweeteners, are permitted individually in diets and beverages. These sweeteners of different classes, used in combination, have been found to possess a synergistic sweetening effect. Whether they also have a synergistic genotoxic effect is unknown. Swiss Albino male mice were exposed to blends of aspartame (3.5, 35, 350 mg/kg body weight) and acesulfame-K (1.5, 15 and 150 mg/kg body weight) by gavage. Bone marrow cells isolated from femora were analyzed for chromosome aberrations. Statistical analysis of the results show that aspartame in combination with acesulfame-K is not significantly genotoxic. (Mukhopadhyay *et al.*, 2000)

This work aimed to study some blood indices of rats as affected by saccharin and the therapeutic action of vitamins C and E. The used adult female Rattus norvegicus albino rats in the present study were weighing 100—120 g. Administration of saccharin at a dose of 35 mg kg^{-1} body weight (b.wt.) day^{-1} for 35 days significantly decreased serum glucose, triglycerides, cholesterol, total protein and albumin values. These decrements were by 20.16%, 22.76%, 44.92%, 20.16% and 40.44%, respectively, compared to control level (p value < 0.01). But it increased levels of kidney function indices. The effect of saccharin was more pronounced on creatinine. Activities of Alanine aminotranferease (ALT), aspartate aminotransferase (AST) and Alkaline phosphatase (ALP) increased significantly following saccharin treatment to rats. Concerning hematoligical parameters, the more obvious changes were observed in the increment of white blood cell (WBC), mean corpuscular volume (MCV) and platelets (PLT) and the decrease in hematocrit, hemoglobin (Hb) and red blood cells (RBCs) count in response to the administration of saccharin. In general, vitamin C or E (150 mg kg^{-1} b.wt. day^{-1} for 35 days) was able to reduce the effects of saccharin intake. Both vitamins, however, generally have beneficial effects in reducing the changes in the studied parameters. (Abdelaziz *et al.*, 2011)

Monosodium phosphate (MSP - E339i), disodium phosphate (DSP - E339ii), and trisodium phosphate (TSPE339iii) are used as antimicrobials, pH control agents (buffers), boiler water additives, cleaners, coagulants, dispersants, leavening agents, stabilizers, emulsifers, sequestrants, texturizers, nutrients, and dietary supplements. The effects of these have been studied on root tips of *Allium cepa* L. Roots of *A. cepa* were treated with a series of concentrations, ranging from 300 to 500 ppm for 24, 48 and 72 h. The results indicated that these food preservatives reduced mitotic division in *A. cepa* when compared with the respective control. Mitotic index values were generally decreased with increasing concentrations and longer treatment times. Additionally, variations in the percentage of mitotic stages were observed. The total percentage of aberrations generally increased with increasing concentrations of these chemicals and longer period of treatment. Different abnormal mitotic figures were observed in all mitotic phases. Among these abnormalities were stickiness, anaphase bridges, C-mitosis and micronuclei. These food additives remarkably depressed the DNA content in the root meristems of *A. cepa.* The interphase nuclear volume (INV) also varied between the untreated (controls) and the treated plants. (Turkoglu Sifa, 2009)

Citric acid is used widely as an acidulant, pH regulator, flavour enhancer, preservative and antioxidant synergist in many foods, like soft drinks, jelly sweet, baked nutrients, jam, marmalade, candy, tinned vegetable and fruit food Clastogenic properties of the food additive citric acid, commonly used as an antioxidant, were analysed in human peripheral blood lymphocytes. Citric acid induced a significant increase of chromosomal aberrations (CAs) at all the concentrations and treatment periods tested. Citric acid significantly decreased mitotic index (MI) at 100 and 200 lg ml-1 concentrations at 24 h, and in all concentrations at 48 h. However, it did not decrease the replication index (RI) significantly. Citric acid also significantly increased sister chromatid exchanges (SCEs) at 100 and 200 lg ml-1 concentrations at 24 h, and in all concentrations at 48 h. This chemical significantly increased the micronuclei frequency (MN) compared to the negative control. It also decreased the cytokinesis-block proliferation index (CBPI), but this result was not statistically significant. (Serkan Yılmaz *et al.*, 2008)

7. Food additives and safety

With the increasing use of processed foods since the 19th century, there has been a great increase in the use of food additives of varying levels of safety. This has led to legislation in many countries regulating their use. For example, boric acid was widely used as a food preservative from the 1870s to the 1920s, but was banned after World War I due to its toxicity, as demonstrated in animal and human studies. During World War II the urgent need for cheap, available food preservatives led to it being used again, but it was finally banned in the 1950s.[2] Such cases led to a general mistrust of food additives, and an application of the precautionary principle led to the conclusion that only additives that are known to be safe should be used in foods. In the USA, this led to the adoption of the Delaney clause, an amendment to the Federal Food, Drug, and Cosmetic Act of 1938, stating that no carcinogenic substances may be used as food additives. However, after the banning of cyclamates in the USA and Britain in 1969, saccharin, the only remaining legal artificial sweetener at the time, was found to cause cancer in rats. Widespread public outcry in the USA, partly communicated to Congress by postage-paid postcards supplied in the packaging of sweetened soft drinks, led to the retention of saccharin despite its violation of the Delaney clause.

In September 2007, research financed by Britain's Food Standards Agency and published online by the British medical journal *The Lancet*, presented evidence that a mix of additives commonly found in children's foods increases the mean level of hyperactivity. The team of researchers concluded that "the finding lends strong support for the case that food additives exacerbate hyperactive behaviors (inattention, impulsivity and overactivity) at least into middle childhood." That study examined the effect of artificial colors and a sodium benzoate preservative, and found both to be problematic for some

Colour	Status worldwide:	Where found:	Possible negative effects:	References
Erythrosine FD&C Red No. 3	Banned for use in cosmetics and external drug, but not food and ingested drugs in the U.S.	Cocktail, canned fruits salads confections dairy products snack foods.	Cancer	*The Washington Post*, February 7, 1990 CBS News, June 3, 2008
Tartrazine (E102) FD&C Yellow No. 5,	Banned in Norway and Austria.	Ice cream Carbonated drinks Fish sticks	Hyperactivity, asthma, skin rashes, and migraine headaches.	UK Food Guide. http://www.ukfoodguide.net/e102.htm. Retrieved 2007 FDA, 2007
Quinoline yellow (E104)	* Banned in Australia, Japan, Norway and the U.S. Restricted to max. permitted levels in U.K.	Soft drinks Ice creams Candies Cosmetics medications	Asthma, rashes and hyperactivity. Potential carcinogen in animals: implicated in bladder and liver cancer. Altered reproduction in animals.	efsa.europa.eu - EFSA updates safety advice on six food colours 091112
Sunset yellow (E110)* Yellow FCF Orange Yellow S	Banned in Norway, Sweden and Finland. Restricted to maximum permitted levels in U.K.	Sweets Snack foods Ice-creams, Yoghurts Drinks	AVOID in allergies & asthma. Cancer – DNA damage, increases tumors in animals. Growth retardation and severe weight loss in animals.	091113 efsa.europa.eu doi:10.1016/S0140-6736(07)
Carmosine (E122)*	Banned in Canada, Japan, Norway, Austria, Sweden and the U.S. Restricted to maximum permitted levels in U.K.	Yoghurts Sweets	DNA damage and tumours in animals.	Food additives *CBC News*. 29 September 2008
Allura red (E129)* FD&C Red No. 3	Banned in Denmark, Belgium, France, Germany, Switzerland, Sweden, Austria and Norway	Carbonated drinks Bubble gum, snacks, Sauces, preserves, Soups, wine, cider, etc.	May worsen or induce asthma, rhinitis (including hayfever), or urticaria (hives).	*UK Food Guide*, a British food additives website. Last retrieved 20 May 2007
Ponceau 4R (E124)* Conchineal	Banned in US, Canada, Norway, Sweden and Japan. Restricted to maximum permitted levels in the UK	Carbonated drinks Ice-creams Confectioneries Desserts	Cancer - DNA damage and tumours in animals. Can produce bad reactions in asthmatics	Food And Drug Administration Compliance Program Guidance Manual p.10
Amaranth (E123) Wine	Banned in the U.S.	Alcoholic drinks Fish roe	May worsen or induce asthma, allergies or hives.	FDA/CFSAN Food Compliance Program: Domestic Food Safety Program
Indigo Carmine (E132)*	Banned in the US, Japan, Australia and Norway. UK use restricted to maximum permitted levels	Ice-creams Sweets Baked goods Confectionery items Biscuits	May cause nausea, vomiting, skin rashes, and brain tumors. DNA damage and tumors in animals.	United States Food and Drug Administration
Brilliant Blue (E133)*	Banned in Austria, Belgium, France, Norway, Sweden, Switzerland and Germany. Restricted to maximum permitted levels in U.K.	Dairy products Sweets Drinks	Hyperactivity and skin rashes. Listed as human carcinogen by the US EPA. Causes DNA damage and tumors in animals	FDA, 1993

Table 1. Colour additives to avoid

children. Further studies are needed to find out whether there are other additives that could have a similar effect, and it is unclear whether some disturbances can also occur in mood and concentration in some adults. In the February 2008 issue of its publication, *AAP Grand Rounds*, the American Academy of Pediatrics concluded that a low-additive diet is a valid intervention for children with ADHD: "Although quite complicated, this was a carefully conducted study in which the investigators went to great lengths to eliminate bias and to rigorously measure outcomes. The results are hard to follow and somewhat inconsistent. For many of the assessments there were small but statistically significant differences of measured behaviors in children who consumed the food additives compared with those who did not. In each case increased hyperactive behaviors were associated with consuming the additives. For those comparisons in which no statistically significant differences were found, there was a trend for more hyperactive behaviors associated with the food additive drink in virtually every assessment. Thus, the overall findings of the study are clear and require that even we skeptics, who have long doubted parental claims of the effects of various foods on the behavior of their children, admit we might have been wrong."

In 2007, Food Standards Australia New Zealand published an official shoppers' guidance with which the concerns of food additives and their labeling are mediated.

There has been significant controversy associated with the risks and benefits of food additives. Some artificial food additives have been linked with cancer, digestive problems, neurological conditions, ADHD, heart disease or obesity. Natural additives may be similarly harmful or be the cause of allergic reactions in certain individuals. For example, safrole was used to flavor root beer until it was shown to be carcinogenic. Due to the application of the Delaney clause, it may not be added to foods, even though it occurs naturally in sassafras and sweet basil.

Blue 1, Blue 2, Red 3, and Yellow 6 are among the food colorings that have been linked to various health risks. Blue 1 is used to color candy, soft drinks, and pastries and there has been some evidence that it may cause cancer. Blue 2 can be found in pet food, soft drinks, and pastries, and has shown to cause brain tumors in mice. Red 3, mainly used in cherries for cocktails has been correlated with thyroid tumors in rats and humans as well. Yellow 6, used in sausages, gelatin, and candy can lead to the attribution of gland and kidney tumors and contains carcinogens, but in minimal amounts.

Some studies have linked some food additives to hyperactivity in children. A recent British study found that children without a history of any hyperactive disorder showed varying degrees of hyperactivity after consuming fruit drinks with various levels of additives. Among those that were studied were: Sodium benzoate (E211), Tartrazine (E102), quinoline yellow (E104), Sunset yellow (E110), Carmosine (E122), Allura red (E129). See tables below for more information.

8. Joint FAO/WHO Expert Committee on Food Additives (JECFA)

The Joint FAO/WHO Expert Committee on Food Additives (JECFA) is an international scientific expert committee that is administered jointly by the Food and Agriculture Organization of the United Nations FAO and the World Health Organization WHO. It has been meeting since 1956, initially to evaluate the safety of food additives. Its work now also includes the evaluation of contaminants, naturally occurring toxicants and residues of veterinary drugs in food.

ii) Food preservatives to avoid:

Sodium benzoate (E211)*		Carbonated drinks Pickles Sauces Certain medicines (even some "natural and homeopathic" medications for kids)	Aggravates asthma and suspected to be a neurotoxin and carcinogen, may cause fetal abnormalities. Worsens hyperactivity	Food Standards Agency issues revised advice on certain artificial colours, 2007
Sulphur Dioxide (E220)*	Not banned anywhere.	Carbonated drinks Dried fruit Juices Cordials Potato products	May induce gastric irritation, nausea, diarrhea, asthma attacks, skin rashes. Destroys vitamin B1. Causes fetal abnormalities and DNA damage in animals.	International Chemical Safety Card 0074
Sodium metabisulphite		Preservative and antioxidant.	May provoke life threatening asthma	http://www.fedupwithfoodadditives.info/factsheets/Factsafeadditives.htm
Potassium nitrate (E249)	Not banned anywhere	Cured meats and canned meat products.	May lower oxygen carrying capacity of blood; may combine with other substances to form nitrosamines that are carcinogens; may negatively effect the adrenal gland.	International Chemical Safety Card 1069
Calcium benzoate (E213)	Not banned anywhere	Drinks, low-sugar products, cereals, meat products.	May temporarily inhibit digestive enzyme function and may deplete levels of the amino acid glycine. AVOID with allergies, hives, & asthma.	http://www.fedupwithfoodadditives.info/factsheets/Factsafeadditives.htm
Calcium sulphite (E226)	In the U.S., sulphites are banned from many foods, including meat	In a vast array of foods-from burgers to biscuits, from frozen mushrooms to horseradish. Used to make old produce look fresh.	May cause bronchial problems, flushing, low blood pressure, tingling, and anaphylactic shock. Avoid them if you suffer from bronchial asthma, cardiovascular or respiratory problems and emphysema.	http://www.fedupwithfoodadditives.info/factsheets/Factsafeadditives.htm
Butylated Hydroxy-anisole (E320) BHA/BHT		Particularly in fat containing foods, confectionery, meats.	BHA/BHT is may be carcinogenic to humans. BHA also interacts with nitrites to form chemicals known to cause changes in the DNA of cells.	doi:10.1021/jm00191a020
Benzoic acid (E210)		Drinks, low sugar products, cereals, meat products.	May temporarily inhibit digestive enzyme function. May deplete glycine levels. AVOID in asthma, or allergies.	International Chemical Safety Card 0103

Table 2. Food preservatives to avoid

iii) Flavourings & sweeteners to avoid:

Monosodium Gluatamate MSG (E621)* **	Not banned anywhere	Processed foods & drinks, soup mixes.	Destroys nerve cells in brain and linked with aggravating or accelerating Huntington's, Alzheimer's and Parkinson's diseases. Causes cancer, DNA damage and fetal abnormalities in animals. Increases hyperactivity.	doi:10.1111/j.1365-2222.2009.03221.x
Aspartame (E951)*	US Air Force pilots are banned from drinking soft drinks containing aspartame.	200 times sweeter than sugar	May cause neurological damage, especially in younger children where brain is still developing. Breaks down in the body to phenylalanine (neurotoxin - may cause seizures), aspartic acid (damages developing brain) and methanol (converts to formaldehyde). Crosses the placental barrier from mother to baby, even in small doses. Implicated in diseases such as MS and Non-Hodgkin's Lymphoma. May contribute to obesity.	FDA Consumer Magazine, 1999
Acesulphame K (E950)*	Not banned anywhere.	200 times sweeter than sugar	Causes cancer in animals. Linked to hypoglycemia, lung tumours, increased cholesterol and leukemia. May contribute to obesity	British Pharmacopoeia Commission Secretariat, 2009
Saccharine (E954)*	Banned in Germany, Spain, Portugal, Hungary, France, Malaysia, Zimbabwe, Fiji, Peru, Israel, Taiwan.	350 times sweeter than sugar	May interfere with blood coagulation, blood sugar levels and digestive function. Causes cancer of the bladder, uterus, ovaries, skin and blood vessels in animals. Linked to DNA damage and congenital abnormalities in animals. May contribute to Obesity.	USDA, 1972
High Fructose Corn Syrup (HFCS)	Not banned anywhere	Carbonated drinks other sweetened drinks (juices) baked goods candies canned fruits jams & jellies dairy products	Obesity Accelerated aging Insulin resistance Diabetes mellitus Complications of diabetes Fatty liver Increased triglycerides Increased uric acid Chronic diarrhea Irritable bowel syndrome Hives	The American Medical Association, 2007

*All of these additives are considered the "Dirty Dozen Food Additives" and are prohibited in the UK for foods marketed for children less than 36 months.
**MSG-intolerant people can develop MSG symptom complex, which is characterized by one of more of the following:

- A burning sensation in the back of the neck, forearms and chest.
- Numbness in the back of the neck, radiating to the arms and back.
- A tingling, warmth and weakness in the face, temples, upper back, neck and arms.
- Facial pressure or tightness, swelling of lips/face
- Chest pain, rapid heartbeat
- Headache, nausea, drowsiness
- Bronchospasm (difficulty breathing) in MSG-intolerant people with asthma.

Table 3. Flavourings & sweeteners to avoid

JECFA has evaluated more than 1500 food additives, approximately 40 contaminants and naturally occurring toxicants, and residues of approximately 90 veterinary drugs. The Committee has also developed principles for the safety assessment of chemicals in food that are consistent with current thinking on risk assessment and take account of recent developments in toxicology and other relevant sciences.

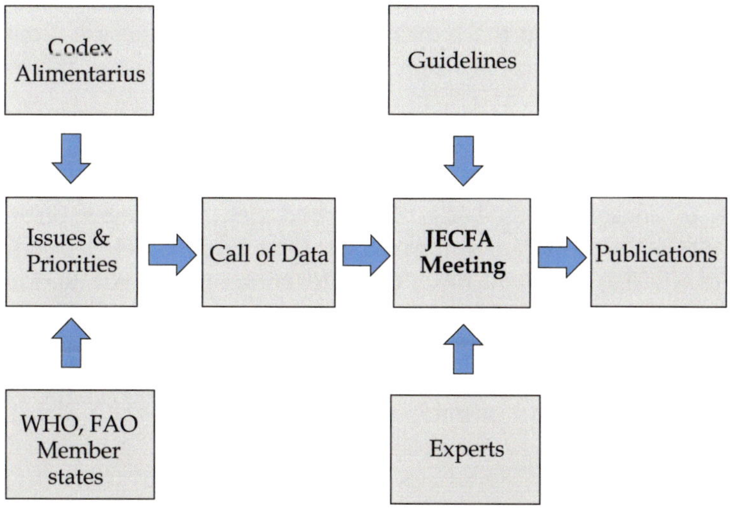

Fig. 3. Showing flow chart which shows the process flow for JECFA

The above flow chart shows the process flow for JECFA.

A brief summary of the purpose, history, and workings of the Joint FAO/WHO Expert Committee on Food Additives.

9. Sites related to JECFA

- JMPR
  The Joint FAO/WHO Meeting on Pesticide Residues is the related expert committee for pesticide residues in food. It is responsible for reviewing and evaluating toxicological residue and analytical aspects of the pesticides under consideration.
- Codex Alimentarius
  The Codex Alimentarius Commission was created in 1963 by FAO and WHO to develop food standards, guidelines and related texts such as codes of practice under the Joint FAO/WHO Food Standards Programme. The main purposes of this Programme are protecting health of the consumers and ensuring fair trade practices in the food trade, and promoting coordination of all food standards work undertaken by international governmental and non-governmental organizations. Codex standards are based on scientific advice as provided by JECFA and JMPR.

10. General information related to chemical risks in food

Chemicals are the building blocks of life and are important for many, if not all, aspect of human metabolism. However, human exposure to chemicals at toxic levels, as well as nutritional imbalances, are known or suspected to be involved in causing cancer, cardiovascular disease, kidney and liver dysfunction, hormonal imbalance, reproductive disorders, birth defects, premature births, immune system suppression, musculoskeletal disease, impeded nervous and sensory system development, mental health problems, urogenital disease, old-age dementia, and learning disabilities. Possibly a significant part of these disorders and diseases can be attributed to chemical exposure, and for many (environmental) chemicals food is the main source of human exposure. Consequently, the protection of our diet from these hazards must be considered one of the essential public health functions of any country.

Under the World Trade Organization's Agreement on Sanitary and Phytosanitary Measures, food traded internationally must comply with Codex Standards that are established to protect health of consumers on basis of a sound risk assessment. Such independent international risk assessments are performed by Joint FAO/WHO Expert Committee on Food Additives (JECFA) and Joint FAO/WHO Meeting on Pesticide Residues (JMPR) as well as ad hoc expert meetings to address specific and emerging issues. The experts estimate a safe level of exposure (acceptable or tolerable daily intake ADI, TDI) and estimate the exposure to chemicals from the diet and from specific foods. Such exposure assessments often are based on national data or international data from the WHO Global Environment Monitoring System - Food Contamination Monitoring and Assessment Programme (GEMS/Food).

Based on the risk assessments provided through these expert meetings, the Codex Alimentarius Commission can recommend specific measures, such as maximum limits in foods to assure that exposure do not exceed the acceptable/tolerable level of intake. Other measure can be the development of 'Codes of Practices' to reduce levels of contaminants in food. Also, levels of use for food additives can be recommended or maximum residue levels for pesticides or veterinary drug residues when applied in accordance with good practices. The scientific advice provided through these expert meetings also often serves directly as the basis for national food safety standards.

11. Conclusion

Food additives preserve the freshness and appeal of food between the times it is manufactured and when it finally reaches the market. Additives may also improve nutritional value of foods and improve their taste, texture, consistency or color. All food additives approved for use in the United States are carefully regulated by federal authorities to ensure that foods are safe to eat and are accurately labeled. Food additives have been used by man since earliest times. Today, food and color additives are more strictly regulated than at any time in history. Additives may be incorporated in foods to maintain product consistency, improve or maintain nutritional value, maintain palatability and wholesomeness provide leavening or control acidity/alkalinity, and/or enhance flavor or impart desired color. The Food and Agriculture Organization (FAO), however, recognizes additives as any substance whose intended use will affect, or may reasonably be expected to affect, the characteristics of any food. FAO law prohibits the use of any additive that has been found to cause cancer in humans or animals. To market a new food or color additive, a manufacturer must first petition the FAO for its approval. FAO regulations require evidence that each substance is safe at its intended levels of use before it may be added to foods. In deciding whether an additive should be approved, the agency considers the composition and properties of the substance, the amount likely to be consumed, its probable long-term effects and various safety factors. Some food additives like boric acid, citric acid and sodium metabisulphite showed mitotoxicity and genotoxicity having potential risks for human health. Thus, all additives are subject to ongoing safety review as scientific understanding and methods of testing continue to improve.

12. Acknowledgements

Authors are thankful to Dr. C. S. Nautiyal, Director, CSIR-National Botanical Research Institute, Lucknow for encouragement. Thanks are also due to Ms. Rekha Singh, Project Assistant, Cytogenetics Laboratory, CSIR-NBRI for help.

13. References

Added use level data, SPET calculations and indication of food categories with highest SPET as used in exposure estimations of flavourings at the 73rd JECFA meeting (2010), Complete Table

Ashida H., Hashimoto T., Tsuji S. , Kanazawa K., Dasno G.I. (2000) Synergistic effects of food colors on the toxicity of 3-amino-1,4-dimethyl-5h-pyrido[4,3-b]indole (Trp-P-1) in primary cultured rat hepatocytes. Journal of nutritional science and vitaminology vol. 46, no3, pp. 130-136

Assessment of technologies for determining cancer risks from the environment. Darby, PA, USA: DIANE publishing. 1981. pp. 177. ISBN 142892437X.

Bucci, Luke (1995). Nutrition applied to injury rehabilitation and sports medicine. Boca Raton: CRC Press. pp. 151. ISBN 0-8493-7913-X.

Codex Alimentarius. "Class Names and the International Numbering System for Food Additives.". http://www.codexalimentarius.net/download/standards/7/CXG_036e.pdf.

Evaluation of the carcinogenic hazards of food additives (Fifth report of the Joint FAO/WHO Expert Committee on Food Additives). WHO Technical Report Series, No. 220, 1961 (out of print).

Evaluation of the toxicity of a number of antimicrobials and antioxidants (Sixth report of the Joint FAO/WHO Expert Committee on Food Additives). WHO Technical Report Series, No. 228, 1962 (out of print).

Evaluation of food additives: specifications for the identity and purity of food additives and their toxicological evaluation: some extraction solvents and certain other substances; and a review of the technological efficacy of some antimicrobial agents. (Fourteenth report of the Joint FAO/WHO Expert Committee on Food Additives). WHO Technical Report Series, No. 462, 1971 (out of print).

Evaluation of food additives: some enzymes, modified starches, and certain other substances: Toxicological evaluations and specifications and a review of the technological efficacy of some antioxidants (Fifteenth report of the Joint

FAO/WHO Expert Committee on Food Additives). WHO Technical Report Series, No. 488, 1972.

Evaluation of certain food additives and the contaminants mercury, lead, and cadmium (Sixteenth report of the Joint FAO/WHO Expert Committee on Food Additives). WHO Technical Report Series, No. 505, 1972, and corrigendum (out of print).

Evaluation of certain food additives (Eighteenth report of the Joint FAO/WHO Expert Committee on Food Additives). WHO Technical Report Series, No. 557, 1974, and corrigendum (out of print).

Evaluation of certain food additives: some food colours, thickening agents, smoke condensates, and certain other substances. (Nineteenth report of the Joint FAO/WHO Expert Committee on Food Additives). WHO Technical Report Series, No. 576, 1975 (out of print).

Evaluation of certain food additives (Twentieth report of the Joint FAO/WHO Expert Committee on Food Additives). WHO Technical Report Series, No. 599, 1976. English français español

Evaluation of certain food additives (Twenty-first report of the Joint FAO/WHO Expert Committee on Food Additives). WHO Technical Report Series, No. 617, 1978.

Evaluation of certain food additives and contaminants (Twenty-second report of the Joint FAO/WHO Expert Committee on Food Additives). WHO Technical Report Series, No. 631, 1978 (out of print).

Evaluation of certain food additives (Twenty-third report of the Joint FAO/WHO Expert Committee on Food Additives). WHO Technical Report Series, No. 648, 1980, and corrigenda.

Evaluation of certain food additives (Twenty-fourth report of the Joint FAO/WHO Expert Committee on Food Additives). WHO Technical Report Series, No. 653, 1980.

Evaluation of certain food additives (Twenty-fifth report of the Joint FAO/WHO Expert Committee on Food Additives). WHO Technical Report Series, No. 669, 1981.

Evaluation of certain food additives and contaminants (Twenty-sixth report of the Joint FAO/WHO Expert Committee on Food Additives). WHO Technical Report Series, No. 683, 1982.

Evaluation of certain food additives and contaminants (Twenty-seventh report of the Joint FAO/WHO Expert Committee on Food Additives). WHO Technical Report Series, No. 696, 1983, and corrigenda (out of print).

Evaluation of certain food additives and contaminants (Twenty-eighth report of the Joint FAO/WHO Expert Committee on Food Additives). WHO Technical Report Series, No. 710, 1984, and corrigendum.

Evaluation of certain food additives and contaminants (Twenty-ninth report of the Joint FAO/WHO Expert Committee on Food Additives). WHO Technical Report Series, No. 733, 1986, and corrigendum.

Evaluation of certain food additives and contaminants (Thirtieth report of the Joint FAO/WHO Expert Committee on Food Additives). WHO Technical Report Series, No. 751, 1987.

Evaluation of certain food additives and contaminants (Thirty-first report of the Joint FAO/WHO Expert Committee on Food Additives). WHO Technical Report Series, No. 759, 1987, and corrigendum.

Evaluation of certain veterinary drug residues in food (Thirty-second report of the Joint FAO/WHO Expert Committee on Food Additives). WHO Technical Report Series, No. 763, 1988.

Evaluation of certain food additives and contaminants (Thirty-third report of the Joint FAO/WHO Expert Committee on Food Additives). WHO Technical Report Series, No. 776, 1989.

Evaluation of certain veterinary drug residues in food (Thirty-fourth report of the Joint FAO/WHO Expert Committee on Food Additives). WHO Technical Report Series, No. 788, 1989.

Evaluation of certain food additives and contaminants (Thirty-fifth report of the Joint FAO/WHO Expert Committee on Food Additives). WHO Technical Report Series, No. 789, 1990, and corrigenda.

Evaluation of certain veterinary drug residues in food (Thirty-sixth report of the Joint FAO/WHO Expert Committee on Food Additives). WHO Technical Report Series, No. 799, 1990.

Evaluation of certain food additives and contaminants (Thirty-seventh report of the Joint FAO/WHO Expert Committee on Food Additives). WHO Technical Report Series, No. 806, 1991, and corrigenda.

Evaluation of certain veterinary drug residues in food (Thirty-eighth report of the Joint FAO/WHO Expert Committee on Food Additives). WHO Technical Report Series, No. 815, 1991.

Evaluation of certain food additives and naturally occurring toxicants (Thirty-ninth report of the Joint FAO/WHO Expert Committee on Food Additives). WHO Technical Report Series No. 828, 1992.

Evaluation of certain veterinary drug residues in food (Fortieth report of the Joint FAO/WHO Expert Committee on Food Additives). WHO Technical Report Series, No. 832, 1993.

Evaluation of certain food additives and contaminants (Forty-first report of the Joint FAO/WHO Expert Committee on Food Additives). WHO Technical Report Series, No. 837, 1993.

Evaluation of certain veterinary drug residues in food (Forty-second report of the Joint FAO/WHO Expert Committee on Food Additives). WHO Technical Report Series, No. 851, 1995.

Evaluation of certain veterinary drug residues in food (Forty-third report of the Joint FAO/WHO Expert Committee on Food Additives). WHO Technical Report Series, No. 855, 1995, and corrigendum.

Evaluation of certain food additives and contaminants (Forty-fourth report of the Joint FAO/WHO Expert Committee on Food Additives). WHO Technical Report Series, No. 859, 1995.

Evaluation of certain veterinary drug residues in food (Forty-fifth report of the Joint FAO/WHO Expert Committee on Food Additives). WHO Technical Report Series, No. 864, 1996.

Evaluation of certain food additives and contaminants (Forty-sixth report of the Joint FAO/WHO Expert Committee on Food Additives). WHO Technical Report Series, No. 868, 1997.

Evaluation of certain veterinary drug residues in food (Forty-seventh report of the Joint FAO/WHO Expert Committee on Food Additives). WHO Technical Report Series, No. 876, 1998.

Evaluation of certain veterinary drug residues in food (Forty-eighth report of the Joint FAO/WHO Expert Committee on Food Additives). WHO Technical Report Series, No. 879, 1998.

Evaluation of certain food additives and contaminants (Forty-ninth report of the Joint FAO/WHO Expert Committee on Food Additives). WHO Technical Report Series, No. 884, 1999.

Evaluation of certain veterinary drug residues in food (Fiftieth report of the Joint FAO/WHO Expert Committee on Food Additives). WHO Technical Report Series, No. 888, 1999.

Evaluation of certain food additives (Fifty-first report of the Joint FAO/WHO Expert Committee on Food Additives). WHO Technical Report Series, No. 891, 2000.

Evaluation of certain veterinary drug residues in food (Fifty-second report of the Joint FAO/WHO Expert Committee on Food Additives). WHO Technical Report Series, No. 893, 2000.

Evaluation of certain food additives and contaminants (Fifty-third report of the Joint FAO/WHO Expert Committee on Food Additives). WHO Technical Report Series, No. 896, 2000.

Evaluation of certain veterinary drug residues in food (Fifty-fourth report of the Joint FAO/WHO Expert Committee on Food Additives). WHO Technical Report Series, No. 900, 2001.

Evaluation of certain food additives and contaminants (Fifty-fifth report of the Joint FAO/WHO Expert Committee on Food Additives). WHO Technical Report Series, No. 901, 2001.

Evaluation of certain mycotoxins (Fifty-sixth report of the Joint FAO/WHO Expert Committee on Food Additives). WHO Technical Report Series, No. 906, 2002.

Evaluation of certain food additives and contaminants (Fifty-seventh report of the Joint FAO/WHO Expert Committee on Food Additives). WHO Technical Report Series, No. 909, 2002.

Evaluation of certain veterinary drug residues in food (Fifty-eighth report of the Joint FAO/WHO Expert Committee on Food Additives). WHO Technical Report Series, No. 911, 2002.

Evaluation of certain food additives (Fifty-ninth report of the Joint FAO/WHO Expert Committee on Food Additives). WHO Technical Report Series, No. 913, 2002.

Evaluation of certain veterinary drug residues in food (Sixtieth report of the Joint FAO/WHO Expert Committee on Food Additives). WHO Technical Report Series No. 918, 2003.

Evaluation of certain food additives and contaminants (Sixty-first report of the Joint FAO/WHO Expert Committee on Food Additives). WHO Technical Report Series, No. 922, 2004.

Evaluation of certain veterinary drug residues in food (Sixty-second report of the Joint FAO/WHO Expert Committee on Food Additives). WHO Technical Report Series, No. 925, 2004.

Evaluation of certain food additives (Sixty-third report of the Joint FAO/WHO Expert Committee on Food Additives). WHO Technical Report Series, No. 928, 2005.

Evaluation of certain food contaminants (Sixty-fourth report of the Joint FAO/WHO Expert Committee on Food Additives). WHO Technical Report Series, No. 930, 2006.

Evaluation of certain food additives (Sixty-fifth report of the Joint FAO/WHO Expert Committee on Food Additives). WHO Technical Report Series, No. 934, 2006.

Evaluation of certain veterinary drug residues in food (Sixty-sixth report of the Joint FAO/WHO Expert Committee on Food Additives). WHO Technical Report Series, No. 939, 2006.

Evaluation of certain food additives and contaminants (Sixty-seventh report of the Joint FAO/WHO Expert Committee on Food Additives). WHO Technical Report Series, No. 940, 2007.

Evaluation of certain food additives and contaminants (Sixty-eighth report of the Joint FAO/WHO Expert Committee on Food Additives). WHO Technical Report Series, No. 947, 2007.

Evaluation of certain food additives (Sixty-ninth report of the Joint FAO/WHO Expert Committee on Food Additives). WHO Technical Report Series, No. 952, 2009.

Evaluation of certain veterinary drug residues in food (Seventieth report of the Joint FAO/WHO Expert Committee on Food Additives). WHO Technical Report Series, No. 954, 2009.

Evaluation of certain food additives (Seventy-first report of the Joint FAO/WHO Expert Committee on Food Additives). WHO Technical Report Series, No. 956, 2010

Evaluation of certain contaminants in food (Seventy-second report of the Joint FAO/WHO Expert Committee on Food Additives). WHO Technical Report Series, No. 959, 2011

Evaluation of certain food additives and contaminants (Seventy-third report of the Joint FAO/WHO Expert Committee on Food Additives). WHO Technical Report Series, No. 960, 2011

Fennema, Owen R. (1996). Food chemistry. New York, N.Y: Marcel Dekker. pp. 827. ISBN 0-8247-9691-8.

Fumiyo Sako, Noriko Kobayashi, Hiroyuki Watabe and Naoyuki Taniguchi (1980) Cytotoxicity of food dyes on cultured fetal rat hepatocytes Toxicology and Applied Pharmacology Volume 54, Issue 2, Pages 285-292

General principles governing the use of food additives (First report of the Joint FAO/WHO Expert Committee on Food Additives). WHO Technical Report Series, No. 129, 1957 (out of print).

International Organization for Standardization. "67.220: Spices and condiments. Food additives".

Ismael Abdelaziz, Abd El Rahiem,A Ashour (2011) Effect of saccharin on albino rats' blood indices and the therapeutic action of vitamins C and E. Hum Exp Toxicol vol. 30 no. 2 129-137)

Karen Lau, W. Graham McLean, Dominic P. Williams, and C. Vyvyan Howard (2006) Synergistic Interactions between Commonly Used Food Additives in a Developmental Neurotoxicity Test. TOXICOLOGICAL SCIENCES 90(1), 178–187

McCann, D; Barrett, A; Cooper, A; Crumpler, D; Dalen, L; Grimshaw, K; Kitchin, E; Lok, K et al. (2007). "Food additives and hyperactive behaviour in 3-year-old and 8/9-year-old children in the community: a randomised, double-blinded, placebo-controlled trial.". Lancet 370 (9598): pp. 1560–7. doi:10.1016/S0140-6736(07)61306-3. PMID 17825405.

Mukherjee A., K. Agarwal and J. Chakrabarti (1992) Genotoxicity studies of the food additive ester gum. Food and Chemical Toxicology Volume 30, Issue 7, Pages 627-630

Mukherjee A. and J. Chakrabarti (1997) In vivo cytogenetic studies on mice exposed to acesulfame-K—A non-nutritive sweetener. Food and Chemical Toxicology Volume 35, Issue 12, Pages 1177-1179

Mukhopadhyay K M., A. Mukherjee and J. Chakrabarti (2000) In vivo cytogenetic studies on blends of aspartame and acesulfame Food and Chemical Toxicology Volume 38, Issue 1, Pages 75-77

Nobakht, M., Fattahi, M., Hoormand, M., Milanian, I., Rahbar, N., Mahmoudian (2000) A study on the teratogenic and cytotoxic effects of safflower extract. M. Journal of Ethnopharmacology Vol. 73 No. 3 pp. 453-459

Pandey Ram Milan and Santosh Upadhyay, (2007) Impact of Food Additives on Mitotic Chromosomes of Vicia faba L. CARYOLOGIA Vol. 60, no. 4: 309-314

Procedures for the testing of intentional food additives to establish their safety for use (Second report of the Joint FAO/WHO Expert Committee on Food Additives). WHO Technical Report Series, No. 144, 1958 (out of print).

Rev. Lyman Abbott (Ed.) (1900). The Outlook (Vol. 64). Outlook Co. pp. 403.

Serkan Yılmaz, Fatma Ünal, Deniz Yüzbaşıoğlu and Hüseyin Aksoy (2008) Clastogenic effects of food additive citric acid in human peripheral lymphocytes. Cytotechnology 56:137–144)

Specifications for identity and purity of food additives (antimicrobial preservatives and antioxidants) (Third report of the Joint FAO/WHO Expert Committee on Food Additives). These specifications were subsequently revised and published as Specifications for identity and purity of food additives, Vol. I. Antimicrobial preservatives and antioxidants, Rome, Food and Agriculture Organization of the United Nations, 1962 (out of print).

Specifications for identity and purity of food additives (food colours) (Fourth report of the Joint FAO/WHO Expert Committee on Food Additives). These specifications were subsequently revised and published as Specifications for identity and purity of food additives, Vol. II. Food colours, Rome, Food and Agriculture Organization of the United Nations, 1963 (out of print).

Specifications for the identity and purity of food additives and their toxicological evaluation: emulsifiers, stabilizers, bleaching and maturing agents (Seventh report of the Joint FAO/WHO Expert Committee on Food Additives). WHO Technical Report Series, No. 281, 1964 (out of print). English, pp. 1-100; English, pp. 101-189;

Specifications for the identity and purity of food additives and their toxicological evaluation: food colours and some antimicrobials and antioxidants (Eighth report of the Joint FAO/WHO Expert Committee on Food Additives). WHO Technical Report Series, No. 309, 1965 (out of print).

Specifications for the identity and purity of food additives and their toxicological evaluation: some antimicrobials, antioxidants, emulsifiers, stabilizers, flour-treatment agents, acids, and bases (Ninth report of the Joint FAO/WHO Expert Committee on Food Additives). WHO Technical Report Series, No. 339, 1966 (out of print).

Specifications for the identity and purity of food additives and their toxicological evaluation: some emulsifiers and stabilizers and certain other substances (Tenth report of the Joint FAO/WHO Expert Committee on Food Additives). WHO Technical Report Series, No. 373, 1967 (out of print).

Specifications for the identity and purity of food additives and their toxicological evaluation: some flavouring substances and non-nutritive sweetening agents (Eleventh report of the Joint FAO/WHO Expert Committee on Food Additives). WHO Technical Report Series, No. 383, 1968 (out of print).

Specifications for the identity and purity of food additives and their toxicological evaluation: some antibiotics (Twelfth report of the Joint FAO/WHO Expert Committee on Food Additives). WHO Technical Report Series, No. 430, 1969 (out of print).

Specifications for the identity and purity of food additives and their toxicological evaluation: some food colours, emulsifiers, stabilizers, anticaking agents, and certain other substances (Thirteenth report of the Joint FAO/WHO Expert Committee on Food Additives). WHO Technical Report Series, No. 445, 1970 (out of print).

Toxicological evaluation of certain food additives with a review of general principles and of specifications (Seventeenth report of the Joint FAO/WHO Expert Committee on Food Additives). WHO Technical Report Series, No. 539, 1974, and corrigendum (out of print).

Turkoglu Si. fa (2009) Genotoxic effects of mono-, di-, and trisodium phosphate on mi¬totic activity, DNA content, and nuclear volume in Allium cepa L. CARYOLOGIA Vol. 62, no. 3: 171-179

U.S. Food and Drug Administration. (1993). Everything Added to Food in the United States. Boca Raton, FL: C.K. Smoley (c/o CRC Press, Inc.).

Wansu Park, Mun Seog Chang, Hocheol Kim, Ho Young Choi, Woong Mo Yang, Do Rim Kim, Eun Hwa Park and Seong Kyu Park (2008) Cytotoxic effect of gallic acid on testicular cell lines with increasing H_2O_2 level in GC-1 spg cells. Toxicology in Vitro Volume 22, Issue 1, Pages 159-163

Online References

http://www.ncbi.nlm.nih.gov/pubmed/17825405?ordinalpos=7&itool=EntrezSystem2.PEntrez.Pubmed.Pubmed_ResultsPanel.Pubmed_DefaultReportPanel.Pubmed_RVDocSum.

http://www.foodstandards.gov.au/newsroom/publications/choosingtherightstuff/.

http://www.webmd.com/diet/features/the-truth-about-seven-common-food-additives

http://www.sixwise.com/newsletters/06/04/05/12-dangerous-food-additives-the-dirty-dozen-food-additives-you-really-need-to-be-aware-of.htm

http://www.iso.org/iso/iso_catalogue/catalogue_ics/catalogue_ics_browse.htm?ICS1=67&ICS2=220.

2

The Safety Assessment of Food Additives by Reproductive and Developmental Toxicity Studies

Cansın Güngörmüş* and Aysun Kılıç*
Hacettepe University/Department of Biology, Ankara
Turkey

1. Introduction

The overall consumption of food additives is 139 lbs/year/person. If the common additives like spices, sugars, salt, honey, pepper, mustard, dextrose etc. are excluded, the consumption decreases to 5 lbs/year. Due to widespread consumption, it is necessary to evaluate the implications for the health of consumers because of the presence of newly synthesized food additives before commence production according to accepted guidelines such as Food and Drug Administration (FDA), U.S. Environmental Protection Agency (EPA) and European Food Safety Authority (EFSA). "Redbook 2000" is one of the revised form of Redbook II guideline published in 1993 by FDA; also defined as "Toxicological Principles for the Safety Assessment of Direct Food Additives and Color Additives Used in Food". This document is a guidance for determining toxicity studies, for designing and reporting the results of toxicity studies, conducting statistical analyses of data, the review of histological data and the submission of this information to FDA. The toxicological testing should provide not only information relevant to the average consumer, but also relevant to those population groups whose pattern of food consumption, physiological or health status may make them vulnerable such as young age, pregnancy and other metabolic disorders. Possible toxicological effects due to additive consumption should be tested especially in reproduction and developmental studies which are designed to evaluate effects on sexuality and fertility of males and females, developing organisms (mortality, structural abnormality and functional deficiencies). Besides, multigenerational reproductive toxicity studies provide information about the effects of a test substance on gonadal function, estrous cycle, mating behavior, lactation and development of the offspring.

For new food additives, a safety evaluation is obtained generally from experimental data derived from investigations in laboratory animals. Although it may be possible to use human data derived from medical use, occupational epidemiology or from volunteers, the obtained data would be limited. Therefore, the likely effects on man can be estimated by intentive extrapolation from laboratory animals. The end points and the indices obtained must provide sufficient information and statistical power to permit FDA to determine whether the additive is associated with changes in reproduction and fertility.

---

* Both authors have equal contribution in the chapter.

This chapter will highlight the possible health effects of newly synthesized food additives on market and focus on their reproductive and developmental toxicity perspectives. Considering that food technology is a complex area in which even the simplest additive interacts with all the others to produce qualified food; we are expecting that this chapter will help to examine food additives that can probably affect the human life.

2. Overview of food additives

Advances in food technology have resulted in an increased number of modified foods and additives in 20th century. An additive is a substance which may intentionally become a component of food or affect its characteristics. There are about 3000 different food additives defined up to date. Food additives may be divided in several groups; although there is some overlap between them. Main six categories of food additives are classified as preservatives, nutritional supplements, flavoring agents, colorings, texturing agents and miscellaneous. According to the functional classes, definitions and technological functions, food additives are summarized in Figure 1 (COABISCO, 2011).

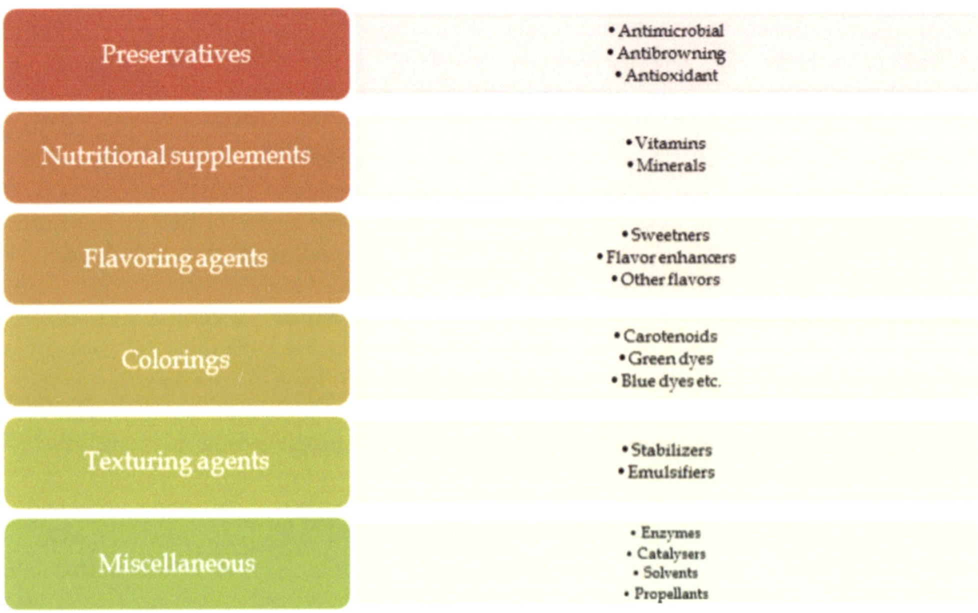

Fig. 1. Six main categories of food additives

Each food additive is assigned a unique E number, which have been assessed for use within the European Union (EU) to inform consumers (Figure 2). E numbers for European countries are all prefixed by "E"; on the other hand non-European countries do not use this prefix. E letter stands for the approval of the food additive in Europe. The numbering scheme fallows that of the International Numbering System (INS) as determined by Codex Alimentarius Committee (Codex Alimentarius, 2009). Though, only a subset of INS additives are approved for use in the EU. The United States Food and Drug Administration listed these items as "generally recognized as safe" or GRAS. As an example, additive E 341 (Tricalcium phosphate) is approved by US so has an "E" prefix and 341 numbering which stands for 340-349 subset known as "phosphates" under Antioxidants and Acidity Regulators group.

Fig. 2. Classification of additives by numeric range

In the EU, to authorize a substance as a food additive, a reasonable case of technological need, no hazard to consumers at level of proposed use and no misguidance to consumers should be demonstrated. To evaluate whether the newly released food additive has an effect on health; The European Commission is required a consultancy from Scientific Committee on Food (SCF). In this context, SCF deals with questions relating to the toxicology and hygiene in the entire food production chain for consumer health and food safety issues. For submission of a new food additive, the evaluation process by the SCF requires administrative, technical, toxicological data and references (European Commission Health & Consumer Protection Directorate, 2011). Among these, toxicological data obtained from experimental studies have a crucial role for consumers' health due to the presence of any additive in food.

During the general toxicological evaluation of food additives, the SCF first issued Guidelines for the Safety Assessment of Food Additives in 1980 (Scientific Committee for Food, 1980). However, new guidance documents have been published as Joint FAO/WHO Expert Committee on Food Additives (JECFA) (IPCS/JECFA, 1987), SCF guideline is still applicable. The aim of toxicological testing should provide sufficient information relevant to average consumer and vulnerable populations such as young age, pregnancy, diabetes, etc. The testing conditions depend on the chemical structure, proposed levels of use in food. The human data is derived from occupational epidemiology, medical use and volunteers but; for newly submitted food additives experimental data is commonly derived from laboratory animals. For evaluation of the safety of food additives, core studies are required such as; metabolism/toxicokinetics, subchronic toxicity, genotoxicity, chronic toxicity, carcinogenicity, reproduction and developmental toxicity. In this chapter, we are going to discuss the significant role of reproductive and developmental toxicity studies for evaluating new food additives.

3. Traditional and newly released food additives

"Toxicological Principles for the Safety Assessment of Food Ingredients" (Redbook 2000) is the new name of Redbook I which was previously published in draft form in 1983. Redbook 2000 provides information about toxicological data of food ingredients which are submitted to Center for Food Safety and Applied Nutrition and Office of Food Additive Safety for industry and other stakeholders. Food and color additives, food contact substances (which are also defined as indirect food additives) and substances classified as generally recognized as safe (GRAS) are the components of food ingredients (U.S. Food and Drug Administration, 2007).

At the end of the toxicological studies data derived from animal studies can be used to extrapolate to give information on human exposure. Therefore, defining the Acceptable Daily Intake (ADI), described as the dose level at which the additive causes effects on the health of the animals, is important. The highest level at which no adverse effect on the health of the animals is observed is called the NOAEL (No-Observed-Adverse-Effect-Level). An ADI is derived by dividing the NOAEL obtained from these studies, by an appropriate 'uncertainty' factor, which is intended to take account of differences between the animals on which the additive was tested and humans, in order to reduce further possibility of risk to humans. This uncertainty factor is commonly 100 (assuming that human beings are 10 times more sensitive than test animals and that the different levels of sensitivity within the human population is in a 10 fold range), but may be as much as 1,000 (if, for example, the toxic effect in animals is found to be particularly severe) or as low as 10 (where it has been found that humans are less likely than animals to be affected, based on actual data on the additive in humans) (Food Safety Authority of Ireland, 2011).

4. Assessment of potential reproductive and developmental toxicity of recently used food additives

Before a new food additive is introduced to the market, it should be tested if it causes any reproductive and developmental toxicity (EFSA, 2010a). To observe the potential effects, multigeneration reproduction studies have to be conducted. Laboratory species such as mouse, rabbit and especially rat are used at least for two generations and one litter per generation (Scientific Committee for Food, 1980). The test substance should be administrated in normal diet.

On the other hand, two laboratory species, usually a rodent and a non-rodent should be used in developmental toxicity studies. The test substance should either be in normal diet or administered by oral gavage during whole gestation period in order to detect the potential toxicological effects. In addition to a multigenerational and/or developmental toxicity study; in order to provide the possible effects after postnatal development and function (such as neurological function and behavior), examinations should be continued from the beginning of embryogenesis through to weaning.

4.1 Reproductive toxicity studies

The aim of a reproductive toxicity study is to ensure data about effects on the sexuality and fertility of males and females. These include reproductive behavior, pregnancy carriage ability, pre-postnatal survival rate, reproductive ability-capacity of the offspring and to examine major target organs for toxicity including reproductive organs in both parents and offspring histopathologically (Scientific Committee for Food, 1980). Besides, multigeneration reproductive toxicity studies ensure information about the effects on gonadal function, estrous cycles, conception, parturition, lactation (Joint FAO/WHO Expert Committee, 2000).

Reproductive studies constitutively target multigenerational studies as a result of human exposure to most food additives and preservatives during the whole lifetime. Studies performed with multigeneration enable researchers to detect any potential effect of a specific additive on each litter per generation. The administration of the test substance to parental and offspring generations should be continuous via the diet.

The end point evaluated in the indices calculated must provide sufficient information and statistical power to permit FDA whether the additive has effects on reproduction and fertility (U.S. Food and Drug Administration, 2007). The minimal reproduction study should consist at least two generations with one litter per generation. The reproduction study generation number should be expanded if the developmental toxicity effect of food additive is observed. A brief summary of the recommended reproduction study design according to FDA was given in Figure 3.

In a basic two generation reproduction and teratology study, the first step is to find the appropriate dose range of the compound in order to conduct the main study. The second step, includes the selection of experimental animal species due to its' life span, body size, breeding conditions, gestation length, high fertility rate etc. 5-9 weeks of aged animals, preferably rats are typically chosen. Each test and control group should include approximately 20 males and 20 pregnant females with uniform weight and age. For the detection of dose-related responses, minimum three doses of the test substance should be used: high, intermediate and low doses. The administration of the test substance may be via diet, drinking water or by gavage. In the first parental group (F_0), males should be administered for the duration of spermatogenesis and epididymal transit before and throughout the mating period. First parental females should be administered for the same length of time as males and through pregnancy to the weaning of the F_{1a} litter. Litters should be exposed throughout their entire lives. A female should be mated with a single randomly selected male from the same dose group until the pregnancy occurs or three weeks have elapsed. Each morning all females should be examined for the presence of vaginal plug or sperm in the vaginal lavage which is considered as "day zero" of gestation. Each animal should be observed twice each day at predefined time intervals. All animals should be weighted before administration, once weekly thereafter and at necropsy. During necropsy, the organs of reproductive system belonging to weanlings and parental animals (males-females) should be examined histopathologically. Uterus and ovaries for females; testis, seminal vesicle, prostate, epididymis for males should be weighted and evaluated separately. Brain, thymus and spleen tissues should also be examined for weanlings. Organ weights should be recorded both as absolute and relative weights. Indices which are the animal number responding to the test substance during conception until weaning period should be calculated for each reproduction study. To evaluate the endpoints of male reproductive toxicity, apart from counting testicular spermatid numbers, minimum 200 sperm per sample from cauda epididymis or proximal vas deferens should be examined. Acquired data from control and test groups of animals should be compared statistically using suitable statistics program.

Fig. 3. A reproduction study design by FDA

According to the recommendations of FDA (U.S. Food and Drug Administration, 2007), a toxicity study should be conducted according to Good Laboratory Practice Regulations. The animals used in the study should be well cared and housed according to the recommendations of Guide for the Care and the Use of Laboratory Animals. All test animals should be categorized according to their species, strain, sex and weight or age.

4.2 Pre- and post-natal developmental toxicity studies

Developmental toxicity studies assess the effects of the test substance on the developing organism including the death, structural abnormalities, altered growth and functional deficiencies (Joint FAO/WHO Expert Committee, 2000).

Lethal, teratogenic and other toxic effects on the embryo and fetus are examined under the consideration of prenatal developmental toxicity studies (Scientific Committee for Food,

1980). Besides, postnatal developmental toxicity studies deal with the observed side effects caused by maternal or maternal milk way exposure.

Embryonic and fetal resorptions, death, fetal weight, sex ratio, external and visceral skeletal morphology are in the context of prenatal developmental toxicity. The major objectives of postnatal developmental studies are physical, functional and behavioral development in animals exposed from embryogenesis through to weaning. These tests include neurological function, behavior during both early postnatal and adulthood phases, the measurement of vaginal opening in female pups, etc.

4.3 Current studies about recently used food additives

All food additives prior to their authorization should be evaluated for their safety by SCF or EFSA. According to the international legislations, all food additives must be kept under continuous observations and must be re-evaluated in the light of new scientific techniques. This part will be an overview of commonly used food additives listed below:

4.3.1 Lutein (E 161b)

Food colors are the first evaluated additives whose data are old. As soon as new available studies are conducted, the results should be renewed with the old ones. This is the reason of the evaluation priority of food color additives.

Lutein (E 161b) is a natural carotenoid dye which is approved as a food additive by the EU. In a 90 day rat study, according to NOAEL (No Observed Adverse Effect Level) of 200 mg/kg bw/day, which was defined by EFSA Panel, there were no developmental toxicity effects observed at dose levels up to 1000 mg/kg bw/day, the highest dose tested. Additionally, available reported data showed no effects on reproductive organs in oral 90-day studies. Due to the fact that lutein is a normal constituent of diet, ADI can be altered. Lutein is found as a non-genotoxic additive; but the absence of multigeneration reproductive toxicity and chronic toxicity/carcinogenicity studies caused EFSA to get the decision for ADI of lutein as 1 mg/kg bw/day with the uncertainty factor of 200 (EFSA, 2010a).

4.3.2 Caramel colours (E 150a,b,c,d)

Caramel colours are colouring substances approved as food additives according to the reactants used in their manufacture. E 150a is Class I Plain Caramel or Caustic Caramel, E 150b is Class II Caustic Sulphite Caramel, E 150c is Class III Ammonia Caramel and E 150d is Class IV Sulphite Ammonnia Caramel. EFSA determined the NOAEL as 30 g/kg bw/day with the uncertainty factor of 100 and ADI as 300 mg/kg bw/day for caramel colours. Up to date, there are no reproductive and/or developmental study of Class I and II caramels. JECFA states that there are only three developmental studies on Class III caramels. For Class III and IV caramels, studies with pregnant CD1 mice showed that, after treatment on days 6-15 of gestation, there were no effects on the number of implantation sites and resorption numbers, maternal-fetal survival and fetal skeletal defects (Morgareidge, 1974a; 1974b). In other studies conducted with pregnant Wistar rats and Dutch-belted rabbits, no treatment-related effects of Class III caramels were seen in the dams and fetal parameters (Morgareidge, 1974a). In the studies which were conducted on Wistar rats for Class IV caramels, no adverse effect was seen on female fertility, litter size, number of implantation sites or sex ratio of the pups (Til and Spanjers, 1973). Available reproduction and developmental studies, although limited, do not reveal any effects of concern. The studies also did not reveal any effects on reproductive organs. Lymphocytopenia is the only concern reported in short term studies with caramel colours (EFSA, 2010b).

4.3.3 Erythrosine (E 127)

Erythrosine is a xanthene-dye which was evaluated by JECFA with the ADI of 0-0.1 mg/kg bw/day. It is used especially for cocktail and candied cherries. In rats (Collins et al., 1993a; 1993b) and rabbits (Burnett et al., 1974) exposed to erythrosine by gavage or in drinking water, on day 0-19 of gestation, it was found neither fetotoxic nor teratogenic. Vivekanandhi et al. conducted a study with Swiss male Albino mice in which 64, 128 and 256 mg/kg bw/day erythrosine were administered daily by gavage resulted in decreased sperm motility and increased sperm abnormalities in dose dependent manner (Vivekanandhi et al., 2006).

4.3.4 Green S (E 142)

Green S is a triarylmethane dye authorized as a food additive in the EU. JECFA has previously established an ADI of 25 mg/kg bw/day in 1975; however, this was re-evaluated and later considered to be 5 mg/kg bw/day by SCF. In a previous study established by rats, the NOAEL of Green S was derived as 500 mg/kg bw/day, based on the results as increased spleen and kidney weight. At the high dose treatment (1000 mg/kg bw/day), there was evidence that the amniotic membranes had a green colouring. However, intra-uterine and post-natal development was not affected. Within the light of these results, the NOAEL is concluded as 1000 mg/kg bw/day for fetal development. In an additional study, 15 male and 15 female Wistar rats were administrated Green S in the diet at dose levels of 250, 500 and 1500 mg/kg bw/day for 13 weeks. When compared with controls, increases in mean body weight were observed in all treatment groups (BIBRA, 1978; Clode et al., 1987). Also, adequate reproduction and embryotoxicity studies including teratology are still requested (EFSA, 2010c).

4.3.5 Amaranth (E 123)

Amaranth (E 123) is an azo dye approved as a food additive in the EU. The ADI of Amaranth was established as 0-0.5 and 0.08 mg/kg bw/day by JECFA and SCF respectively. There are several studies which examined the reproductive and developmental toxicity of Amaranth. When all of these studies are taken into account, NOAELs for Amaranth is as follows: mouse 100 mg/kg bw/day, rat 15 mg/kg bw/day and rabbit 15 mg/kg bw/day (EFSA, 2010d). According to the results of Shtenberg and Gavrilenko, oral Amaranth exposure at doses of 1.5 and 15 mg/kg bw/day for 12-14 months in parental generation causes a significantly higher percentage of unsuccessful pregnancies with no live born born pups and increases percentages of stillborns in rats (Shtenberg and Gavrilenko, 1970). In the study, 1.5 and 15 mg/kg bw/day administration of Amaranth in drinking water was reported to cause increased rate of post-implantation death with increased mortality at post-partum in both treatment groups and a higher incidence of stillbirth at 15 mg/kg bw/day group rats. Khera et al. administrated Amaranth to Wistar rats at dose levels of 15, 30, 100, 200 mg/kg bw/day on Days 0-18 of gestation, either by gavage or in diet (Khera et al., 1974). According to the results, all dams were killed on day 19. The Panel considered that the NOAEL of this study is 200 mg/kg bw/day Amaranth, which is the highest dose tested.

4.3.6 Brilliant Blue FCF (E 133)

Brilliant Blue FCF (E 133) is another commonly used triarylmethane dye which is authorized as a food additive by the EU. Both JECFA and SCF established an ADI of 12.5 mg/kg bw/day. Depending on the newly conducted long-term studies, the ADI was revised to 10 mg/kg bw/day by SCI in 1984. The NOAEL was assigned as 2500 mg/kg bw/day by JECFA with uncertainty factor of 200. Afterwards, SCF established Brilliant Blue NOAEL as 1073 mg/kg bw/day in male and female rats with uncertainty factor of 100 (EFSA, 2010e). Among few chronic toxicity studies, the lowest NOAEL came from the most recent toxicity study (IRCD, 1981; Borzelleca, 1990). The Panel agreed with the authors and considered that the new NOAEL as 631 mg/kg bw/day. With uncertainty factor 100, the new established ADI for Brilliant Blue is 6 mg/kg bw/day. Data available up to date shows Brilliant Blue is poorly absorbed by the body and it is mainly excreted as unchanged in faeces. In addition, there have been several rat studies up to date; however in none of the studies treatment-related abnormalities were observed.

4.3.7 Curcumin (E 100)

Curcumin (E 100) is a dicinnamoylmethane dye consisting of three principal colouring components. It is also approved by the EU for the use as a food additive. JECFA allocated an ADI dose of 0-3 mg/kg bw/day and a NOAEL dose of 250-320 mg/kg bw/day with uncertainty factor of 100 depending on the results of a multigeneration study which is conducted by Ganiger et al. In the study, rats were fed with Curcumin for 24 weeks at doses of 250-320 mg/kg bw/day and 960-100 mg/kg bw/day (Ganiger et al., 2007). In the high dose group, there was a decrease in body weight gain. Garg conducted a multigeneration study in Wistar rats according to OECD Testing guideline administering curcumin (Garg, 1974). Rats were fed with diets containing 0, 1500, 3000, and 10000 mg/kg bw/day curcumin. At the end of the study, it was reported that there was a dose-related decrease on body weight gain in the dams of the parental generation during days 10-15 of gestation. However, no other effects were observed. According to its' chemical composition, Curcumin is a rapidly metabolized dye which is later excreted with faeces (EFSA, 2010f).

4.3.8 Canthaxanthin (E 161 g)

Canthaxanthin is a carotenoid pigment which is authorized by the EU as a food additive. It is mainly composed of all-*trans* β-carotene-4,4'-dione with other minor carotenoids. The ADI dose of canthaxanthin is established as 0.03 mg/kg bw/day. There is no data reporting adverse effects of xanthaxanthin on reproductive system or on the developing fetus in high doses up to 1000mg/kg bw/day in rats and in high doses up to 400 mg/kg bw/day in rabbits. Hoffmann-La Roche reported that there were no adverse effects on fertility, litter size, the number of young weaned and their weights after 0 (placebo) and 0.1% canthaxanthin exposure in rats (Hoffmann - La, 1990). In a three-generation reproduction study by Buser, male and female rats were fed with diet including 0, 250, 500 and 1000 mg/kg bw/day canthaxanthin (Buser, 1987). The results indicated that there were no treatment-related effects in reproductive system. The EFSA Panel evaluated the highest dose tested of 1000 mg/kg bw/day as NOAEL for reproductive system, embryotoxicity and teratogenicity (EFSA, 2010g).

4.3.9 Aspartame

Nowadays, aspartame (APM) is the most commonly used artificial sweetener in the world (Hazardous Substances Data Bank, 2005). APM is approved by both FDA and the EU for the use in all foods (FDA, 1996; EFSA, 2006h). According to long term studies, ADI of APM is 2.5-5 mg/kg bw/day (Butchko et al., 2002). Long term carcinogenicity bioassays performed on rat and mice indicated that APM is a high effective carcinogenic additive causing lymphomas, leukemias and neoplastic lesions in females and Schwannomas in males (Soffritti et al., 2006). However, recently conducted lifespan studies with Swiss mice at dose levels of 0, 2000, 8000, 16000, 32000 ppm resulted that it does not affect the daily feed consumptions, mean body weights and the survival of males-females (Soffritti et al., 2010).

4.3.10 Paraben

Parabens have wide range of use in different industrial areas. One of them is its' use in food ingredients as an anti-microbial agent (Hossani et al., 2000). Parabens are lately reported to act as xenoestrogens which are a class of endocrine disruptors due to the lengths of their alkyl side chains (Okubo et al., 2001). It is known that parabens have an effect on reproductive tissues, induce aberrant estrogenic signaling in cells, cause changes in the expression patterns of multiple genes in rat fetal reproductive system (Naciff et al., 2003). Despite the potential health effects, parabens are approved as food additives by the EU with ADI dose of 0-10 mg/kg bw/day (Ishiwatari et al., 2007) and with NOAEL as 1000 mg/kg bw/day (Boberg et al., 2010). In a developmental study conducted by Thuy et. al, during juvenile-peripubertal period, female rats were administered with methyl-, ethyl-, propyl-, isopropyl-, butyl- and isobutylparabens at doses of 62.5, 250, 1000 mg/kg bw/day (Thuy et al., 2010). Their results showed that in the highest dose group there was a significant delay in the date of vaginal opening, a decrease in length of estrous cycle, morphological changes in the uterus and increased number of cystic follicles in ovaries. Additionally, after 10, 100, 1000 mg/kg bw/day polyparaben treatment, decreases in daily testis sperm production and in serum testosterone levels in a dose-dependent manner were observed in all doses (Oishi, 2002).

4.3.11 Coriander essential oil

Coriander essential oil is obtained by steam distillation of the dried fruits (seeds) of Coriandrum sativum L. In the food industry, coriander oil is used as a flavoring agent and adjuvant. Coriander oil is both approved for the use as food additive by FDA and The EU (Vollmuth et al., 1990). While maternal NOAEL of coriander oil was determined as 250 mg/kg bw/day, developmental NOAEL was established as 500 mg/kg bw/day (FFHPVC, 2002). Vollmuth et al. administered 250, 500, 100 mg/kg bw/day coriander oil to pregnant pregnant Crl CD rats (7 day before cohabitation, during gestation, 4 day post-parturition). At the highest dose significant decreases in gestation index, length of gestation, viability of pups and litter size were noted.

4.3.12 Allyl isothiocyanate (AITC)

Allyl isothiocyanate (AITC) is used both as a food additive and a flavouring agent. It can occur naturally in certain vegetables such as cabbage, mustard and horseradish (EFSA 2010i). EFSA Panel (2010) established the ADI dose as 0.02 mg/kw bw/day with an uncertainty factor of 500 (EFSA, 2010i). The Panel regarded a Low Observed Adverse Effect Level (LOAEL) as 12 mg/kw bw/day rather than the NOAEL in order to cover uncertainties resulting from extrapolation. Oral doses of AITC up to 18.5, 23.8 and 12.3 mg/kg bw/day did not cause any developmental toxicity in pregnant rats, hamsters and rabbits and may be fetotoxic to mouse at doses higher than 6 mg/kg bw/day without any teratogenic effects (EFSA, 2010i).

4.3.13 Tricalcium phosphate (E 341)

Tricalcium phosphate (E 341) is a commonly used flavor preservative, anti-caking, stabilizing and anti-souring food additive. JECFA reported the ADI dose of E 341 as 70 mg/kg bw/day (JECFA, 2001). In a recently published study, Wistar rats were treated with 175 and 350 mg/kg bw/day E 341 during gestation days 0-20. Decrease in placental weights and skeletal morphometry in fetus were observed in both doses (Güngörmüş et al., 2010). There was also a decrease in trans-umblical cord lengths in treatment group. According to these results, it was concluded that rat prenatal development during gestation is sensitive to E 341 exposure.

5. Alternative methods

To support and explore in more depth of results obtained from fundamental studies; immunotoxicity, allergenicity, neurotoxicity, genotoxicity, human volunteer studies, predictive mechanistic and special studies may also be helpful. Another recent method for developmental toxicity testing is *in vitro* studies, which are not based on the use of animals. *In vitro* studies provide sufficient data by using cellular and subcellular systems to predict the mechanisms involved in early stage of development. As a result of ethical issues about animal use, *in vitro* testing mechanisms will come into prominence in upcoming years. Over the past few decades, one of the most notable revolutionized technologies is the application of nanotechnology in food sector. Therefore, in this part, concerns and health implications on the application of nanotechnology in food will be discussed. The possible effects on reproductive system due to consumption of foods involving nanoparticles are evaluated.

5.1 Approaches of *in vitro* studies for assessing food additives

Commonly, the toxicological risk to humans from exposure to an individual chemical is evaluated using animal data from long-term or acute *in vivo* toxicity studies. Over the last 20 years, there has been a clear tendency for increased use of *in vitro* methods in toxicology as supplements to animal tests. Although such studies have previously been considered during the hazard characterization of many compounds they generally have had no direct influence on the calculation of ADI values. *In vitro* studies may be a useful perspective in bridging the gap between a test species and the human situation, thereby providing a more scientific basis for the use of a specific data. *In vitro* testing systems are increasingly becoming an essential tools as part of integrated toxicology testing strategy and scientific progress in the fields of cellular and molecular biology. These studies are used in a wide range of processes including the determination of ADI, for reporting suggestions in safety approaches, metabolism pathways of specific compounds. According to the limited number of toxicological studies, *in vitro* applications are useful for the prospective toxicological classification and characterization of food additives. *In vitro* studies have the potential for calculation and detection of both inter-species and inter-individual variability in toxicokinetics and toxicodynamics of food additives (Walton, 1999).

5.2 Nanoparticle based applications of food additives in food industry

Public interest in the subject of nanotechnology in the food industry is growing. It opens up many new possibilities which are of interest to the food industry. Nanofood market potential was predicted as 20.4 billion US dollars for the year 2010 (IFST, 2006). More than 200 companies worldwide are already believed to be involved in this sector, especially in the USA, Japan and China. Several companies are investigating encapsulation technology for the delivery of active ingredients in food products (e.g. flavouring agents, vitamins, fatty acids). Nanotechnology in the food industry is a sensitive subject. Manufacturers fear a blanket rejection of products containing nanomaterials, similar to what has happened with genetic engineering. Food products naturally contain nano-sized ingredients. These are different from synthetically manufactured nanomaterials. Food proteins can be mentioned as examples of natural nanostructures whose size can vary between several hundred nanometers such as milk proteins and casein.

With encapsulation (Figure 4), in which active agents and substances can be encapsulated in nanostructured materials, the purpose is to enhance solubility (e.g. of colouring agents), facilitate controlled release (e.g. only in certain parts of the alimentary tract, for instance in order to prevent the bad taste of an ingredient which in itself is beneficial such as omega-3 fatty acids in fish oils), improve bioavailability, i.e. the amount of a nutritional ingredient which is actually absorbed by the body (e.g. vitamins, minerals), protect micronutrients and bioactive compounds during manufacture, storage and retail. The most important nanostructured materials are currently nano-capsules (micelles, liposomes) and nanoemulsions (Greßler et al., 2010). Nanocarrier systems can be used to mask the unpleasant tastes and flavours of ingredients and additives such as fish oils, to protect the encapsulated ingredients from degradation during processing and storage, as well as to improve dispersion of water-insoluble food ingredients. However, current studies on the application of nanoencapsulation mainly address its potential for target delivery of active ingredients of functional food and nutraceuticals (Hsieh and Ofori, 2007).

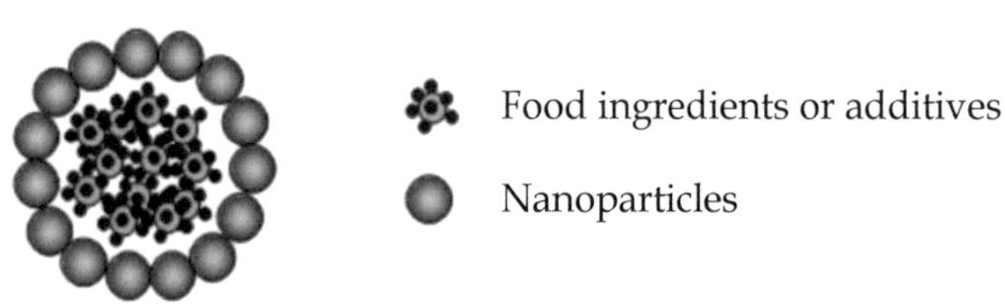

Fig. 4. Schematic diagram of nanoencapsulation (Centre for Food Safety, Hong Kong, 2010)

Experimental data demonstrated that the distribution of nanoparticles after oral administration is dependent upon particle size. Smaller-sized nanoparticles have a more widespread tissue distribution in organs like kidney, liver, lungs and brain while the bigger particles (28 nm and 58 nm) remain almost solely inside the gastrointestinal tract (The Government of the Hong Kong Centre for Food Safety Food and Environmental Hygiene Department, 2010). Studies have been performed on the ability of nanoparticles to penetrate the placental barrier. There is also information that certain nanomaterial (C60 fullerene) can pass across the placenta. However, due to the inconsistent results of some *in vitro* and animal studies, no general conclusion on the penetration power of nanoparticles across the placental barrier can be made. There is no information on whether nanomaterials are transferred into milk (Tsuchiya et al., 1996; EFSA, 2010j). Because nanoparticle food industry is a recently developing field, the reproductive and developmental toxicity studies are rare. In one of the few studies conducted by Durnev et al., silicon crystal 2-5 nm nanoparticles in the form of 1-5 µ granules in water suspension were injected intraperitoneally in a single dose to male F_1(CBA×C57Bl/6) mice or to outbred albino rats on days 1, 7, and 14 of gestation (Durnev et al., 2010). It was reported that injection of 50 mg/kg dose of silicon crystal nanoparticles reduced body weight gain in pregnant rats and newborn rats at different stages of the experiment, but had no effect on other parameters of physical development of rat progeny and caused no teratogenic effects. In a recent study it was also reported that nanosized silicon materials are generally nontoxic and biodegradable (Fucikova et al., 2011). Nowadays, in comparison to other materials currently used in medicine applications, nanosized Si based materials are the only one showing complete biodegrability and nontoxicity without any significant inflammatory reactions.

6. Discussion

Food additives are natural or manufactured substances, which are added to foods for restoring colors lost during processing, providing sweetness, preventing deterioration during storage and guarding with preservatives against food poisoning. A food additive is defined as a substance not normally consumed as a food in itself and not normally used as a characteristic ingredient of food whether or not it has nutritive value, the intentional addition of which to food for a technological purpose in the manufacture, processing, preparation, treatment, packaging, transport or storage of such food results, or may be reasonably expected to result, in it or its by-products becoming directly or indirectly a component of such foods (Food Safety Authority of Ireland, 2011).

All food additives undergo a safety assessment that may be used in the manufacture or preparation of foodstuffs in the European Union. Up to 2002, this safety assessment was carried out by the EU SCF but since 2003, the responsibilities of the SCF have been taken by EFSA.

The safety evaluation of a food additive involves examination of the chemical structure and characteristics, including its specifications, its impurities and potential breakdown products. Toxicological data is essential to identify and characterize the possible health hazards of an additive and to allow extrapolation of the findings in animals and other test systems to humans. In these studies, the additive is administered to laboratory animals.

Such tests are designed to give information on any possible effects from short-term or long-term exposure to the additive, including whether it may have any potential to cause cancer (carcinogenicity), or to affect male or female reproduction or the development of the embryo or the fetus if consumed by a pregnant woman (reproductive or developmental toxicity). Other effects include the genotoxicity potential of the compound; which is the ability to cause the development of cancer or adverse effects in future generations.

This chapter is an overview of the developmental and reproductive studies conducted by the administration of commonly used food additives. Recent studies show that there is a lot of concern about the safety of food additives in toxicological manner. With the increasing amount of progression in food additive industry, people who need more information about popular additives admit to being confused with the possible health effects. There are both safety evaluations and regulations of newly released food additives. As a matter of fact, it is inevitable to use food additives with the increasing demand of high-quality food. Excessive parental exposure of food additive throughout lifespans makes reproductive and developmental endpoints remarkable for investigating. This chapter strived to focus on the fine prints of toxicological assessments, especially the possible reproductive and developmental effects of food additives and to give perspectives for new approaches in the evaluations of food additives in concordance with the improvement of food industry.

7. Acknowledgements

The authors would like to thank Prof. Dr. Dürdane Kolankaya and Prof. Dr. M. Turan Akay for their improving suggestions and advices as supervisors.

8. References

British Industrial Biological Research Association (BIBRA). (1978). Report No 196/1/78. Short term toxicity of Green S in rats. Clode, SA, Gaunt, IF, Hendy, RJ, Cottrell, RC, Gangolli, SD.

Boberg, J., Camilla, T., Christiansen, S. & Ulla., Hass. (2010). Possible endocrine disrupting effects of parabens and their metabolites. *Reproductive Toxicology,* Vol.30, pp.301-312.

Borzelleca, J.F., Depukat, K. & Hallagan, J.B. (1990). Lifetime toxicity/carcinogenicity studies of FD & C Blue No. 1 (Brilliant blue FCF) in rats and mice. *Food and Chemical Toxicology.* Vol.28, pp.221-234.

Burnett, C.M., Agersborg, H., Borzelleca, J., Eagle, E., Ebert, A., Pierce, E., Kirschman, J. & Scala, R. (1974). Teratogenic studies with certified colors in rats and rabbits. *Toxicology and Applied Pharmacology.* Vol.25, pp.121-128.

Buser, S. (1987). Canthaxanthin in a three-generation study in rats. Unpublished Report No. HLR 138/86755 of Huntingdon Research Centre Ltd. Submitted by Hoffmann-La Roche & Co., Basel, Switzerland.

Butchko, H.H., Stargel, W.W., Comer, C.P., Mayhew, D.A., Benninger, C., Blackburn, G.L., De Sonneville, L.M.J., Geha, R.F., Herteley, Z., Kostner, A., Leon, A.S., Liepa, G.U., McMartin, K.E., Mendnhall, C.L., Munro, I.C., Novotny, E.J., Renwick, A., Gi

Schiffman, S.S., Schomer, D.L., Shaywitz, B.A., Spiers, P.A., Tephly, J.A. & Trefz, F.K. (2002). Intake of aspartame vs the acceptable daily intake. *Regulative Toxicology and Pharmacology*. Vol.35, pp.13-16.

CAOBISCO Website. (2011). www.caobisco.com/doc_uploads/legislation/XOT4-EN.pdf, Belgium.

Clode, S.A., Gaunt, I.F., Hendy, R.J., Contrell, R.C. & Gangolli, S.D. (1987). Short-term toxicity of Green S in rats. *Food and Chemical Toxicology*. Vol.25, pp.969-975.

Codex Alimentarius (2009). "Noms de Categorie et Systeme International de Numerotation des Additifs Alimentaries"

Collins, T.F., Black, T.N., O'Donnell, M.W.J., Shackelford, M.E. & Bulhack, P. (1993a). Teratogenic potential of FD & C Red no. 3 when given in drinking water. *Food and Cosmetics Toxicology*. Vol.31, pp.161-167.

Collins, T.F., Black, T.N. & Ruggles, D.I. (1993b). Teratogenic potential of FD&C Red No. 3 when given by gavage. *Toxicology and Industrial Health*. Vol.9, No.4, pp.605-16.

Durnev, A.D., Solomina, A.S., Daugel-Dauge, N.O., Zhanataev, A.K., Shreder, E.D., Nemova, E.P., Shreder, O.V., Veligura, V.A., Osminkina, L.A., Timoshenko, V.Y. & Seredenin, S.B. (2010) Evaluation of Genotoxicity and Reproductive Toxicity of Silicon Nanocrystals. *Bulletin of Experimental Biology and Medicine*. Vol.149, No.4, pp.445-449.

EFSA. (2010). Scientific Opinion on the safety of allyl isothiocyanate for the proposed uses as a food additive. *EFSA Journal*. Vol.8, No.12, pp.1943.

EFSA. (2010). Scientific Opinion on the re-evaluation of Amaranth (E 123) as a food additive. *EFSA Journal*. Vol.8, No.7, pp.1649.

EFSA. (2010). Scientific Opinion on the re-evaluation of Brilliant Blue FCF (E 133) as a food additive. *EFSA Journal*. Vol.8, No.11, pp.1853.

EFSA. (2010). Scientific Opinion on the re-evaluation of Canthaxanthin (E 161 g) as a food additive. *EFSA Journal*. Vol.8, No.10, pp.1852.

EFSA Scientific opinion on the re-evaluation of Caramel colours (E 150 a,b,c,d) as food additives (2011) EFSA Journal, Vol.9, No.3, pp.2004.

EFSA. (2010). Scientific Opinion on the re-evaluation of Curcumin (E 100) as a food additive. *EFSA Journal*. Vol.8, No.9, pp.1679.

EFSA. (2010). Scientific Opinion on the re-evaluation of Green S (E 142) as a food additive. *EFSA Journal*. Vol.8, No.11, pp.1851.

EFSA. (2010). Scientific opinion on the re-evaluation of Lutein (E161b) as a food additive. *EFSA Journal*. Vol.8, No.7, pp.1678.

EFSA. (2009). Scientific Opinion of the Scientific Committee on the Potential Risks Arising from Nanoscience and Nanotechnologies on Food and Feed Safety.

European Commission Health & Consumer Protection Directorate-General. (2011). Guidance on Submissions for Food Additive Evaluations by the Scientific Committee on Food.

European Food Safety Authority (EFSA). (2006). Opinion of the Scientific Panel on Food Additives, Flavourings, Processing Aids and Materials in contact with Food (AFC) on a request from the Commission related to a new long-term carcinogenicity study on aspartame. *EFSA Journal*. Vol.356, pp.1-44.

FFHPVC. (2002). The terpene consortium test plan for terpenoid tertiary alcohols and related esters. In: Report Submitted to the EPA Under the HPV Challenge Program, The Flavor and Fragrance High Production Volume Consortia, Washington, DC.

Food and Drug Administration (FDA). (1996). Food additives permitted for direct addition to food for human consumption: Aspartame. *Federal Regulations.* Vol.61, pp.33654-33656.

Food Safety Authority of Ireland. (2011) Guidance on Food Additives, Revision 1.

Fucikova, A., Valenta, J., Pelant, I., Kusova, K. & Brezina, V. (2011). Nanocrystalline silicon in biological studies. *Physica Status Solidi.* Vol.8, No.3, pp.1093–1096.

Ganiger, S., Malleshappa, H.N., Krishnappa, H., Rajahekhar, G.V., Ramakrishna, R. & Sullivan, F. (2007). A two generation reproductive toxicity study with curcumin, turmeric yellow, in Wistar rats. *Food and Chemical Toxicology.* Vol.45, pp.64-69.

Garg, S.K. (1974) Effects of Curcuma longa on fertility. *Planta medica.* Vol.26, pp.225-227.

Greßler, S., Gazsó, A., Simkó, M., Nentwich, M., Fiedeler, U. (2010). Nanoparticles and nanostructured materials in the food industry. *Nano Trust Dossiers.* No.004en.

Güngörmüş, C., Kılıç, A., Akay, M.T. & Kolankaya, D. (2010). The effects of maternal exposure to food additive E341 (tricalcium phosphate) on foetal development of rats. *Environmental Toxicology and Pharmacology.* Vol.29, pp.111–116.

Hazardous Substances Data Bank. (2005). TOXNET: Toxicological Data Network. Available: http://www.toxnet.net.nlm.nih.gov

Hoffmann-La, R. (1990). Canthaxanthin. Unpublished report submitted to WHO by F. Hoffmann - La Roche & Co., Basle, Switzerland (as referred to by JECFA).

Hossaini, A., Larsen, J.J. & Larsen, J.C. (2000) Lack of oestrogenic effects of food preservatives (parabens) in uterotrophic assays. *Food and Chemical Toxicology.* Vol.38, pp.319–23.

Hsieh, Y.H.P. & Ofori, J.A. (2007). Innovation in food technology for health. *Asia Pacific Journal of Clinical Nutrition.* Vol.16, No.Supplemental1, pp.65-73.

Institute of Food Science & Technology (IFST). (2006) Nanotechnology Information Statement.

IPCS/JECFA. (1987) Environmental Health Criteria 70: Principles for the Safety Assessment of Food Additives and Contaminants in Food. World Health Organization, Geneva.

International Research & Development Corporation (IRCD). (1981). Longt-term dietary toxicity/carcinogenicity study in rats. Unpublished report.

Ishiwatari, S., Suzuki, T., Hitomi, T., Yoshino, T., Matsukuma, S. and Tsuji, T. (2007) Effects of methyl paraben on skin keratinocytes. *Journal of Applied Toxicology.* Vol.27, pp.1–9.

JECFA. (2001). http://www.inchem.org/documents/jecfa/jeceval/jec 2306.htm.

Joint FAO/WHO Expert Committee. (2000). Guidelines for the preparation of toxicological working papers on food additives. Geneva.

Khera, K.S., Przybylski, W. & McKilnley, W.P. (1974). Implantation and embryonic survival in rats treated with Amaranth during gestation. *Food and Cosmetics Toxicology.* Vol.12, pp.507-510.

Morgareidge, K. (1974a). Teratologic evaluation of FDA 71-82 (caramel, bakers and confectioners)in mice, rats and rabbits. Unpublished report No. PB 234-870 from

Food and Drug Research Laboratories Inc., Waverly, NY; USA (as referred to by JECFA, 1987).

Morgareidge, K., (1974b). Teratologic evaluation of FDA 71-83 (caramel beverage) in mice, rats and rabbits. Unpublished report No. PB 234-867 from Food and Drug Research Laboratories Inc., Waverly, NY; USA.

Naciff, J.M., Overmann, G.J., Torontali, S.M., Carr, G.J., Tiesman, J.P. & Richardson, B.D. (2003) Gene expression profile induced by 17 alpha-ethynyl estradiol in the prepubertal female reproductive system of the rat. *Toxicological Sciences* Vol.72, pp.314–30.

Oishi, S. (2002) Effects of propyl paraben on the male reproductive system. Food and Chemical Toxicology. Vol.40, No.12, pp.1807–1813.

Okubo, T., Yokoyama, Y., Kano, K. & Kano, I. (2001). ER-dependent estrogenic activity of parabens assessed by proliferation of human breast cancer MCF-7 cells and expression of ERalpha and PR. *Food and Chemical Toxicology*. Vol.39, pp.1225–1232.

Scientific Committee for Food. (1980). Guidelines for the Safety Assessment of Food Additives. Reports of the Scientific Committee for Food (Tenth series). Commission of the European Communities, Luxembourg EUR 6892.

Shtenberg, A.I. & Gavrilenko, E.V. (1970). Effects of Amaranth food dye on reproductive function and progeny development in experiments with albino rats. *Voprosy Pitaniia*. Vol.29, pp.66.

Soffritti, M., Belpoggi, F., Degli Esposti, D., Lambertini, L., Tibaldi, E. & Rigano, A. (2006). First experimental demonstration of the multipotential carcinogenic effects of aspartame administered in the feen to Spraguq-Dawley rats. *Environmental Health Perspective* Vol.114, pp.379-385.

Soffritti, M., Belpoggi, F., Manservigi, M., Tibaldi, E., Lauriola, M., Falcioni, L. & Bua, L. (2010). Aspartame administered in feed, beginning prenatally through life span, induces cancers of the live and lung in male swiss mice. *American Journal of Industrial Medicine*. Vol.53, No.12, pp.1197-1206.

The Government of the Hong Kong Centre for Food Safety Food and Environmental Hygiene Department Special Administrative Region. (2010). Nanotechnology and Food Safety, Risk Assessment Studies Report No. 41.

Thuy, T.B., Vo, Y.M.Y., Kyung-Chul, C. & Eui-Bae, J. (2010) Potential estrogenic effect(s) of parabens at the prepubertal stage of a postnatal female rat model. *Reproductive Toxicology*. Vol.29, pp.306–316.

Til, H.P. & Spanjers, M. (1973). Reproduction study in rats with six different Ammonia Caramels. Unpublished report No. R 4068 from Centraal Instuut voor Voedingsonderzoek (CIVO/TNO), Zeist, The Netherlands.

Tsuchiya, T., Oguri, I., Yamakoshi, Y.N. & Miyata, N. (1996). Novel harmful effects of fullerene on mouse embryos in vitro and in vivo. *FEBS Letters* Vol.393, No.1, pp.139-45.

U.S. Food and Drug Administration. (2007). Toxicological Principles for the Safety Assessment of Food Ingredients, Redbook 2000.

Vivekanandhi, J., Devi, C.P.A., Jayaraman, K. & Raghavan, L. (2006). Effect of Erythrosine on testicular function in mice. *Toxicology International*. Vol.13, pp.119-125.

Vollmuth, T.A., Bennett, M.B., Hoberman, A.M. & Christian, M.S. (1990). An evaluation of food flavoring ingredients using an in vivo reproductive and developmental toxicity screening test. *Teratology*. Vol.41, pp.597–598.

Walton, K., Walker, R., Van de Sandt, J.J.M., Castell, J.V., Knapp, A.G.A.A., Kozianowski, G., Roberfroid, M. & Schilter, B. (1999). The application of in vitro data in the derivation of the acceptable daily intake of food additives. *Food and Chemical Toxicology*. Vol.37, pp.1175-1197.

3

Production and Functional Properties of Dairy Products Containing Lactophorin and Lactadherin

Mizuho Inagaki[1], Xijier[2], Yoshitaka Nakamura[3], Takeshi Takahashi[3], Tomio Yabe[1,2], Toyoko Nakagomi[4], Osamu Nakagomi[4] and Yoshihiro Kanamaru[1,2]
[1]*Department of Applied Life Science, Gifu University*
[2]*United Graduate School of Agricultural science, Gifu University*
[3]*Food Science Institute, Division of Research and Development, Meiji Co., Ltd.*
[4]*Department of Molecular Microbiology and Immunology, Graduate School of Biomedical Sciences and Global Center of Excellence, Nagasaki University*
Japan

1. Introduction

In this chapter, we introduce the possible protective utilization of cow milk proteins, lactophorin (LP) and lactadherin (also known as periodic acid Schiff 6/7 (PAS6/7)), against human rotavirus (HRV) gastroenteritis.

Milk is the natural food of the newborn mammal, and it is endowed with protective components against pathogens, such as antibodies. Our previous studies have demonstrated that the 2 proteins, LP and lactadherin, exhibit potent inhibitory activity against HRV. HRV is the single most important etiologic agent of severe gastroenteritis in infants and young children. To determine whether cow's milk could serve as a protective food additive effective against HRV infection, this chapter discusses the potential utilizations of LP and lactadherin from normal cow's milk to protect against HRV gastroenteritis, focusing in particular on sweet whey, a byproduct of industrial-scale cheese manufacturing.

2. Rotavirus gastroenteritis

Infectious gastroenteritis is distinguished between bacterial and viral origin, depending on pathogenesis. Rotavirus, adenovirus, and norovirus are well-known infectious gastroenteritis pathogens of viral origin.

HRV was first discovered by Ruth Bishop et al. in 1973, and was recognized as a major cause of childhood diarrheal morbidity and mortality worldwide (Bishop et al., 1973; Bishop, 2009). The virus is transmitted by the fecal-oral route. It infects the enterocytes of the villi of the small intestine and causes gastroenteritis. The incubation period of rotavirus infection is 2-4 days, and once diarrhea occurs, recovery usually requires approximately 1 week. By the age of 5 years, nearly every child in the world has been infected with rotavirus at least once (Velázquez et al., 1996). The estimated annual incidence of rotavirus gastroenteritis is approximately 114 million episodes requiring home care and 600,000 deaths in children worldwide (Dennehy, 2008). More than 85% of these deaths occur in developing countries, South Asia, and sub-Saharan Africa (Naghipour et al., 2008; Centers for Disease Control and Prevention, 2011). In the absence of vaccination, rotavirus gastroenteritis has been estimated to cause 87,000 hospitalizations in Europe (Soriano-Gabarro et al., 2006), 55,000–70,000 hospitalizations in the USA (Parashar et al., 2006), and 78,000 hospitalizations in Japan (Nakagomi et al., 2005) among children below 5 years of age. Thus, rotavirus gastroenteritis causes large human costs in developing countries and large public medical burdens in developed countries.

In general, vaccination is the most effective method for protection against viral diseases. To reduce the aforementioned global burden posed by rotavirus gastroenteritis, the 2 oral rotavirus vaccines Rotarix® (GlaxoSmithKline Biologicals, Rixensart, Belgium) and RotaTeq® (Merck and Co., Whitehouse station, NJ) have been licensed for use in more than 100 countries worldwide (Tate et al., 2010). Large-scale trials in Europe and North and Latin America demonstrated that these vaccines are safe and effictive (Ciarlet & Schödel, 2009; O'Ryan & Linhares, 2009). Clinical studies are ongoing in Asia and Africa to assess the safety and efficacy of the vaccines in these populations (Zaman et al., 2010; Armah et al., 2010). However, to reduce the risk of intussusceptions, the first doses of both vaccines are strictly limited between the age of 6-15 weeks, and full doses of vaccines need to be completed by ages 6-8 months (Cortese et al., 2009). Therefore, prophylactic options against HRV infection are needed.

Young mammals depend on passive immunity obtained *via* breast-feeding for resistance against infectious diseases, because their immature immune systems cannot produce antibodies immediately after birth. The mother is able to produce antibodies against infectious agents, and they are passively transmitted to the offspring *via* milk.

It has been proposed that passive protection against HRV infection could be achieved by using immunoglobulin G (IgG) from the colostrum of cows hyper-immunized with rotavirus (Ebina et al., 1992; Sarker et al., 1998). Unfortunately, the clinical use of bovine colostrum from hyper-immunized cows has been limited because of difficulties in large-scale production. Recently, skimmed and concentrated bovine late colostrum (SCBLC) obtained from normal cows at 6-7 days after parturition exhibited high potency in inhibiting human rotaviral replication *in vitro* and *in vivo* (Inagaki et al., 2010a), indicating that SCBLC is likely to play an alternative role to colostrum of cows hyper-immunized with rotavirus.

Furthermore, studies of milk components exhibiting inhibitory activity against rotavirus have also been reported. For example, supplemental dietary whey protein concentrate (WPC) (Wolber et al., 2005; Pérez-Cano et al., 2008) and macromolecular bovine whey protein fraction (MMWP) (Kvistgaard et al., 2004; Bojsen et al., 2007) exhibited protective efficacy against simian rotavirus and murine rotavirus infection *in vivo*. The inhibitory mechanism of WPC remains to be elucidated, whereas Bojsen et al. found that mucin 1 and immunoglobulins were the major rotavirus inhibitors in MMWP (2007).

3. Inhibitory activity of LP and lactadherin against HRV infection

Recently, we identified LP16 (16 kDa LP fragment) and bovine lactadherin (PAS6/7) as human rotavirus inhibitors in bovine milk (Inagaki et al., 2010b). In this section, we will describe the anti-viral properties of these components.

3.1 Anti-HRV activity of LP

LP was initially found to be a glycoprotein in the heat- and acid-stable proteose peptone (PP) fraction and was referred to as PP component 3 (Girardet et al., 1996). LP is found in bovine, cameline (Girardet et al., 2000), caprine (Sørensen et al., 1997), and ovine milk (Sørensen et al., 1997), but not in human milk (Sørensen et al., 1997).

LP is present at an average concentration of 0.3 g/l in normal bovine milk (Koletzko et al., 2005). LP consists of 2 major glycopeptides; 28 kDa (LP28) and 18 kDa (LP18) (Girardet et al., 1996). LP28 contains 5 partial phosphorylation sites (Ser_{29}, Ser_{34}, Ser_{38}, Ser_{40}, and Ser_{46}), 3 *O*-glycosylation sites (Thr_{16}, Thr_{60}, (Kjeldsen et al., 2003), and Thr_{86}), and 1 *N*-glycosylation site (Asn_{77}) (Girardet & Linden, 1996). It exists in various molecular forms formed via posttranslational modification (Kanno, 1989a, 1989b). LP18 has an amino acid sequence corresponding to the 54-135 C-terminal portion of LP28, and this sequence is thought to occur as a proteolytic degradation product of LP28 (Girardet & Linden, 1996). Thus far, emulsification and inhibition of lipolytic activity have been reported as the characters and functions of LP (Kanno, 1989a; Girardet et al., 1993). Recently, LP has been found to stimulate immunoglobulin production in human hybridoma cells and human peripheral blood lymphocytes (Sugahara et al., 2005). However, its biological function remains unclear.

The inhibitory activity of LP against HRV infection was identified as follows. Previously, Kanamaru et al. (1999) reported that high-M_r glycoprotein fraction (F1) from cow milk whey potently inhibited HRV infection *in vitro*. They reported that F1 formed a complex with various proteins but failed to identify the inhibitory entity in F1. Ten years later, LP was identified as one of the inhibitory components of HRV replication in F1 (Inagaki et al., 2010b). In brief, F1 was initially heated at 95°C for 30 min, rendering milk antibodies inert, and then subjected to ammonium sulfate fractionation. The component with a molecular size of 16 kDa, found in a certain fraction from ammonium sulfate fractionation, exhibited inhibitory activity against HRV replication. Sequencing analysis of this substance resulted in the first 7 N-terminal amino acid residues of ILKEKHL, which is consistent with the sequence of residues 69-75 of bovine LP. Thus, LP16 exhibited a strong inhibitory activity against HRV replication.

Furthermore, a preliminary experiment revealed that LP28 and LP18 potently inhibited HRV infection, suggesting that the consensus structure of LP28 and LP18 (i.e., sequence of residues 54-135 C-terminal portion of LP) was involved in their inhibitory activities (Inagaki et al., unpublished observation). Further studies are in progress for detailed elucidation of the HRV inhibitory mechanism of LP.

3.2 Anti-HRV activity of lactadherin

Lactadherin is a major milk fat globule membrane component in milk. Lactadherin in bovine milk is also known as PAS6/7.

Lactadherin consists of 2 N-terminal epidermal growth facter (EGF)-like domains followed by 2 repeated C domains with homology to the C1 and C2 domains of blood clotting factors V and VIII (Mather, 2000). Interestingly, lactadherin has first EGF-like domain containing glycosylation sites, whereas human lactadherin has defects in this domain (Mather, 2000).

Lactadherin binds to integrins αvβ3 (Taylor et al., 1997; Andersen et al., 2000; Hanayama et al., 2002) and αvβ5 (Andersen et al., 2000), which are expressed by endothelial cells. However, the physiological function of lactadherin in milk is little known.

The inhibitory activity against HRV infection of human lactadherin was first identified by Yolken et al. (1992). Furthermore, a previous clinical study indicated a correlation between human lactadherin in breast milk and morbidity due to rotavirus gastroenteritis in young children (Newburg et al., 1998). These reports led us to investigate non-immunoglobulin component(s) of rotavirus inhibitor in bovine milk.

One report indicated that bovine lactadherin did not have anti-HRV activity (Kvistgaard et al., 2004). The study was performed using the human Wa strain rotavirus infected to Caco-2 cells and a short-term (1 h) incubation of cells with lactadherin. On the contrary, the inhibitory activity of bovine lactadherin against HRV infection was reported by Inagaki et al. (2010b). The study was performed using the human MO strain rotavirus infected to MA104 cells and demonstrated that long-term (22 h) incubation of cells with lactadherin resulted in significant antiviral effects. The reasons for the inconsistent results are unclear because of distinct experimental conditions. However, it has been reported that lactadherin binds to MA104 cells via integrin αvβ3 (Taylor et al., 1997; Andersen et al., 2000; Hanayama et al., 2002), which is known as one of the cell receptors for rotavirus (Guerrero et al., 2000). Therefore, the interaction between lactadherin and cell surface components is likely important for its antiviral activity. Thus, the inhibitory mechanisms of lactadherin remain controversial.

4. Utilization of sweet whey proteins against HRV gastroenteritis

Sweet whey is manufactured as a byproduct of cheese production. Thus, its production increases as the consumption of cheese expands. However, its routine disposal will become a significant problem in the dairy industry. Therefore, the extended utilization of sweet whey should be pursued. Based on the above findings that LP and lactadherin exhibit inhibitory activities against HRV infection, we attemped to investigate the potential utilization of sweet whey as a protective food additive against HRV gastroenteritis.

4.1 Microfiltration retentate fraction (MFRF) from sweet whey

To concentrate LP and lactadherin, we attempted to examine the presence of both inhibitory components in sweet whey, which was produced during cheese manufacturing. As shown in Fig. 1, the concentrate was collected as the MFRF. Then, it was pasteurized by a high-temperature short-time method sterilization (HTST) method consisting of heating at 72°C for 15 s, followed by spray drying (referred to as Dried MFRF).

Fig. 2A shows the result of two-dimensional electrophoresis of Dried MFRF stained with Coomassie Brilliant Blue. Dried MFRF contains α-lactalbumin (α-LA, Mw: 14,100 Da) and β-lactoglobulin (β-LG, Mw: 18,200 Da) as the major protein components. The existence of LP and lactadherin could be observed, although not as obviously as the major components, indicating that the inhibitory components appeared to be contained in Dried MFRF. When immunochemical detection using the specific monoclonal antiserum for each protein (Aoki et al., 1994) was performed, as shown in Fig. 2B and 2C, LP and lactadherin could certainly be detected in Dried MFRF.

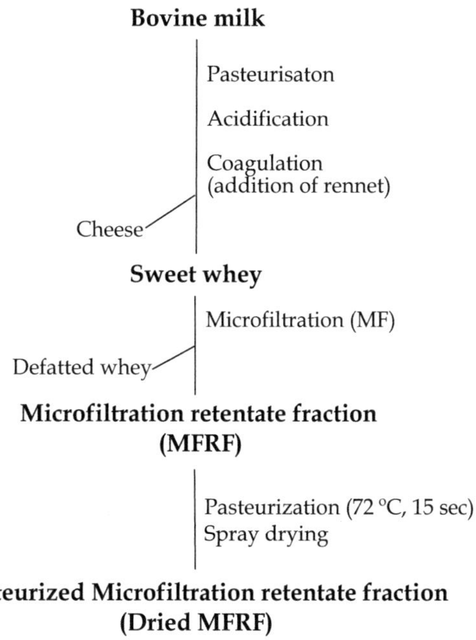

Fig. 1. Flow chart for the production of Dried MFRF

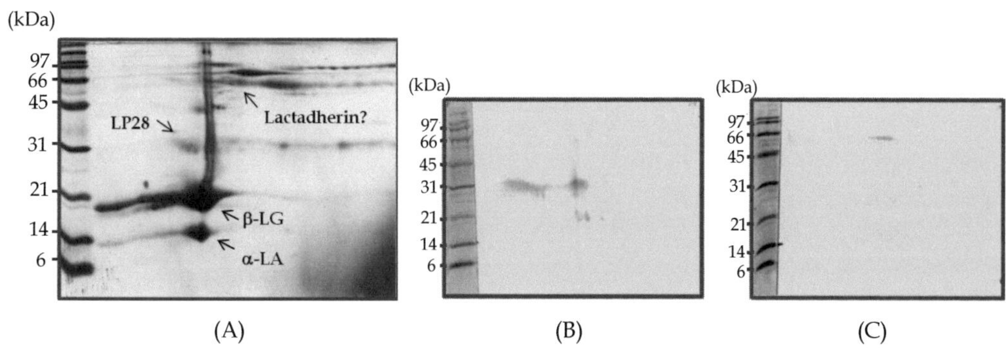

Fig. 2. Dried MFRF contains LP and Lactadherin. A. Two-dimensional protein profile of Dried MFRF. The horizontal dimension was isoelectric focusing (pI, 3.0-10.0), and the second dimension was 15% polyacrylamide gel electrophoresis (PAGE). The gel was stained with Coomassie Brilliant Blue. The molecular weights of the standards (broad range, Bio-Rad) are indicated in kDa on the left. B. Immunochemical detection of LP. Two-dimensional PAGE was performed as in panel A, and then samples were transferred onto a polyvinylidene difluoride (PVDF) membrane and immunostained for LP using the monoclonal anti-LP 1C10 primary antibody (Aoki et al., 1994), followed by horse-radish peroxidase (HRP)-conjugated goat anti-mouse IgG secondary antibody. The molecular weights of the standards (broad range, Bio-Rad) are indicated in kDa on the left. C. Immunochemical detection of lactadherin. Two-dimensional PAGE and western blotting were performed as in panel B. Samples were immunostained for lactadherin with the monoclonal anti-lactadherin 3F12 primary antibody (Aoki et al., 1994), followed by HRP-conjugated goat anti-mouse IgG secondary antibody. The molecular weights of the standards (broad range, Bio-Rad) are indicated in kDa on the left

4.2 Inhibitory activity of Dried MFRF against HRV infection

Next, we investigated the inhibitory activity of Dried MFRF against HRV infection. A replication inhibition (neutralization) assay for HRV was performed using MA104 cells (African rhesus monkey kidney cell line) following a procedure described previously (Inagaki et al., 2010b) with slight modifications. Our previously published focus reduction assay for rotaviral infection was performed using a suspension of MA104 cells, and a preincubated virus/milk sample mixture was incubated further for 22 h with the cells before fixation. In this study, a confluence monolayer of MA104 cells was established in wells of a glass slide, and a virus/milk sample mixture was inoculated for 1 h and removed from the monolayer before further advancing the viral infection to exclude the influence of the milk sample on MA104 cells by prolonged incubation.

As shown in Fig. 3, MFRF was found to potently inhibit the replication of HRV MO strain (serotype G3P[8]) with an MIC of 3.1 μg/ml. Furthermore, even after pasteurization by HTST method, the neutralizing activity of Dried MFRF remained, with an MIC of 4.7 μg/ml. This activity might also be attributed largely to the heat-resistant character of LP. The colostrums whey from the hyper-immunization of pregnant cows with human rotavirus (rotawhey) was used as a positive control. Rotawhey contains a high level of specific anti-human rotavirus antibodies, and it exhibited a robust inhibitory activity, with an MIC of 0.012 μg/ml (Fig. 3). Bovine lactoferrin also exhibited an inhibitory activity, although weak, with an MIC of 180 μg/ml (Fig. 3). The MIC value of Dried MFRF indicated that it has great potential as a protective food additive against HRV infection.

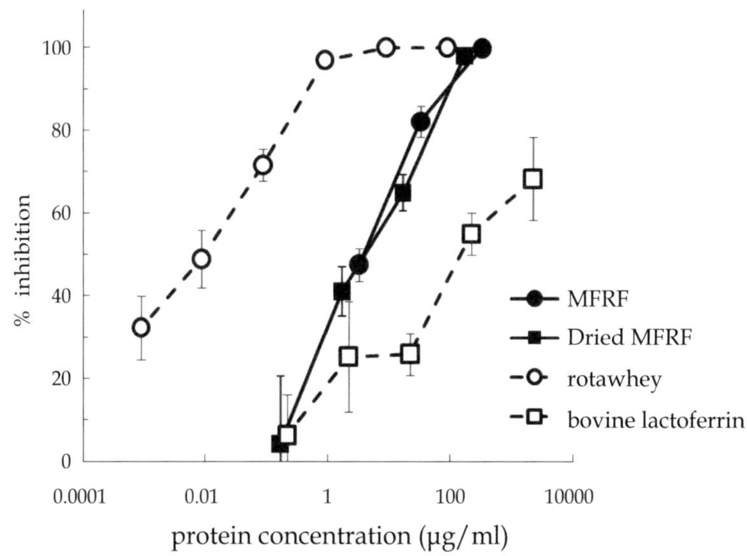

Fig. 3. MFRF exhibits inhibitory activity after pasteurization treatment

MA104 cells were plated into the wells of a 24-well heavy Teflon (HT)-coated slide (AR Brown, Tokyo) and grown to full confluence. A suspension containing infectious virus at a titer of 1 × 10^5 - 1 × 10^6 fluorescent cell focus-forming units (FCFU)/ml was treated with 20 μg/ml trypsin (Sigma-Aldrich, St. Louis, MO) for 30 min at 37°C. After appropriate dilution with Eagle's minimum essential medium (E-MEM) containing 2% fetal calf serum to give a titer of approximately 10^3 FCFU per 100 μl, aliquots were mixed with equal volumes (100 μl) of one-half serially diluted samples in microtubes for 1 h at 37ºC. The diluted mixtures (20 μl/well) were added to the confluent monolayer of MA104 cells. The control produced approximately 100 infected foci per well without the test samples of milk. The cells were further cultured for 1 h at 37°C in an atmosphere of 5% CO_2. After removal of the inoculums, the cells were washed once with E-MEM to remove unbound virus, followed by incubation at 37°C in an atmosphere of 5% CO_2. After 17 h of incubation, the cells were fixed with cold methanol for 10 min. Infected cells were detected by an indirect immunofluorescence assay using the PO-13 monoclonal anti-pigeon rotavirus antibody (Minamoto et al., 1993) and fluorescein isothiocyanate-conjugated goat anti-mouse IgG serum. The foci numbers of infected cells were measured by observation of fluorescence microscopy. Neutralizing activity was expressed as the percentage reduction in the foci numbers of infected cells as compared with infected cells without milk sample. The minimum inhibitory concentration (MIC), the minimum concentration inducing a 50% reduction in infected cells, was calculated for each sample from a logarithmic regression of the concentration-dependent percentage focus reduction. The inhibitory activity of each sample is expressed as a percentage of infected cells as compared to control cells (100%). The experiments were performed in triplicate at least 3 times, and representative results for each sample are given as the mean (SD).

4.3 Analysis of the protective components in Dried MFRF

To verify and further characterize the effective components of Dried MFRF regarding protection against HRV infection, we attempted to fractionate Dried MFRF by size exclusion chromatography on Sephacryl S-500 HR. As shown in Fig. 4, 3 fractions were collected according to the elution pattern of Dried MFRF.

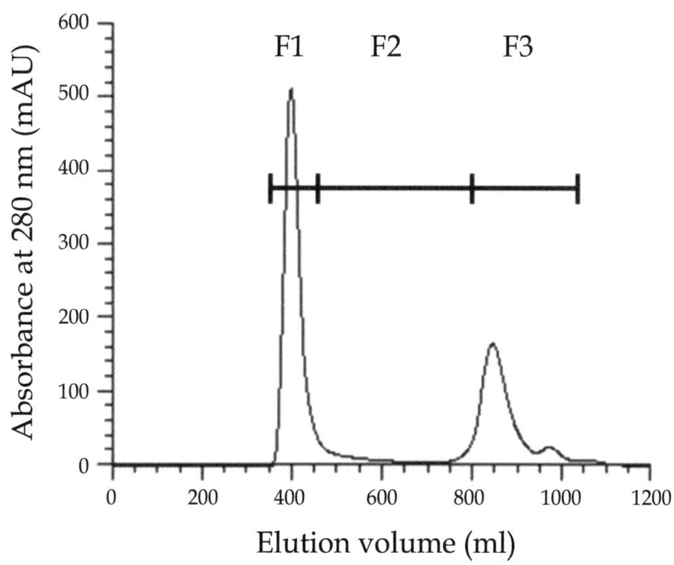

Fig. 4. Fractionation of Dried MFRF by Size Exclusion Chromatography on Sephacryl S-500 HR

The column (60 × 5.0 cm, GE Healthcare UK Ltd., Little Chalfont, UK) was equilibrated with 50 mM Tris-HCl buffer (pH 8.0) containing 0.15 M NaCl, 2 mM EDTA, and 0.02% NaN_3. Dried MFRF was dissolved in elution buffer at a concentration of 5 mg/ml, and 30 ml were added to a Sephacryl S-500 HR column. The flow rate was 10 ml/min. Eluted fractions were freeze-dried after dialysis against distilled water.

Next, to investigate the protein components in the fractions, we attempted to resolve the fractions by two-dimensional PAGE. The results are shown in Fig. 5. We confirmed by immunoblot analysis that only F1 contained lactadherin (result not shown). LP was mainly detected in F2 and slightly present in F1. Although 2 major whey proteins, α-LA and β-LG, were detected in each of the three fractions, the vast majority of them detected in F3. Lactadherin and LP28 were present as minor components in F1 and F2, respectively. α-LA and β-LG were present as major components in F3.

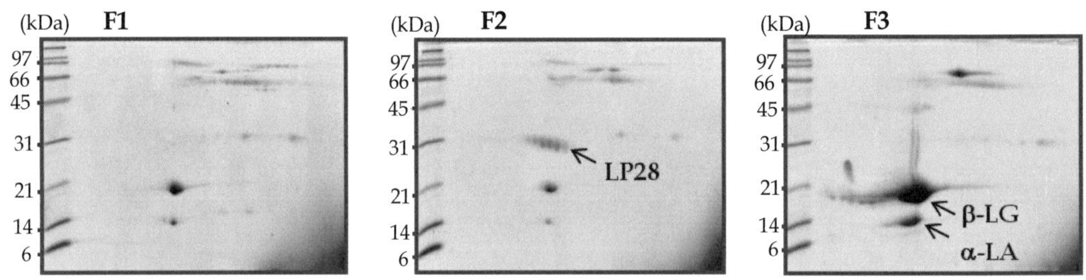

Fig. 5. Protein profiles of size exclusion chromatography fractions. Two-dimensional PAGE profiles of each fractions. The horizontal dimension was isoelectric focusing (pI, 3.0-10.0), and the second dimension was 15% PAGE. The gel was stained with Coomassie Brilliant Blue. The molecular weights of the standard (broad standard, Bio-Rad) are indicated in kDa on the left

As mentioned above, IgG was identified as a rotavirus inhibitor in bovine milk (Ebina et al., 1992; Sarker et al., 1998). Accordingly, to address the contribution of IgG to the anti-HRV activity of Dried MFRF, we attempted to separate IgG in each fraction from other components by using affinity chromatography on a HiTrap Protein G HP column (5 ml, GE Healthcare UK Ltd.). A typical elution pattern of F3 is shown in Fig. 6. The bound fraction was IgG, and the unbound fractions from each fraction were collected as F1′, F2′, and F3′. We found that IgG was removed from F2 and F3, although a small portion remained, as shown in Fig. 6. Conversely, we did not observe the elution of IgG from F1 (results not shown). These results indicated that IgG might represent a minor component in Dried MFRF. In this manner, we obtained 4 fractions: F1′ (fraction containing lactadherin), F2′ (fraction containing LP28), F3′ (fraction containing α-LA and β-LG as major components) and IgG collected from F3.

F3 was dissolved in 20 mM sodium phosphate buffer (pH 7.0) at a concentration of 1 mg/ml. The column was equilibrated with the same buffer, and the column was connected with and controlled by the ÄKTA prime system (GE Healthcare UK Ltd.). The flow rate was 2 ml/min. Proteins were monitored at 280 nm (solid line). The unbound fraction was collected as F3′. The bound fraction, IgG, eluted with a step 100% elution buffer (0.1 M glycine-HCl, pH 2.7) (dotted line). The eluted IgG fractions were neutralized with 1 M Tris-HCl (pH 9.0). Each fraction was freeze-dried after dialysis against distilled water.

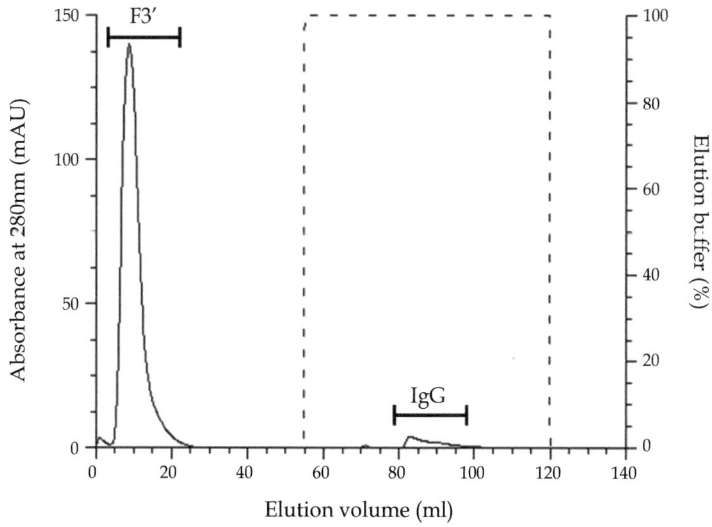

Fig. 6. Fractionation of F3 by Affinity Chromatography on HiTrap Protein G HP column

In the following focus reduction assay, IgG exhibited inhibitory activity against HRV MO infection, with an MIC of 0.27 μg/ml (Fig. 7). F2′ exhibited similar inhibitory activity level as IgG, with an MIC of 0.32 μg/ml (Fig. 7). F1′ exhibited slightly weaker inhibitory activity than did F2′, with an MIC of 1.2 μg/ml (Fig. 7). Although F3, before proteinG affinity chromatography, exhibited a strong inhibitory activity (result not shown), F3′ lost this activity after chromatography, resulting in an MIC of 20,000 μg/ml (Fig. 7). Taken together, the inhibitory components in Dried MFRF should include at least lactadherin, LP, and IgG. These components exhibited very similar activity, although the former 2 could not be

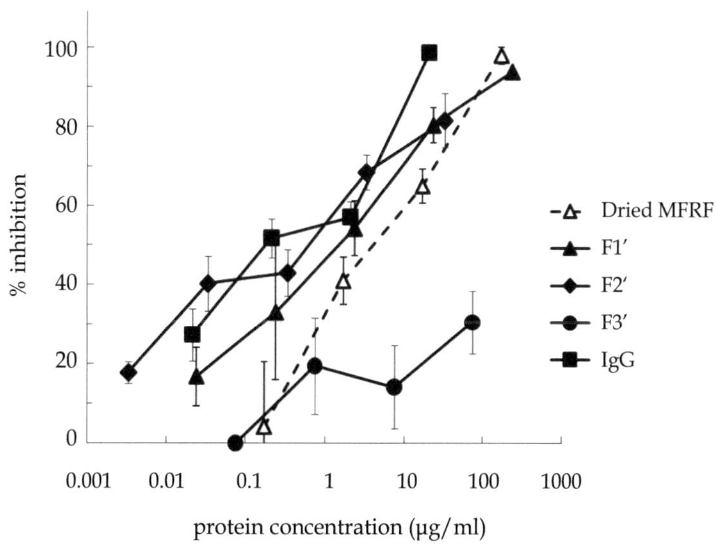

Fig. 7. The Fractions Obtained by Protein G Affinity Chromatography and Their *in Vitro* Inhibitory Activity against HRV MO Strain

purified from Dried MFRF in this study. Our previous study demonstrated that the MICs of lactadherin and LP16 in pure form were 0.016 and 1.8 µg/ml, respectively (Inagaki et al., 2010b). As the precise content of these active components in Dried MFRF is at present not clear, we conclude that their contribution to the inhibitory efficacy against HRV infection of Dried MFRF is likely to be comparable.

Inhibitory activity was determined as described in the legend to Fig. 3. The inhibitory activity of each sample was expressed as the percent decrease in foci numbers of infected cells as compared to the foci numbers of control cells, which were treated with PBS in place of the milk sample (100%). The experiments were performed in triplicate at least 3 times, and representative results for each sample are given as the mean (SD).

4.4 Inhibitory activity of Dried MFRF against various types of HRV

Furthermore, we investigated the protective efficacy of Dried MFRF against other types of HRV besides the MO strain. Rotavirus has two independent serotypes (G and P types), and they are defined by VP7 and VP4, respectively. Epidemiological studies on rotavirus showed that strains with G-types of G1, G2, G3, and G4 and those with P-types of P[4] and P[8] are the most prevalent causes of rotavirus gastroenteritis in humans (Gentsch et al., 2005; Santos and Hoshino, 2005; McDonald et al., 2009). Furthermore, the rotavirus G/P-type distribution varies from year-to-year (O'Ryan, 2009). As shown in Fig. 8, Dried MFRF also exhibited inhibitory activity against the Wa strain (serotype G1P[8]) and the Hochi strain (serotype G4P[8]), with MICs of 2.8 and 3.2 µg/ml, respectively. Therefore, Dried MFRF can be concluded to have potential as a protective food additive against several serotypes of HRV.

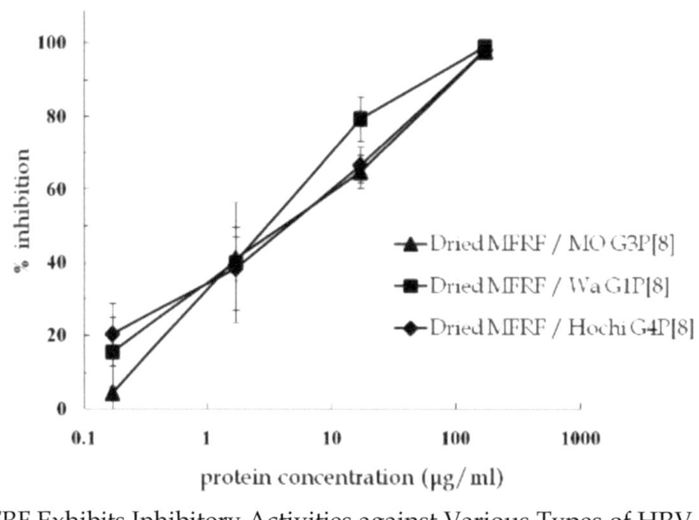

Fig. 8. Dried MFRF Exhibits Inhibitory Activities against Various Types of HRV

Inhibitory activity was determined as described in the legend to Fig. 3. The inhibitory activity of each sample was expressed as the percent decrease in foci numbers of infected cells as compared to the foci numbers of control cells, which were treated with PBS in place of the milk sample (100%). The experiments were performed in triplicate at least 3 times, and representative results for each sample are given as the mean (SD).

4.5 Protective efficacy of Dried MFRF against HRV-induced diarrhea in suckling mice

Finally, we investigated whether a single administration of Dried MFRF exhibits prophylactic efficacy against HRV-induced diarrhea *in vivo*. As shown in Fig. 9, in the PBS group, 10 of the 11 mice developed diarrhea 48 h post inoculation (hpi), and all mice recovered from diarrhea by 96 hpi. In the Dried MFRF (2.5 mg) group, only 2 of 16 mice developed diarrhea at 48 hpi, and all mice recovered from symptoms by 72 hpi. In the Dried MFRF (1.0 mg) group, 4 of 11 mice developed diarrhea at 48 hpi, and all mice recovered by 72 hpi. This result clearly indicated that Dried MFRF is a promising candidate for a prophylactic food additive against HRV infection.

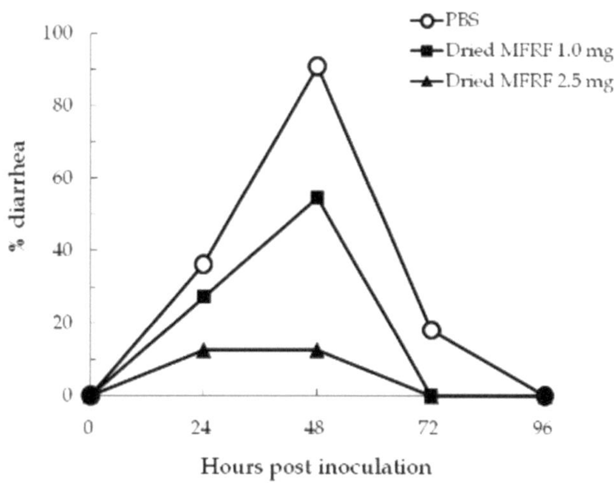

Fig. 9. Dried MFRF Exhibits Preventive Efficacy against HRV-Induced Diarrhea in Suckling Mice

Pregnant BALB/c mice were purchased from Japan SLC (Hamamatsu, Japan). Litters of 5-day-old mice were orally administered with PBS (n = 11), 1.0 mg of Dried MFRF (n = 11), or 2.5 mg of Dried MFRF (n = 16) for 60 min before inoculation with 2.5 $\times 10^5$ FCFU of the HRV MO strain. Stools were examined daily to assess diarrhea for 4 days after viral inoculation. Liquid-like mucous yellow stool was considered diarrhea.

5. Summary

Milk contains essential components for child growth. In this chapter, we introduced the inhibitory activity of LP and lactadherin against HRV infection, and examined the possibility of MFRF, which is obtained as a byproduct of cheese manufacturing, as an alternative therapeutic option against HRV gastroenteritis.

Dried MFRF exhibited inhibitory activity against several types of HRV *in vitro*. Furthermore, we demonstrated that prophylactic oral administration of Dried MFRF once before inoculation of HRV prevented the development of diarrhea in suckling mice *in vivo*. Finally, we concluded that Dried MFRF contained LP, lactadherin, and IgG as rotavirus inhibitors. As the anti-HRV activity of LP was not affected by heating at 95°C for 30 min (Inagaki et al., result not shown), the anti-HRV activity of MFRF would be stable to partial heat sterilization.

Recently, it was reported that one-third of all pediatric rotavirus gastroenteritis patients are children between 3 and 6 years of age, an age group outside the primary target of rotavirus vaccine in Japan (Ito et al., 2011; Nakanishi et al., 2009). Thus, these epidemiological studies indicated the need for not only vaccination but also alternative preventive procedures against HRV infection. In conclusion, Dried MFRF, in which the non-immunoglobulin components including LP and lactadherin are concentrated, is a promising candidate prophylactic food additive against HRV infection. Dried MFRF was also found to be a potent inhibitor of several types of bovine rotavirus derived from field breeds (Inagaki et al., results not shown). Rotavirus gastroenteritis is an important issue in livestock animals as well. Taken together, Dried MFRF is very useful as a protective food additive against rotaviral infection.

6. Acknowledgements

We thank Dr. Nobuyuki Minamoto and Dr. Makoto Sugiyama for kindly providing the PO-13 monoclonal antibody, and Dr. Tsukasa Matsuda for kindly providing the 1C10 and 3F12 monoclonal antibodies. We also express our thanks to Mr. Kengo Kishita and Mr. Tomohiro Katsura for their technical assistance. This research was supported by the Program for Promotion of Basic and Applied Researches for Innovations in Bio-Oriented Industry.

7. References

Andersen, M. H., Graversen, H., Fedosov, S. N., Petersen, T. E., & Rasmussen, J. T. (2000). Functional analyses of two cellular binding domains of bovine lactadherin. *Biochemistry*, Vol. 39, No. 20, (April 2000), pp. 6200–6206, ISSN 1520-4995

Aoki, N., Kuroda, H., Urabe, M., Taniguchi, Y., Adachi, T., Nakamura, R., & Matsuda, T. (1994). Production and characterization of monoclonal antibodies directed against bovine milk fat globule membrane (MFGM). *Biochimica et biophysica acta*, Vol. 1199, No. 1, (January 1994), pp. 87-95, ISSN 006-3002

Armah, G. E., Sow, S. O., Breiman, R. F., Dallas, M. J., Tapia, M. D., Feikin, D. R., Binka, F. N., Steele, A. D., Laserson, K. F., Ansah, N. A., Levine, M. M., Lewis, K., Coia, M. L., Attah-Poku, M., Ojwando, J., Rivers, S. B., Victor, J. C., Nyambane, G., Hodgson, A., Schödel, F., Ciarlet, M., & Neuzil, K. M. (2010). Efficacy of pentavalent rotavirus vaccine against severe rotavirus gastroenteritis in infants in developing countries in sub-Saharan Africa: a randomised, double-blind, placebo-controlled trial. *The Lancet*, Vol. 376, No. 9741, (August 2010), pp. 606-614, ISSN 0099-5355

Bishop, R. F., Davidson, G. P., Holmes, I. H., & Ruck, B. J. (1973). Virus particles in epithelial cells of duodenal mucosa from children with acute non-bacterial gastroenteritis. *The Lancet,* Vol. 302, No. 7841, (December 1973), pp. 1281-1283, ISSN 0099-5355

Bishop, R. F. (2009). Discovery of rotavirus: Implications for child health. *Journal of gastroenterology and hepatology,* Vol. 24, No. s3, (October 2009), pp. S81-S85, ISSN 1440-1746

Bojsen, A., Buesa, J., Montava, R., Kvistgaard, A. S., Kongsbak, M. B., Petersen, T. E., Heegaard, C. W., & Rasmussen, J. T. (2007). Inhibitory activities of bovine macromolecular whey proteins on rotavirus infections in vitro and in vivo. *Journal of dairy science,* Vol. 90, No. 1, (January 2007), pp. 66-74, ISSN 0022-0302

Centers for Disease Control and Prevention (CDC) (2011). Rotavirus surveillance --- worldwide, 2009. *MMRW. Morbidity and mortality weekly report*, Vol. 60, No. 16, (April 2009), pp. 514-516, ISSN 1545-861X

Cortese, M. M., Parashar, U. D., & Centers for Disease Control and Prevention (CDC) (2009). Prevention of rotavirus gastroenteritis among infants and children: recommendations of the Advisory Committee on Immunization Practices (ACIP). *MMWR. Recommendations and reports: Morbidity and mortality weekly report. Recommendations and reports / Centers for Disease Control*, Vol. 58, No. RR-2, (February 2009), pp. 1-25, ISSN 1545-8601

Ciarlet, M., & Schödel, F. (2009). Development of a rotavirus vaccine: clinical safety, immunogenicity, and efficacy of the pentavalent rotavirus vaccine, RotaTeq. *Vaccine*, Vol. 27, No. suppl 6, (December 2009), pp. G72-G81, ISSN 0264-410X

Dennehy, P. H. (2008). Rotavirus vaccine: an overview. *Clinical microbiology reviews*, Vol. 21, No. 1, (January 2008), pp. 198-208, ISSN 0893-8512

Ebina, T., Ohta, M., Kanamaru, Y., Yamamoto-Osumi, Y., & Baba, K. (1992). Passive immunizations of suckling mice and infants with bovine colostrum containing antibodies to human rotavirus. *Journal of medical virology*, Vol. 38, No. 2, (October 1992), pp. 117-123, ISSN 0146-6615

Guerrero, C. A., Méndez, E., Zárate, S., Isa, P., López, S., & Arias, C. F. (2000). Integrin alpha(v)beta(3) mediates rotavirus cell entry. *Proceeding of the National Academy of Sciences of the United States of America*, Vol. 97, No. 26, (December, 2000), pp. 11644-14649, ISSN 1091-6490

Gentsch, J. R., Laird, A.R., Bielfelt, B., Griffin, D. D., Banyai, K., Ramachandran, M., Jain, V., Cunliffe, N. A., Nakagomi, O., Kirkwood, C. D., Fischer T. K., Parashar U. D., Bresee J. S., Jiang, B., and Glass, R. I. (2005). Serotype diversity and reassortment between human and animal rotavirus strains: implications for rotavirus vaccine programs. *The Journal of infectious diseases*, Vol. 192, No. Suppl 1, (September 2005) pp. 146-159. ISSN 0022-1899

Girardet, J. M., Linden, G., Loye, S., Courthaudon, J. L., & Lorient, D. (1993). Study of mechanism of lipolysis inhibition by bovine milk proteose-peptone component 3. *Journal of dairy science*, Vol. 76, No. 8, (August 1993), pp. 2156-2163, ISSN 0022-0302

Girardet, J. M., & Linden, G. (1996). PP3 component of bovine milk: a phosphorylated whey glycoprotein. *The Journal of dairy research*, Vol. 63, No. 2. (May 1996), pp. 333-350, ISSN 0022-0299

Girardet, J. M., Saulnier, F., Gaillard, J. L., Ramet, J. P., & Humbert, G. (2000). Camel (*Camelus dromedarius*) milk PP3: evidence for an insertion in the amino-terminal sequence of the camel milk whey protein. *Biochemistry and cell biology*, Vol. 78, No. 1, (January 2000), pp. 19-26, ISSN 0829-8211

Hanayama, R., Tanaka, M., Miwa, K., Shinohara, A., Iwamatsu, A., & Nagata, S. (2002). Identification of a factor that links apoptotic cells to phagocytes. *Nature*, Vol. 417, No.6885, (May 2002), pp. 182-187, ISSN 1476-4687

Inagaki, M., Yamamoto, M., Xijier, Cairangzhouma, Uchida, K., Yamaguchi, H., Kawasaki, M., Yamashita, K., Yabe, T., & Kanamaru, Y. (2010a). *In vitro* and *in vivo* evaluation of the efficacy of bovine colostrum against human rotavirus infection. *Bioscience, biotechnology, and biochemistry*, Vol. 74, No. 3, (March 2010), pp. 680-682, ISSN 0916-8451

Inagaki, M., Nagai, S., Yabe, T., Nagaoka, S., Minamoto, N., Takahashi, T., Matsuda, T., Nakagomi, O., Nakagomi, T., Ebina, T., & Kanamaru, Y. (2010b). The bovine lactophorin C-terminal fragment and PAS6/7 were both potent in the inhibition of human rotavirus replication in cultured epithelial cells and the prevention of experimental gastroenteritis. *Bioscience, biotechnology, and biochemistry,* Vol. 74, No. 7, (July 2010), pp. 1386-1390, ISSN 0916-8451

Ito, H., Trabe, O., Katsumi, Y., Matsui, F., Kidowaki, S., Mibayashi, A., Nakagomi, T., & Nakagomi, O. (2011). The incidence and direct medical cost of hospitalization due to rotavirus gastroenteritis in Kyoto, Japan, as estimated from a retrospective hospital study. *Vaccine,* in press, (August 2011), ISSN 1873-2518

Kanamaru, Y., Etoh, M., Song, X-G., Mikogami, T., Hayakawa, H., Ebina, T., & Minamoto, N. (1999). A high-Mr glycoprotein fraction from cow's milk potent in inhibiting replication of human rotavirus in vitro. *Bioscience, biotechnology, and biochemistry,* Vol. 63, No. 1, (January 1999), pp. 246-249, ISSN 0916-8451

Kanno, C. (1989a). Purification and separation of multiple forms of lactophorin from bovine milk whey and their immunological and electrophoretic properties. *Journal of dairy science,* Vol. 72, No. 4, (April 1989), pp. 883-891, ISSN 0022-0302

Kanno, C. (1989b). Characterization of multiple forms of lactophorin isolated from bovine milk whey. *Journal of dairy science,* Vol. 72, No. 7, (July 1989), pp. 1732-1739, ISSN 0022-0302

Kjeldsen, F., Haselmann, K. F., Budnik, B. A., Sørensen, E. S., & Zubarev, R. A. (2003). Complete characterization of posttranslational modification sites in the bovine milk protein PP3 by tandem mass spectrometry with electron capture dissociation as the last stage. *Analytical chemistry,* Vol. 75, No. 10, (May 2003), pp. 2355-2361, ISSN 0003-2700

Koletzko, B., Baker. S., Cleghorn, G., Neto, U. F., Gopalan, S., Hernell, O., Hock, Q. S., Jirapinyo, P., Lonnerdal, B., Pencharz, P., Pzyrembel, H., Ramirez-Mayans, J., Shamir, R., Turck, D., Yamashiro, Y., & Zong-Yi, D. (2005). Global standard for the composition of infant formula: recommendations of an ESPGHAN coordinated international expert group. *Journal of pediatric gastroenterology and nutrition,* Vol. 41, No. 5, (November 2005), pp. 584-599, ISSN 1536-4801

Kvistgaard, A. S., Pallesen, L. T., Arias, C. F., López, S., Petersen, T. E., Heegaard, C. W., & Rasmussen, J. T. (2004). Inhibitory effects of human and bovine milk constituents on rotavirus infections. *Journal of dairy science,* Vol. 87, No. 12, (December 2004), pp. 4088-4096, ISSN 0022-0302

McDonald, S. M., Matthijnssens, J., McAllen, J. K., Hine, E., Overton, L., Wang, S., Lemey, P., Zeller, M., Van Ranst, M., Spiro, D. J., & Patton, J. T. (2009). Evolutionary dynamics of human rotaviruses: balancing reassortment with preferred genome constellations. *PLoS Pathogens,* Vol. 5, No. 10, (October 2009), pp. e1000634, ISSN 1553-7374

Mather, I. H. (2000). A review and proposed nomenclature for major proteins of the milk-fat globule membrane. *Journal of dairy science,* Vol. 83, No. 2, (February 2000), pp. 203-247, ISSN 0022-0302

Minamoto, N., Sugimoto, O., Yokota, M., Tomita, M., Goto, H., Sugiyama, M., & Kinjo, T. (1993). Antigenic analysis of avian rotavirus VP6 using monoclonal antibodies. *Archives of virology,* Vol. 131, No. 3-4, (February 1993), pp. 293-305, ISSN 0304-8608

Nakanishi, K., Tsugawa, T., Honma, S., Nakata, S., Tatsumi, M., Yoto, Y., & Tsutsumi, H. (2009). Detection of enteric viruses in rectal swabs from children with acute gastroenteritis attending the pediatric outpatient clinics in Sapporo, Japan. *Journal of Clinical Virology: the official publication of the Pan American Society for Clinical Virology*, Vol. 46, No. 1, (September 2009), pp. 94-97, ISSN 1873-5967

Nakagomi, T., Nakagomi, O., Takahashi, Y., Enoki, M., Suzuki, T., & Kilgore, P. E. (2005). Incidence and burden of rotavirus gastroenteritis in Japan, as estimated from a prospective sentinel hospital study. *The Journal of infectious diseases,* Vol. 192, No. Suppl 1, (September 2005), pp. S106-S110, ISSN 0022-1899

Naghipour, M., Nakagomi, T., and Nakagomi, O. (2008). Issues with reducing the rotavirus-associated mortality by vaccination in developing countries. *Vaccine,* Vol. 26, No. 26, (June 2008) pp. 3236-3241. ISSN 1873-2518

Newburg, D. S., Peterson, J. A., Ruiz-Palacios, G. M., Matson, D. O., Morrow, A. L., Shults, J., Guerrero, M. L., Chaturvedi, P., Newburg, S. O., Scallan, C. D. , Taylor, M. R., Ceriani, R. L., & Pickering, L. K. (1998). Role of human-milk lactadherin in protection against symptomatic rotavirus infection. *Lancet,* Vol. 351, No. 9110, (April 1998), pp. 1160-1164, ISSN 0140-6736

O'Ryan, M. (2009). The ever-changing landscape of rotavirus serotypes. *The Pediatric infectious disease journal,* Vol. 28, No. 3 Suppl, (March 2009), pp. S60-62, ISSN 1532-0987

O'Ryan, M., & Linhares, A. C. (2009). Update on rotarix: an oral human trotavirus vaccine. *Expert review of vaccines,* Vol. 8, No. 12, (December 2009), pp. 1627-1641, ISSN 1744-8395

Parashar, U. D., Alexander, J. P., & Glass, R. I. (2006). Prevention of rotavirus gastroenteritis among infants and children. Recommendations of the Advisory Committee on Immunization Practices (ACIP). *MMWR. Recommendations and reports: Morbidity and mortality weekly report. Recommendations and reports / Centers for Disease Control,* Vol. 55, No. RR-12, (August 2006), pp. 1-13, ISSN 1545-8601

Pérez-Cano, F. J., Marín-Gallén, S., Castell, M., Rodríguez-Palmero, M., Rivero, M., Castellote, C., & Franch, A. (2008). Supplementing suckling rats with whey protein concentrate modulates the immune response and ameliorates rat rotavirus-induced diarrhea. *The Journal of nutrition,* Vol. 138, No. 12, (December 2008), pp. 2392-2398, ISSN 0022-3166

Santos, N., & Hoshino, Y. (2005). Global distribution of rotavirus serotypes/genotypes and its implication for the development and implementation of an effective rotavirus vaccine. *Reviews in medical virology,* Vol. 15, No. 1, (January/February 2005) pp. 29-56. ISSN 1099-1654

Sarker, S. A., Casswall, T. H., Mahalanabis, D., Alam, N. H., Albert, M. J., Brüssow,.H., Fuchs, G. J., & Hammerström, L. (1998). Successful treatment of rotavirus diarrhea in children with immunoglobulin from immunized bovine colostrum. *The Pediatric infectious disease journal,* Vol. 17, No. 12, (December 1998), pp. 1149-1154, ISSN 0891-3668

Soriano-Gabarro, M., Mrukowicz, J., Vesikari, T., & Verstraeten, T. (2006). Burden of rotavirus disease in European Union countries. *The Pediatric infectious disease journal*, Vol. 25, No. 1 Suppl, (January 2006), pp. S7–S11, ISSN 1532-0987

Sørensen, E. S., Rasmussen, L. K., Møller, L., & Petersen, T. E. (1997). The localization and multimeric nature of component PP3 in bovine milk: purification and characterization of PP3 from caprine and ovine milks. *Journal of dairy science,* Vol. 80, No. 12 (December 1997), pp. 3176-3681, ISSN 0022-0302

Tate, J. E., Patel, M. M., Steele, A. D., Gentsch, J. R., Payne, D. C., Cortese, M. M., Nakagomi, O., Cunliffle, N. A., Jiang, B., Neuzil, K. M., de Oliveira, L. H., Glass, R. I., & Parashar, U. D. (2010). Global impact of rotavirus vaccines. *Expert review of vaccines,* Vol. 9, No. 4, (April 2010), pp. 395-407, ISSN 1744-8395

Taylor, M. R., Couto, J. R., Scallan, C. D., Ceriani, R. L., & Peterson, J. A. (1997). Lactadherin (formerly BA46), a membrane-associated glycoprotein expressed in human milk and breast carcinomas, promotes ArgGlyAsp (RGD)-dependent cell adhesion. *DNA and cell biology,* Vol. 16, No. 7, (July 1997), pp. 861–869, ISSN 1044-5498

Wolber, F. M., Broomfield, A. M., Fray, L., Cross, M. L., & Dey, D. (2005). Supplemental dietary whey protein concentrate reduces rotavirus-induced disease symptoms in suckling mice. *The Journal of nutrition,* Vol. 135, No. 6, (January 2005), pp. 1470-1474, ISSN 1541-6100

Yolken, R. H., Peterson, J. A., Vonderfecht, S. L., Fouts, E. T., Midthum, K., & Newburg, D. S. (1992). Human milk mucin inhibits rotavirus replication and prevents experimental gastroenteritis. *The Journal of clinical investigation,* Vol. 90, No. 5, (November 1992), pp. 1984-1991, ISSN 0021-9738

Zaman, K., Dang, D. A., Victor, J. C., Shin, S., Yunus, M., Dallas, M. J., Podder, G., Vu, D. T., Le, T. P., Luby, S. P., Le, H. T., Coia, M. L., Lewis, K., Rivers, S. B., Sack, D. A., Schödel, F., Steele, A. D., Neuzil, K. M., & Ciarlet, M. (2010). Efficacy of pentavalent rotavirus vaccine against severe rotavirus gastroenteritis in infants in developing countries in Asia: a randomised, double-blind, placebo-controlled trial. *Lancet*, Vol. 376, No. 9741, (August 2010), pp. 615-623, ISSN 1474-547X

4

Emerging Preservation Methods for Fruit Juices and Beverages

H. P. Vasantha Rupasinghe and Li Juan Yu
Nova Scotia Agricultural College
Canada

1. Introduction

This chapter provide a review of traditional and non-traditional food preservation approaches including physical methods (non-thermal pasteurization), chemical methods (natural food preservatives) and their combinations for extension of the shelf life of fruit juices and beverages.

Traditionally, the shelf-life stability of juices has been achieved by thermal processing. Low temperature long time (LTLT) and high temperature short time (HTST) treatments are the most commonly used techniques for juice pasteurization. However, thermal pasteurization tends to reduce the product quality and freshness. Therefore, some non-thermal pasteurization methods have been proposed during the last couple of decades, including high hydrostatic pressure (HHP), high pressure homogenization (HPH), pulsed electric field (PEF), and ultrasound (US). These emerging techniques seem to have the potential to provide "fresh-like" and safe fruit juices with prolonged shelf-life. Some of these techniques have already been commercialized. Some are still in the research or pilot scale. The first part of the chapter will give an update of these emerging non-thermal techniques.

Apart from thermal pasteurization, some chemical preservatives are also widely used for the extension of the shelf-life of fruit juices and beverages. Two of the most commonly used preservatives are potassium sorbate and sodium benzoate. However, consumer demand for natural origin, safe and environmental friendly food preservatives has been increasing since 1990s. Natural antimicrobials such as bacteriocins, organic acids, essential oils and phenolic compounds have shown considerable promise for use in some food products. The second part of the chapter will comprise of applications of these natural antimicrobials in fruit juice preservation.

From scientific literature, it is apparent that some individual non-thermal methods as well as natural antimicrobials are effective to inactivate microorganisms or reduce the *log* colony forming units (CFU) while not adversely affecting the sensory and nutritional quality. Moreover, the combination of these techniques could also provide synergistic effects on prolonging the shelf-life of fruit juices and beverages and potentially could become replacements for traditional pasteurization methods. The third part of the chapter will provide recent progresses of these combined techniques in fruit juice shelf-life extension.

2. Traditional thermal pasteurization

Thermal processing is the most widely used technology for pasteurization of fruit juices and beverages. Juice pasteurization is based on a 5-*log* reduction of the most resistant microorganisms of public health significance (USFDA 2001). The process could be accomplished by different time-temperature combinations.

2.1 Low temperature long time (LTLT)

Fruit juice has been traditionally pasteurized by batch heating at 63-65°C for relatively long time (D'Amico et al. 2006). This method has been replaced by high temperature short time treatment due to the undesirable quality changes during this process.

2.2 High temperature short time (HTST)

HTST treatment could minimise those undesirable quality changes made by batch heating due to the much less duration of heat treatment. Currently, HTST pasteurization is the most commonly used method for heat treatment of fruit juice. For example, orange juice is processed by HTST at 90 to 95°C for 15 to 30 s (Braddock 1999). And apple juice is treated by HTST at 77 to 88°C for 25 to 30 s (Moyer & Aitken 1980).

3. Non-traditional method

Thermal processing has been proven to be effective for preservation of fruit juice and beverages. However, thermal treatment tends to reduce the product quality and freshness. Nowadays, consumer demand for natural, healthy and convenient food products is fast growing, which leads to the innovation of novel food preservation technologies. Based on the literature, these novel technologies can be generally divided into physical methods (mostly non-thermal methods) and chemical approaches.

3.1 Physical methods (Non-thermal pasteurization)

Some non-thermal pasteurization methods have been proposed during the last couple of decades, including high hydrostatic pressure (HHP), high pressure homogenization (HPH), pulsed electric field (PEF), and ultrasound (US). These emerging techniques seem to have the potential to provide "fresh-like" and safe fruit juices with prolonged shelf-life.

3.1.1 High hydrostatic pressure (HHP)

High hydrostatic pressure (HHP) processing uses pressures up to 1000 MPa, with or without heat, to inactivate harmful microorganisms in food products (Ramaswamy et al. 2005). High hydrostatic pressure has traditionally been used in non-food areas such as ceramic and steel production. The application of HHP in food area started from 1900s when Hite and other researchers applied HHP on the preservation of milk, fruits and vegetables. However, it takes a long time for the commercial products to emerge in the market. In 1990, the first HHP processed fruit jams were sold in the Japanese market. Subsequently, HHP processed commercial products including fruit juices and beverages, vegetable products, among others, have been produced in North America, Europe, Australia, and Asia (Balasubramaniam et al. 2008).

Generally, there are two **principles** that govern the behaviour of foods under pressure: the Le Chatelier-Braun principle and the Isostatic principle. The Le Chatelier-Braun principle indicates that any phenomenon (such as phase transition, change in molecular configuration, chemical reaction, etc.) accompanied by a decrease in volume is enhanced by the increase in applied pressure. The isostatic principle means that the distribution of pressure into the sample is uniform and instantaneous. Thus, the process time is independent of sample size and shape (Ramaswamy et al. 2005).

HPP is proven to meet the FDA requirement of a 5-*log* reduction of microorganisms in fruit juices and beverages without sacrificing the sensory and nutritional attributes of fresh fruits (San Martín et al. 2002). Compared with thermal processing, HHP has many **advantages**. It can provide safe product with reduced processing time. It can maintain maximum fresh-like flavor and taste in the product due to the lower processing temperatures. Moreover, it is environmentally friendly since it requires only electrical energy and no waste by-products generated (Ramaswamy et al. 2005, Toepfl et al. 2006). Due to these advantages, HHP has been widely used in food product preservation including fruit and beverages in the areas of microbial inactivation (Table 1) and shelf-life extension (Table 2).

Foods	Microorganism	Treatment parameters	*Log* reduction	Sources
Orange juice	*Escherichia coli* O157: H7	550 MPa, 30°C, 5 min	6	Linton et al. 1999
Apple jam	*Listeria monocytogenes*	200 MPa, 20°C, 5 min	2.8	Préstamo et al. 1999
Apple juice	*Escherichia coli* 29055	400 MPa, 25°C	>5	Ramaswamy et al. 2003
Apple juice	*Escherichia coli, Listeria innocua, Salmonella*	545 MPa, 1min	5	Avure Technologies
Orange juice	*Escherichia coli, Listeria innocua*	241 MPa, 3 min	5	Guerrero-Beltran et al. 2011

Table 1. Examples of HHP inactivation of microorganisms in fruit products

Foods	Treatment parameters	Storage conditions	Quality changes	Sources
Blueberry juice	200 MPa, 15 min	Tested right after treatment	Total phenolic and anthocyanin content increased, whereas no changes in antioxidant capacity, pH, °Brix and Colors	Barba et al. 2011
Blueberry juice	400-600 MPa, 15 min	Tested right after treatment	Total phenolic and anthocyanin content increased; no changes in pH, °Brix and Colors; but antioxidant capacity decreased	Barba et al. 2011
Blood orange juice	400-600 MPa, 15 min	4°C for 10 days	93.4% retention rate of anthocyanin; 85% retention rate of ascorbic acid	Torres et al. 2011

Table 2. Examples of HHP effect on quality attributes of fruit products

3.1.2 Pulsed electric field (PEF)

Pulsed electric field processing (PEF) applies short bursts of high voltage electricity for microbial inactivation and causes no or minimum effect on food quality attributes. Briefly, the foods being treated by PEF are placed between two electrodes, usually at room temperature. The applied high voltage results in an electric field that causes microbial inactivation. The applied high voltage is usually in the order of 20-80 kV for microseconds. The common types of electrical field waveform applied include exponentially decaying and square wave (Knorr et al. 1994, Zhang et al. 1995, Barbosa-Cánovas et al. 1999).

The **principles** of PEF processing have been explained by several theories including the trans-membrane potential theory, electromechanical compression theory and the osmotic imbalance theory. One of the most accepted theories is associated with the electroporation of cell membranes. It is generally believed that electric fields induce structural changes in the membranes of microbial cells based on generation of pores of the cell membrane, leading consequently to microbial destruction and inactivation (Tsong 1991, Barbosa-Cánovas et al. 1999).

Compared with thermal processing, PEF processing has many **advantages**. It can preserve the original sensory and nutritional characteristics of foods due to the very short processing time and low processing temperatures. Energy savings for PEF processing are also important compared with conventional thermal processing. Moreover, it is environmentally friendly with no waste generated (Toepfl et al. 2006). Due to these advantages, PEF processing has been widely used in food product preservation including fruit and beverages in the areas of microbial inactivation (Table 3) and shelf-life extension (Table 4).

Foods	Microorganism	Treatment parameters	*Log* reduction	Sources
Apple juice	*E. coli* 8739, *Escherichia coli* O157: H7	30kV/cm, 172µs, <35°C	5	Evrendilek et al., 1999
Cranberry juice	Total aerobic count, molds, yeasts	40kV/cm, 150µs, <25°C	4	Jin & Zhang 1999
Orange juice	*Listeria innocua*	30kV/cm, 12µs, 54°C	6.0	McDonald et al. 2000
Apple cider	*Escherichia coli* O157: H7	90kV/cm, 20µs, 42°C	5.9	Iu et al. 2001
Orange juice	*Salmonella typhimurium*	90kV/cm, 100µs, 55°C	5.9	Liang et al., 2002
Apple juice	*Escherichia coli*	34kV/cm, 7.68µs, 55°C	6.2	Heinz et al. 2003
Grape juice	*Escherichia coli*	34kV/cm, 7.68µs, 55°C	6.4	Heinz et al. 2003
Cherry juice,	*Penicillum expansum*	34kV/cm, 163µs, 21°C	100% inaction of spore germination	Evrendilek et al. 2008
Peach nectar	*Penicillum expansum*	34kV/cm, 163µs, 21°C	100% inaction of spore germination	Evrendilek et al. 2008
Apricot nectar	*Penicillum expansum*	34kV/cm, 163µs, 21°C	100% inaction of spore germination	Evrendilek et al. 2008

Table 3. Examples of PEF inactivation of microorganisms in fruit products

Foods	Treatment parameters	Storage conditions	Quality changes	Sources
Cranberry juice	40kV/cm, 150μs, <25°C	4C for 14 days	No changes in color and volatile profile	Jin & Zhang 1999
Apple Juice	35kV/cm, 94μs,	4, 22 and 37°C for 36 days	No changes in color and Vitamin C	Evrendilek et al. 2000
Apple cider	35kV/cm, 94μs,	4, 22 and 37°C for 14 days	No changes in color and Vitamin C	Evrendilek et al. 2000
Orange juice	40kV/cm, 97ms, 45°C	4°C for 196 days	PEF-processed juice retained more ascorbic acid, flavor, and color than thermally processed juice	Min et al. 2003
Citrus juice	28kV/cm, 100μs, <34°C	Tested right after treatment	No changes in pH, Brix, electric conductivity, viscosity, non-enzymatic browning index (NEBI), hydroxymethylfurfural (HMF), color, organic acid content, and volatile flavour compounds	Cserhalmi, et al. 2006
Apple juice	36kV/cm, 190μs, <34°C	Tested right after treatment	PEF preserved better the pH than HTST.	Charles-Rodriguez et al. 2007
Apple juice	35kV/cm, 6.4ms	Tested right after treatment	PEF preserved better the pH, total acidity, phenolics content, and volatile compounds than HTST	Aguilar-Rosas et al. 2007
Water melon juice	35kV/cm, 50μs, <40°C	Tested right after treatment	113% of Lycopene content, 72% of vitamin C and 100% of antioxidant capacity retention were obtained	Oms-Oliu et al. 2009

Table 4. Examples of PEF effect on quality attributes of fruit products

3.1.3 Ultrasound (US)

Power ultrasound (US) has emerged as a potential non-thermal technique for preservation of food products over the last decade. Compared with diagnostic ultrasound, power US uses a lower frequency range of 20 to 100 kHz and a higher sound intensity of 10 to 1000 W/cm^2 (Baumann et al. 2005).

The **principle** of ultrasonic processing could be explained as follows: Firstly, the ultrasonic transducers convert electrical energy to sound energy. Secondly, when the ultrasonic waves propagate in liquid, small bubbles will be formed and collapsed thousands of times per second. This rapid collapse of the bubbles (cavitation) results in high localized temperatures and pressure, causing breakdown of cell walls, disruption of cell membranes and damage of DNA (Manvell, 1997, Knorr et al. 2004, O'Donnell et al. 2010).

The **application** of high power ultrasound in the food industry has been widely investigated. To meet the FDA requirement of a 5-*log* reduction of microorganisms, a combination of sonication with mild heat treatment and /or pressure is essential (Baumann et al. 2005, D'Amico et al. 2006, Ugarte-Romero et al. 2006, Salleh-Mack and Roberts 2007, Tiwari et al. 2009). Several works have been done to examine the

Foods	Microorganism	Treatment parameters	*Log* reduction	Sources
Carrot juice	*E. coli* K12	19.3 kHz, 700-800 W, 1 min, 60°C	2.5	Zenker et al. 2003
Apple cider	*Listeria monocytogenes*	48 kHz, 600 W, 3 min, 25°C	1-2	Rodgers & Ryser 2004
Apple cider	*Escherichia coli* O157: H7	48 kHz, 600 W, 5 min, 25°C	1-2	Rodgers & Ryser 2004
Apple cider	*Listeria monocytogenes*	20 kHz, 750 W, 5 min, 0.46 W/mL, 60°C	5	Baumann et al. 2005
Apple cider	*Escherichia coli* O157: H7	20 kHz, 150 W, 18 min, 118 W/cm2, 57°C	6	D' Amico et al. 2006
Orange juice	Total mesophilic aerobes	500 kHz, 240 W, 15 min, 60°C	3.4	Valero et al. 2007
Apple juice	*Alicyclobacillus acidoterrestris*	24 kHz, 300 W, 60 min	80%	Yuan et al. 2009
Orange juice	Aerobic mesophilic count (AMC)	20kHz, 500W, 8 min, 89.25μm, 10°C	1.38	Gomez-Lopez 2010
Orange juice	Yeast and mold counts (YMC)	20 kHz, 500 W, 8 min, 89.25 μm, 10°C	0.56	Gomez-Lopez 2010

Table 5. Examples of US inactivation of microorganisms in fruit products

Foods	Treatment parameters	Storage conditions	Quality changes	Sources
Carrot juice	19.3 kHz, 700-800 W, 1 min, 60°C	4°C for 35 days	Improvement in surface color stability and L-ascorbic acid retention	Zenker et al. 2003
Apple cider	20 kHz, 750 W, 4 min, 0.46 W/mL, 60°C	Tested right after treatment	Titratable acidity, pH, and °Brix of the cider were not affected.	Ugarte-Romero et al. 2006
Orange juice	500 kHz, 240 W, 15 min, 60°C	5°C for 14 days	No detrimental effects on limonin content, brown pigments and colour	Valero et al. 2007
Orange juice	20 kHz, 1500 W, 10 min, 32-38°C	10C for 30 days	No significant changes in °Brix and titratable acidity. Significant changes in juice pH, colour, non-enzymatic browning, cloud value and ascorbic acid	Tiwari et al. 2009
Orange juice	20 kHz, 500 W, 8 min, 89.25μm, 10°C	4°C for 10 days	Color and ascorbic acid content were affected during storage	Gomez-Lopez 2010

Table 6. Examples of US effect on quality attributes of fruit products

effectiveness of ultrasound on inactivation of microorganisms in fruit juices (Table 5). A few studies have been conducted to examine the effect of ultrasound on quality of US-treated fruit juices (Table 6).

Except HHP, PEF and power US, other non-thermal techniques including high pressure homogenization (HPH), membrane filtration and UV-light, among others, are also being investigated.

3.1.4 Ultraviolet light

Ultraviolet light (UV-light) technology utilizes radiation with the electro-magnetic spectrum in the range of 100 to 400 nanometers, between visible light and x-rays. It could be further divided into UV-A (320–400 nm), UV-B (280–320 nm) and UV-C (200–280 nm). UV-C is known to have biocidal effects and destroys microorganisms by degrading their cell walls and DNA (Ngadi et al. 2003). Therefore, UV-C could be used for the inactivation of microorganisms such as bacteria, yeasts, moulds, among others (Bintsis, Litopoulou-Tzanetaki, & Robinson, 2000). The amount of cell damage depends on the type of medium, microorganisms and the applied UV dose (Ngadi et al. 2003). For fruit juice and beverage processing, the wavelength of 254 nm is widely used (Guerrero-Beltrán & Barbosa-Cánovas, 2004).

As a non-thermal preservation method, UV-C treatment takes the **advantages** of no toxic or significant non-toxic by-products being formed during the treatment, very little energy being required when compared to thermal pasteurization processes, and maximum aroma and color of the treated fruits being maintained (Tran & Farid, 2004).

UV-C treatment has been successfully **applied** to reduce the microbial load in different fruit juices and nectars. Under suitable treatment conditions, more than 5-*log* reduction of some pathogenic microorganism, such as *E. coli*, in fruit juices could be achieved (Guerrero-Beltrán and Barbosa-Cánovas 2004, Keyser et al. 2008). The minimum treatment condition for clear apple juice was under UV dosage of 230 J L^{-1}, whereas higher UV dosage levels would be needed for cloudy juices such as orange juice and tropical juices (Keyser et al. 2008).

3.1.5 High pressure homogenization (HPH)

High pressure homogenization (HPH) is considered to be one of the most promising non-thermal technologies proposed for preservation of fruit juice and beverages. The primary mechanisms of HPH has been identified as a combination of spatial pressure and velocity gradients, turbulence, impingement, cavitation and viscous shear, which leads to the microbial cell disruption and food constituent modification during the HPH process. HPH has shown its ability to increase the safety and shelf-life of fruit juices including orange juice (Lacroix et al. 2005; Tahiri et al. 2006; Welti-Chanes et al. 2009), apple juice (Kumar et al. 2009; Pathanibul et al. 2009) and apricot juice (Patrignani et al. 2009). The effectiveness of the treatment depends on many parameters including processing factors such as pressure, temperature, number of passes and medium factors such as type of juice and microorganisms. For example, up to 350 MPa processing pressure was required to achieve an equivalent 5-*log* inactivation of *L. Innocua*; however, less pressure is required for *E. coli* (> 250 Mpa) in carrot juices (Pathanibul et al. 2009). Another instance is that a higher reduction of *Saccharomyces cerevisiae* 635 was observed in carrot juice (5-*log* reduction) than in apricot juice. (3-*log* reduction) with a pressure level of 100 MPa for up to 8 passes (Patrignani et al. 2009).

3.1.6 Membrane filtration

Ultrafiltration (UF) and microfiltration (MF) are the most commonly used membrane filtration techniques for fruit juice processing. They have been used commercially for the clarification of fruit juices. Through this processing, a "pasteurized" product could be produced with flavours better than thermally treated products (Tallarico et al. 1998; Ortega-Rivas et al. 1998; Cassano et al. 2003). The effectiveness of the treatment depends on many parameters including processing factors such as types of membrane, pore size, transmembrane pressure and medium factors such as type of juice and microorganism. For example, an ultrafiltration (UF) unit, with polysulphone membranes of 10 kDa and 50 kDa pore sizes and trans-membrane pressures of up to 155 kPa, were used to treat apple juices. Results indicated that pH, acid content, and soluble solids did not change but presented less variability for the smaller pore membrane treatment. Relative colour changes were observed for both membranes, which was more detectable for the larger pore membrane treatment (Zarate-Rodriguez et al. 2001). Another application example was to use an ultrafiltration membrane of 15 kDa pore size to filter carrot and citrus juices. Then the clarified juices could be further processed by reverse osmosis and osmotic distillation (Cassano et al. 2003).

3.2 Chemical methods (natural antimicrobials)

Apart from physical methods, some chemical preservatives are widely used for the shelf-life extension of fruit juices and beverages. The most commonly used preservatives are potassium sorbate and sodium benzoate. However, consumer demand for natural origin, safe and environmental friendly food preservatives is increasing. Natural antimicrobials such as *bacteriocins, lactoperoxidase*, herb leaves and oils, spices, chitozan and organic acids have shown feasibility for use in some food products (Gould 2001, Corbo et al. 2009). Some of them have been considered as Generally Recognized As Safe (GRAS) additives in foods. Table 7 lists some natural antimicrobials and their status for GRAS.

3.2.1 Bacteriocins

Bacteriocins are series of antimicrobial peptides which are readily degraded by proteolytic enzymes in the human body. Among them, nisin is the most commonly used food preservative and the GRAS additives permitted by the Food Additive Status List (USFDA, 2006). Apart from dairy, it has been used to preserve fruit and vegetable juices (Yuste & Fung 2004, Settanni & Corsetti, 2008).

3.2.2 Lactoperoxidase

Lactoperoxidase is an enzyme that is widely distributed in colostrum, raw milk and other body liquid. It is an oxidoreductase and catalyses the oxidation of thiocyanate with the consumption of H_2O_2, to produce intermediate products with antibacterial properties (Corbo et al. 2009). These products have been indicated to be bactericidal for some spoilage and pathogenic microorganisms and yeasts (Gould 2001). Not much information had been found on the application of lactoperoxidase in fruit juices. Until recently, it was used for the preservation of tomato juice and mongo fruits (Touch et al. 2004, Le Nguyen et al. 2005).

3.2.3 Herb, spice and flavor oils

Some herbs and spices contain essential oils, which are natural antimicrobials. The main elements of these antimicrobials are phenolic compounds, including caffeic, cinnamic, ferulic and gallic acids, oleuropein, thymol and eugenol (Gould 2001).

Name	Origin	GRAS status
Bay leaves	Plant	Yes
Chitozan	Animal	No
Cinamon	Plant	Yes
Clove	Plant	Yes
Coriander	Plant	No
Garlic	Plant	Yes
Lactoperoxidase	Animal	No
Lemon (peel, balm, grass)	Plant	Yes
Lime	Plant	Yes
Nisin	Microorganism	Yes
Onion	Plant	Yes
Rosemary	Plant	Yes
Sage	Plant	Yes
Thyme	Plant	Yes

* Revised from USFDA (2006): Food Additive Status List

Table 7. Selected natural antimicrobials and their status for GRAS additives*

Among them, sage (*Salvia officinalis*), rosemary (*Rosemarinus officinalis*), clove (*Eugenia aromatica*), coriander (*Coriandrum sativum*), garlic (*Allium sativum*) and onion (*Allium cepa*)) were listed as potential antimicrobials for food use (Deans and Ritchie 1987). The oils of bay leaves, cinnamon, clove and thyme were also proven to be highly effective for food pathogenic microorganisms including *Campylobacter jejuni, Salmonella enteritidis, Escherichia coli, Staphylococcus aureus* and *Listeria monocytogenes* (Smith-Palmer et al. 1998). It is believed that Gram-positive bacteria were more sensitive to inhibition by plant essential oils than the Gram-negative bacteria.

Cinnamon as an antimicrobial agent has been used in apple juice (Yuste and Fung 2004; Friedman et al. 2004), apple cider (Ceylan et al. 2004) and fresh-cut apple slices (Muthuswamy et al. 2008). Ground cinnamon (0.3%) could inhibit the growth of *Staphylococcus aureus, Y. enterocolitica* and *Salmonella typhimurium* in apple juice (Yuste and Fung 2004), whereas oils of cinnamon leaf or bark inactivated *Salmonella enterica* and *E. coli* O157:H7 in apple juice (Friedman et al.2004). Ethanol extract of cinnamon bark (1% to 2% w/v) and cinnamic aldehyde (2 mM) could reduce *E. coli* O157:H7 and *L. innocua in vitro*. Ethanol extract of cinnamon bark (1% w/v) reduced significantly the aerobic growth of bacteria inoculated in fresh-cut apples during storage at 6°C up to 12 days. It was also found that cinnamic aldehyde has greater antimicrobial activity than potassium sorbate (Muthuswamy et al. 2008).

Citrus fruits extracts have also been applied successfully to fruits and vegetables (Fisher & Phillips, 2008). For example, lemon extract was applied for the inhibition of some spoilage microorganisms, such as *Bacillus licheniformis, Lactobacillus spp., Pichia subpelliculosa, Saccharomyces cerevisiae and Candida lusitaniae*, the minimum inhibition concentration is 100 to 150 ppm (Conte et al. 2007). The growth of pathogenic bacteria, *Escherichia coli* O157:H7, *Listeria innocua* and the food spoilage fungus, *Penicillium chrysogenum* were suppressed by three phenolic compounds (catechin, chlorogenic acid and phloridzin) at 25 mM but the growth of food spoilage yeast *Saccharomyces cerevisiae* was inhibited only by chlorogenic acid and phloridzin (Muthuswamy and Rupasinghe, 2007). Vanillin, the predominant phenolic compound present in vanilla beans, has shown a concentration dependent response and the minimal inhibitory concentration (MIC) of 6 to 18 mM for pathogenic and spoilage microorganisms (Rupasinghe et al., 2006).

3.2.4 Chitozan

Chitosan is a modified, natural carbohydrate polymer derived by deacetylation of chitin [poly-β-(1 → 4)-N-acetyl-D-glucosamine] (No & Meyers, 1995). It is widely produced from crab, shrimp and crawfish, with different deacetylation grades and molecular weights which contribute to different functionalities (No et al., 2007).

Chitosan has attracted attention as a potential food preservative of natural origin due to its antimicrobial activity against a wide range of microorganisms (Sagoo and others 2002). The principles of the antimicrobial activity of chitosan could be explained by several hypotheses. One hypothesis is that the positively charged chitosan molecules could interact with the negatively charged microbial cell membranes, which would affect the cell permeability and lead to the leakage of intracellular compounds (Fang et al., 1994). Another hypothesis is that the interaction of diffused hydrolysis substances with microbial DNA could lead to the inhibition of the mRNA and protein synthesis of the microorganisms (Sudarshan et al., 1992).

A limited works have been done to assess the antimicrobial properties of chitosan in fruit juices (Roller and Covill 1999; Rhoades and Roller 2000). Chitosan glutamate was reported to be an effective preservative against spoilage yeasts in apple juice. Chitosan glutamate in apple juice from 0.1 to 5 g/L inhibited the growth of all spoilage yeasts at 25°C. The most sensitive strain, *Z. bailii*, was completely inactivated by chitosan at 0.1 and 0.4 g/L for 32-day of storage at 25°C. The most resistant strain, *S. ludwigii*, required 5 g/L of chitosan for complete inactivation and for maintaining yeast-free conditions in apple juice for 14 days at 25°C (Roller and Covill 1999). Another study by Rhoades and Roller (2000) showed that 0.3 g/L of Chitosan eliminated all the yeasts in pasteurized apple-elderflower juice during a 13-day of storage at 7°C. However, the total bacterial counts and the lactic acid bacterial counts increased slower than the control (Rhoades and Roller 2000).

Foods	Microorganism	Method used	*Log* reduction	Sources
Apple cider	*Escherichia coli* O157:H7	PEF (90 kV/cm, 20 µs, 42°C) with nisin or cinamon	6-8	Iu et al. 2001
Orange juice	Natural microflora	PEF (80 kV/cm, 44°C) with Nisin (100 U/ml)	6	Hodgins et al. 2002
Orange juice	*Escherichia coli* O157:H7	PEF (90 kV/cm, 100 µs, 55°C) with nisin and/or lysozyme	> 7	Liang et al. 2002
Apple cider	*Listeria monocytogenes*	US (48 kHz, 600 W, 5 min, 25°C) with copper, sodium hypochlorite	5	Rodgers and Ryser 2004
Apple cider	*Escherichia coli* O157: H7	US (20 kHz, 150 W, 18 min, 118 W/cm2) with heat (57°C)	> 7	D' Amico et al. 2006
Carrot juice	*Listeria innocua*	HPH (350 MPa) with nisin (10 IU/ml)	> 5	Pathanibul et al. 2009
Strawberry juice	*Escherichia coli* O157: H7 and *Salmonella Enteritidis*	PEF (35 KV/cm, 4 µs, <40°C) with 0.05% cinnamon bark oil or 0.5% citric acid	>5	Mosqueda-Melgar et al. 2008
Apple juice	*Escherichia coli* O157: H7 and *Salmonella Enteritidis*	PEF (35 KV/cm, 4 µs, <40°C) with 0.1% cinnamon bark oil or 1.5% citric acid	>5	Mosqueda-Melgar et al. 2008
Pear Juice	*Escherichia coli* O157: H7 and *Salmonella Enteritidis*	PEF (35 KV/cm, 4 µs, <40°C) with 0.1% cinnamon bark oil or 1.5% citric acid	>5	Mosqueda-Melgar et al. 2008
Strawberry juice	*Escherichia coli* O157: H7 and *Salmonella Enteritidis*	PEF (18.6 KV/cm, 150 µs, 55°C) with 750 ppm sodium benzoate and 350 ppm potassium sorbate	5.11	Gurtler et al. 2011
Strawberry juice	*Escherichia coli* O157: H7 and *Salmonella Enteritidis*	PEF (18.6 KV/cm, 150 µs, 55°C) with 750 ppm sodium benzoate, 350 ppm potassium sorbate and 2.7% citric acid	6.95	Gurtler et al. 2011

Table 8. Preservation of fruit juices by combination methods

Chitosan has been approved as a food additive in Japan in 1983 and in Korea in 1995. However, it is so far not a GRAS approved food additive by the FDA. As long as receiving the FDA approval for GRAS status, Chitosan as a food additive and its applications in food systems will certainly have a brighter future.

3.3 Combination of physical and chemical methods

It is proved that some individual non-thermal methods as well as natural antimicrobials are effective in inactivating microorganisms and at the same time do not adversely affect the sensory and nutritional quality of the fruit juice and other products. Moreover, the combination of these techniques could provide synergistic effects on prolonging the fruit juice shelf-life and potentially as replacement for traditional pasteurization methods. Table 8 and Table 9 list some examples of recent progresses in these combined techniques for the microbial inactivation (Table 8) and shelf-life extension (Table 9) of fruit juices and beverages.

Foods	Method used	Storage conditions	Quality changes	Sources
Apple cider	US (20 kHz, 750W, 4 min, 0.46W/mL) with heat (60°C)	Tested right after treatment	Titratable acidity, pH, and °Brix of the cider were not affected.	Ugarte-Romero et al. 2006
Apple juice	UV (5.3 J/cm^2) with PEF (34 kV/cm, 98 μs) or with US (20 kHz, 750 W, 5 bar, 43°C)	Tested right after treatment	No significant impact on non-enzymatic browning, total phenolics, antioxidant activity, flavor, color and odor of the juices.	Caminiti et al. 2011
Cranberry juice	UV (5.3 J/cm^2) with PEF (34 kV/cm, 98 μs) or with US (20 kHz, 750 W, 5 bar, 43°C)	Tested right after treatment	No significant impact on non-enzymatic browning, total phenolics, antioxidant activity, flavor, color and odor of the juices.	Caminiti et al. 2011

Table 9. Examples of combined preservation methods on quality attributes of fruit products

4. Conclusions

Non-thermal processing is a promising and useful approach for fruit juice and beverage preservation. The products based on these techniques show many advantages such as the retention of sensorial qualities and nutritional values over traditional thermal processing. However, among these non-thermal techniques, only high pressure processing has been adopted by the food industry so far. Additional pilot-scale testing may require for these non-thermal preservation methods to become a real alternative for thermal processing.

Similarly, the application of natural antimicrobial compounds in fruit juice and beverages is in the laboratory scale. But the potential benefits of these compounds would lead to a fast growth of scale-up and commercial application in food industry.

More practically, the combination of non-thermal techniques and natural antimicrobial compounds would be the future trends for fruit juice and beverages preservation, due to the proven records for effective inhibition of microorganisms and shelf-life extension of fruit juices and beverages.

5. References

Aguilar-Rosas, S.F., Ballinas-Casarrubias, M.L., Nevarez-Moorillon, G.V., Martin-Belloso, O. & Ortega-Rivas, E. (2007). Thermal and pulsed electric fields pasteurization of apple juice: Effects on physicochemical properties and flavour compounds, *J Food Eng* 83(1): 41-46.

Avure Technologies, www.avure.com/archive/documents/.../juice-and-beverage-data.pdf

Balasubramaniam, V. M., Farkas, D. & Turek, E. J. (2008). Preserving food through high pressure processing, *Food Technology*, 11: 33-38.

Barba, F.J., Esteve, M.J. & Frigola, A. (2011). Physicochemical and nutritional characteristics of blueberry juice after high pressure processing, *Food Res Int.* In press.

Barbosa-Cánovas, G.V., Góngora-Nieto, M.M., Pothakamury, U.R. & Swanson, B.G. (1999). *Preservation of foods with pulsed electric fields*, Academic Press, San Diego, pp. 4-47, 108-180.

Baumann, A.R., Martin, S.E. & Feng, H. (2005). Power ultrasound treatment of *listeria monocytogenes* in apple cider, *J Food Prot* 68(11): 2333-2340.

Bintsis, T., Litopoulou-Tzanetaki, E. & Robinson, R.K., (2000). Existing and potential applications of ultraviolet light in the food industry- a critical review, *Journal of Science, Food and Agriculture.* 80: 637-645.

Braddock, R. J. (1999). Single-strength orange juice and concentrates, *in* R. J. Braddock (Ed.), *Handbook of citrus by-products and processing technology,* Wiley, New York, pp. 53-83.

Caminiti, I.M., Noci, F., Munoz, A., Whyte, P., Morgan, D.J., Cronin, D.A. & Lyng, J.G. (2011). Impact of selected combinations of non-thermal processing technologies on the quality of an apple and cranberry juice blend, *Food Chem* 124(4): 1387-1392.

Cassano, A., Drioli, E., Galaverna, G., Marchelli, R., Di Silvestro, G., & Cagnasso, P. (2003). Clarification and concentration of citrus and carrot juice by integrated membrane process, *Journal of Food Engineering* 57: 153-163.

Ceylan, E., Fung, D.Y.C. & Sabah, J.R. (2004). Antimicrobial activity and synergistic effect of cinnamon with sodium benzoate or potassium sorbate in controlling *Escherichia coli* O157:H7 in apple juice, *J. Food Sci.* 69, 102–106.

Charles-Rodríguez, A.V., Nevárez-Moorillón, G.V., Zhang, Q.H. & Ortega-Rivas E. (2007). Comparison of thermal processing and pulsed electric fields treatment in pasteurization of apple juice, *Food Bioprod Process* 85(2C): 93-97.

Conte, A., Speranza, B., Sinigaglia, M. & Del Nobile, M.A. (2007). Effect of lemon extract on foodborne microorganisms, *Journal of Food Protection* 70: 1896–1900.

Corbo, M. R., Bevilacqua, A., Campaniello, D., D'Amato, D., Speranza, B. & Sinigaglia, M. (2009). Prolonging microbial shelf life of foods through the use of natural compounds and non-thermal approaches - A review, *International Journal of Food Science and Technology* 44(2): 223-241.

Cserhalmi, Z., Sass-Kiss, A., Tóth-Markus, M. & Lechner, N. (2006). Study of pulsed electric field treated citrus juices, *Innovative Food Science and Emerging Technologies* 7(1-2): 49-54.

D'Amico, D. J., Silk, T.M., Wu, J., Guo, M. (2006). Inactivation of microorganisms in milk and apple cider treated with ultrasound, J Food Prot 69(3): 556-563.

Deans, S.G. & Ritchie, G. (1987). Antibacterial properties of plant essential oils, *International Journal of Food Microbiology* 5: 165-180.

Evrendilek, G., Zhang, Q. H. & Richter, E. (1999). Inactivation of *Escherichia coli* O157:H7 and *Escherichia coli* 8739 in apple juice by pulsed electric fields, *Journal of Food Protection* 62: 793–796.

Evrendilek, G.A., Tok, F.M., Soylu, E.M. & Soylu, S. (2008). Inactivation of *Penicillium expansum* expansum in sour cherry juice, peach and apricot nectars by pulsed electric fields, *Food Microbiology* 25 (5): 662-667.

Evrendilek, G.A., Jin, Z.T., Ruhlman, K.T., Qiu, X., Zhang, Q.H. & Richter, E.R. (2000). Microbial safety and shelf-life of apple juice and cider processed by bench and pilot scale PEF systems, *Innovative Food Science and Emerging Technologies* 1(1): 77-86.

Fang, S.W., Li,C.F. & Shih, D.Y.C. (1994). Antifungal activity of chitosan and its preservative effect on low-sugar candied kumquat, *J Food Prot* 57: 136–140.

Fisher, K. & Phillips, C. (2008). Potential antimicrobial uses of essential oils in food: is citrus the answer? *Trends in Food Science and Technology* 19: 156–164.

Friedman, M., Henika, P. R., Levin, C. E. & Mandrell, R. E. (2004). Antibacterial activities of plant essential oils and their components against *Escherichia coli* O157:H7 and *Salmonella enterica* in apple juice, *Journal of Agricultural and Food Chemistry* 52(19): 6042-6048.

Gould, G. W. (2001). Symposium on 'Nutritional effects of new processing technologies' – New processing technologies: an overview, *Proceedings of the Nutrition Society* 60: 463-474.

Gómez-López, V.M., Orsolani, L., Martínez-Yépez, A. & Tapia, M.S. (2010). Microbiological and sensory quality of sonicated calcium-added orange juice, *LWT - Food Science and Technology* 43(5): 808-813.

Guerrero-Beltran, J. A., & Barbosa-Canovas, G. (2004). Review: advantages and limitations on processing foods by UV light, *Food Science and Technology International* 10: 137-147.

Guerrero-Beltran, J.A., Barbosa-Canovas, G.V. & Welti-Chanes, J. (2011). High hydrostatic pressure effect on natural microflora, *Saccharomyces cerevisiae, Escherichia coli*, and *Listeria Innocua* in navel orange juice, *International Journal of Food Engineering* 7 (1): Article number 14.

Gurtler, J.B., Bailey, R.B., Geveke, D.J., Zhang, H.Q. (2011). Pulsed electric field inactivation of *E. coli* O157:H7 and non-pathogenic surrogate *E. coli* in strawberry juice as influenced by sodium benzoate, potassium sorbate, and citric acid. Food Control, 22(10): 1689-1694.

Heinz, V., Toepfl, S., & Knorr, D. (2003). Impact of temperature on lethality and energy efficiency of apple juice pasteurization by pulsed electric fields treatment, *Innovative Food Science and Emerging Technologies* 4 (2): 167-175.

Hodgins, A.M., Mittal, G.S. & Griffiths, M.W. (2002). Pasteurization of fresh orange juice using low-energy pulsed electrical field, *J Food Sci* 67(6): 2294-2299.

Iu, J., Mittal, G. S., & Griffiths, M. W. (2001). Reductions in levels of *Escherichia coli* O157:H7 in apple cider by pulsed electric fields, *Journal of Food Protection* 64: 964-969.

Jin, Z.T. & Zhang, Q.H. (1999). Pulsed electric field inactivation of microorganisms and preservation of quality of cranberry juice, *Journal of Food Processing and Preservation* 23 (6): 481-497.

Keyser, M., Muller, I. A., Cilliers, F. P., Nel, W., & Gouws, P. A. (2008). Ultraviolet radiation as a non-thermal treatment for the inactivation of microorganisms in fruit juice, *Innovative Food Science and Emerging Technologies* 9(3): 348-354.

Knorr, D., Geulen, M., Grahl, T. & Sitzmann, W. (1994). Food application of high electric Field pulses, *Trends in Food Science and Technology* 5: 71-75.

Knorr, D., Zenker M., Heinz V. & Lee D. (2004). Applications and potential of ultrasonics in food processing, *Trends in Food Science and Technology* 15(5): 261-266.

Kumar, S., Thippareddi, H., Subbiah, J., Zivanovic, S., Davinson, P.M., & Harte, F., (2009). Inactivation of *Escherichia coli* K-12 in apple juice using combination of high pressure homogenization and chitosan. Journal of Food Science, 74, 8-14.

Lacroix, N., Fliss I., & Makhlouf J. (2005). Inactivation of pectin methylesterase and stabilizationof opalescence in orange juice by dynamic high pressure. Food Res Int 38(5):569-576.

Le Nguyen, D.D., Ducamp, M.N., Dornier, M., Montet, D. & Loiseau, G. (2005). Effect of the lactoperoxidase system against three major causal agents of disease in mangoes. Journal of Food Protection, 68, 1497–1500.

Liang, Z., Mittal, G. S., and Griffiths, M.W. (2002). Inactivation of Salmonella Typhimurium in orange juice containing antimicrobial agents by pulsed electric field. Journal of Food Protection. 65: 1081–1087.

Linton, M., McClements, J. M. J., and Patterson, M. F. (1999). Inactivation of *Escherichia coli* O157:H7 in orange juice using a combination of high pressure and mild heat, J. Food Protect., 62(3): 277–279.

Manvell, C. (1997). Minimal processing of food. Food Science and Technology Today, 11, 107–111.

McDonald, C. J., Lloyd, S. W., Vitale, M. A., Petersson, K., and Inning, F. (2000). Effects of pulsed electric fields on microorganisms in orange juice using electric fields strengths of 30 and 50kV/cm, Journal of Food Science. 5: 984–989.

Min, S., Jin, Z.T., Min, S.K., Yeom, H., Zhang, Q.H. (2003). Commercial-scale pulsed electric field processing of orange juice. *Journal of Food Science*, 68 (4): 1265-1271.

Mosqueda-Melgar, J., Elez-Martínez, P., Raybaudi-Massilia, R. M., & Martín-Belloso, O. (2008). Effects of pulsed electric fields on pathogenic microorganisms of major concern in fluid foods: A review. Critical Reviews in Food Science and Nutrition, 48(8), 747-759.

Moyer, J. C. & Aitken, H. C. (1980). Apple juice. In P. E. Nelson & D. K. Tressler (Eds.), Fruit and vegetable juice processing technology (pp. 212-267). Westport: AVI.

Muthuswamy, S. & Rupasinghe, H.P.V. 2007. Fruit phenolics as antimicrobial agents: selective antimicrobial effect of catechin, chlorogenic acid, and phloridzin. J. Agric. Food Environ. 5,81-85.

Muthuswamy, S., Rupasinghe, H. P. V. & Stratton, G. W. (2007). Antimicrobial effect of cinnamon bark extract on *Escherichia coli* O157:H7, *Listeria innocua* and fresh-cut apple slices. Journal of Food Safety, 28(4), 534-549.

Ngadi, M.O., J.P. Smith and B. Cayouette, (2003). Kinetics of ultraviolate light inactivation of *Escherichia coli* O157:H7 in liquid foods. J. Sci. Food Agri., 83: 1551-1555.

No, H.K. & Meyers, S.P. (1995). Preparation and characterization of chitin and chitosan - a review. Journal of Aquatic and Food Protection Technology, 4, 27–52.

No, H.K., Meyers, S.P., Prinyawiwatkul, W. & Xu, Z. (2007). Application of chitosan for improvement of quality and shelf life of foods - a review. Journal of Food Science, 72, 87-100.

O'Donnell, C.P., Tiwari, B.K., Bourke, P., & Cullen, P.J. (2010). Effect of ultrasonic processing on food enzymes of industrial importance. Trends in Food Science and Technology 21(7): 358-367.

Oms-Oliu, G., Odriozola-Serrano I., Soliva-Fortuny R., & Martín-Belloso O. (2009). Effects of high-intensity pulsed electric field processing conditions on lycopene, vitamin C and antioxidant capacity of watermelon juice. Food Chem 115(4): 1312-9.

Ortega-Rivas, E., Zárate-Rodríguez, E., & Barbosa-Cánovas, G. V. (1998). Apple juice pasteurization using ultrafiltration and pulsed electric fields. Food and Bioproducts Processing, 76C, 193–198.

Pathanibul, T., Taylor, M., Davidson, P. M. & Harte, F. (2009). Inactivation of *Escherichia coli and Listeria innocua* in apple and carrot juices using high pressure homogenization and nisin. International Journal of Food Microbiology, 129: 316-320.

Patrignani, F., Vannini L., Kamdem, S. L. S., Lanciotti R., Guerzoni, M. E. (2009). Effect of high pressure homogenization on *Saccharomyces cerevisiae* inactivation and physico-chemical features in apricot and carrot juices. International Journal of Food Microbiology, 136: 26-31.

Préstamo, G., Sanz, P. D., Fonberg-Broczek, M., & Arroyo, G. (1999). High-pressure response of fruit jams contaminated with *Listeria monocytogenes*, Lett. Appl. Microbiol., 28:313–316.

Ramaswamy, H.S., Riahi, E., & Idziak, E. (2003). High-pressure destruction kinetics of *E. coli* (29055) in apple juice, J Food Sci 68(5), 1750-1756.

Ramaswamy, H.S., Chen, C. & Marcotte, M. (2005). Novel processing technologies for food preservation, in Barrett, D. M., Somogyi, L. P. & Ramaswamy, H (ed.), Processing fruits: science and technology, Boca Raton, FL, USA, CRC Press, 211-214.

Rhoades, J. & Roller, S. (2000). Antimicrobial actions of degraded and native Chitosan against spoilage organisms in laboratory media and foods. Appl Environ Microbiol 66(1), 80-86.

Rodgers, S. L. & Ryser, E.T. (2004). Reduction of microbial pathogens during apple cider production using sodium hypochlorite, copper ion, and sonication. J Food Prot 67(4):766-771.

Roller, S. & Covill, N. (1999). The antifungal properties of chitosan in laboratory media and apple juice. Int J Food Microbiol 47, 67-77.

Rupasinghe, HPV, Boulter-Bitzer, J, Ahn, T, and Odumeru, JA. (2006). Vanillin inhibits pathogenic and spoilage microorganisms in vitro and aerobic microbial growth on fresh-cut apples. Food Res. Intern. 39,575-580.

Sagoo, S., Board, R. & Roller, S. (2002). Chitosan inhibits growth of spoilage micro-organisms in chilled pork products. FoodMicrobiol 19:175–182.

Salleh-Mack, S.Z. & Roberts, J.S. (2007). Ultrasound pasteurization: The effects of temperature, soluble solids, organic acids and pH on the inactivation of escherichia coli ATCC 25922. Ultrason Sonochem 14(3):323-329.

San Martín, MF, Barbosa-Cánovas GV, Swanson BG. (2002). Food processing by high hydrostatic pressure. Crit Rev Food Sci Nutr., 42(6): 627-645.

Settanni, L. & Corsetti, A. (2008). Application of bacteriocins in vegetable food biopreservation. International Journal of Food Microbiology, 121, 123–138.

Smith-Palmer, A., Stewart, J., & Fyfe, L. (1998). Antimicrobial properties of plant essential oils and essences against five important food-borne pathogens. Letters in Applied Microbiology, 26(2), 118-122.

Sudarshan NR, Hoover DG, Knorr D. (1992). Antibacterial action of chitosan. Food Biotechnol 6:257–272.

Tahiri, I., Makhlouf, J., Paquin, P. & Fliss, I. (2006). Inactivation of food spoilage bacteria and *Escherichia coli* O157:H7 in phosphate buffer and orange juice using dynamic high pressure. Food Res Int 39(1):98-105.

Tallarico, P., Todisco, S. & Drioli, E. (1998). Use of ultrafiltration in the preventing of orange juice bitterness and its effects on the aroma compounds distribution. Agro Food Industry Hi-Tech 9(3): 32-36.

Tiwari, B. K., Muthukumarappan, K., O'Donnell, C. P., & Cullen, P. J. (2009). Inactivation kinetics of pectin methylesterase and cloud retention in sonicated orange juice. *Innovative Food Science and Emerging Technologies,* 10(2): 166-171.

Tran, M.T.T. & Farid, M. (2004). Ultraviolet treatment of orange juice. Innovative Food Science and Emerging Technologies, 5(4): 495-502.

Tsong, T.Y. (1991). Electroporation of cell membranes. Biophys. J., 60, 297-306.

Toepfl, S., Mathys, A., Heinz, V., & Knorr, D. (2006). Review: Potential of high hydrostatic pressure and pulsed electric fields for energy efficient and environmentally friendly food processing. Food Reviews International, 22, 405–423.

Torres, B., Tiwari, B.K., Patras, A., Cullen, P.J., Brunton, N., ODonnell, C.P. (2011). Stability of anthocyanins and ascorbic acid of high pressure processed blood orange juice during storage. Innovative Food Science and Emerging Technologies 12(2), 93-97.

Touch, V., Hayakawa, S., Yamada, S. & Kaneko, S. (2004). Effects of a lactoperoxidase-thiocyanate-hydrogen peroxide system on Salmonella enteritidis in animal or vegetable foods. International Journal of Food Microbiology, 93, 175–183.

Ugarte-Romero, E., Feng, H., Martin, S.E., Cadwallader, K.R., Robinson, SJ. (2006). Inactivation of *Escherichia coli* with power ultrasound in apple cider. J Food Sci., 71(2): E102-108.

U.S. Food and Drug Administration. (2001). Hazard analysis and critical control point (HACCP); procedures for the safe and sanitary processing and importing of juices; final rule, *Fed. Regist* 66: 6138-6202.

U.S. Food and Drug Administration. (2006). Food Additive Status List. Available at: http://www.cfsan.fda.gov/~dms/opa-appa.html. Accessed 6/06/2011.

Valero, M., Recrosio, N., Saura, D., Muñoz, N., Martí, N., & Lizama, V. (2007). Effects of ultrasonic treatments in orange juice processing. J Food Eng 80(2):509-516.

Welti-Chanes, J., Ochoa-Velasco, C.E., & Guerrero-Beltrán, J.A. (2009). High-pressure homogenization of orange juice to inactivate pectinmethylesterase. Innovative Food Science and Emerging Technologies, 10, 457-462.

Yuan, Y., Hu, Y., Yue, T., Chen, T., Lo, Y.M. (2009). Effect of ultrasonic treatments on thermoacidophilic *Alicyclobacillus acidoterrestris* in apple juice. J Food Process Preserv 33(3): 370-383.

Yuste, J. & Fung, D.Y. (2004). Inactivation of Salmonella typhimurium and *Escherichia coli* O157:H7 in apple juice by a combination of nisin and cinnamon. Journal of Food Protection, 67, 371–377.

Zárate-Rodríguez, E., Ortega-Rivas, E., Barbosa-Cánovas, G.V. (2001). Effect of membrane pore size on quality of ultrafiltered apple juice. International Journal of Food Science and Technology 36(6): 663-667.

Zenker, M., Hienz, V., & Knorr, D. (2003). Application of ultrasound-assisted thermal processing for preservation and quality retention of liquid foods. Journal of food Protection, 66 (9), 1642–1649.

Zhang, Q.H., Barbosa-Cánovas, G.V., and Swanson, B.G. (1995). Engineering aspects of pulsed electric field pasteurization. J Food Eng., 25(2), 261-281.

5

Biosynthesis, Purification and Biotechnological Use of Exopolysaccharides Produced by Lactic Acid Bacteria

María Laura Werning[1], Sara Notararigo[1], Montserrat Nácher[2], Pilar Fernández de Palencia[1], Rosa Aznar[2,3] and Paloma López[1]
[1]*Departamento de Microbiología Molecular y Biología de las Infecciones, Centro de Investigaciones Biológicas (CSIC),*
[2]*Departamento de Biotecnología, Instituto de Agroquímica y Tecnología de Alimentos (CSIC),*
[3]*Departamento de Microbiología y Ecología, Universitat de València*
Spain

1. Introduction

Polysaccharides have been used traditionally by the food industry for their viscosifying, emulsifying and biothickening properties and more recently for manufacture of functional food due to their prebiotic and immunomodulating properties.

Bacteria can synthesize cytoplasmic storage polysaccharides (e.g. glycogen), cell wall structural polysaccharides such as peptidoglycan, and lipoteichoic acids of gram-positive bacteria, and the lipopolysaccharides anchored in the outer membrane of gram-negative bacteria. In addition, some bacteria can secrete polysaccharide layers on their surface, which together with a few glycoproteins, constitute the glycocalyx. These exocellular polymers comprise the capsular polysaccharides, which form a cohesive layer or capsule covalently linked to the cell surface, and the exopolysaccharides (EPS), which form a slime layer loosely attached to the cell surface or secreted into the environment (Brock, 2008). The physiological role of these molecules are not yet clearly understood, although it is generally recognized that exocellular polysaccharides are not normally used as energy and carbon sources by the producing microorganism. They can serve for a variety of functions including cell recognition and interaction, adherence to surfaces and biofilm formation.

The majority of the polysaccharides used as additives by the food industry such as pectin, cellulose and alginate are obtained from plants and algae. However, other biopolymers like xanthan and gellan, also used as bio-thickeners, are synthesized by gram-negative bacteria. Furthermore, lactic acid bacteria (LAB) producing EPS are used mainly in the dairy industry for improvement of the rheological properties of fermented products as well as for the manufacture of functional food.

The taste/texture benefits of the EPS produced by LAB in fermented foods are well established, because these organisms produce polymers that improve the rheological properties of dairy products. When they are added to food, polysaccharides show functions as thickeners, stabilizers, emulsifiers, gelling agents, and water binding agents (Kimmel et al., 1998). They also contribute to preservation, and enhance the organoleptic characteristics of milk and dairy products such as flavour and aroma (Macedo et al., 2002). More recently, these bio-molecules have been regarded as health promoters due to their role as prebiotics and/or the immunomodulatory properties linked to their structure. As a result, a number of studies are in progress in order to characterize the unmapped diversity of the EPS produced by LAB, since they are considered food-grade organisms.

In this chapter, we shall review the current knowledge pertaining to the EPS synthesized by LAB, from biogenesis to application, detailing their nature and structure. Moreover, the methods most frequently used for the production and purification of these biopolymers will be presented.

2. Composition, structure and classification of EPS

The EPS synthesized by LAB vary greatly in their chemical composition, structure and molecular weight. According to their chemical composition, EPS are classified into heteropolysaccharides (HePS) and homopolysaccharides (HoPS).

HePS are constructed of a backbone of repeated subunits that are linear or branched, with variable molecular masses (up to 10^6 Da). Each one of these subunits can contain between three and eight different monosaccharides and frequently has a range of different linkage patterns. The monosaccharides are present as the α- or β-anomer in the pyranose or furanose form and D-glucose, D-galactose and L-rhamnose are the most frequently encountered. In few cases, N-acetylglucosamine, manose, fucose, glucuronic acid and non-carbohydrate substituents (phosphate, acetyl and glycerol) are also present (de Vuyst & Degeest, 1999; de Vuyst et al., 2001).

Different strains of LAB isolated from dairy products, cereals and alcoholic beverages synthesize HePS. These belong to the genera *Lactococcus* (*L. lactis* subsp. *cremoris, L. lactis* subsp. *lactis*), *Lactobacillus* (*Lb. acidophilus, Lb. delbrueckii* subsp. *bulgaricus, Lb. casei, Lb. sakei, Lb. rhamnosus, Lb. helveticus*), *Streptococcus* (*S. thermophilus, S. macedonicus*) and *Leuconostoc* (*Lc. mesenteroides*) (Montersino et al., 2008; Mozzi et al., 2006; Van der Meulen et al., 2007).

HoPS are composed of repeated units that contain only one type of monosaccharide: D-glucopyranose (glucans) or D-fructopyranose (fructans). These polysaccharides usually display high molecular masses (up to 10^7 Da), and have different degrees and types of branching, linking sites and chain length. Based on their structure, the fructans can be divided into two groups: (i) inulins (linked β-2,1) and (ii) levans (linked β-2,6), both are synthesized by different species of the genera *Leuconostoc, Lactobacillus, Streptococcus* and *Weissella.*

Glucans can be classified into α- and β-D-glucans. The former are more widely found in LAB and they are produced by strains belonging to the genera *Lactobacillus, Leuconostoc* and *Streptococcus.* According to the linkages in the main chain, the α-glucans are subdivided into dextrans (α-1,6), mutans (α-1,3), glucans (α-1,2), reuterans (α-1,4) and alternans (α-1,3 and α-1,6) (Figure 1). These polymers may have side-chain branches that involve others α-linkages different from the main chain. For example, the dextrans produced by various LAB such as *Leuconostoc mesenteriodes* and Lactobacilli, may have branches with α-1,2, α-1,3 or α-1,4 linkages. The dextran most widely used by the industry is a polysaccharide containing 95% α-1,6 and 5% α-1,3 linkages synthesized by *Lc. mesenteroides* NRRL B-512F (Korakli & Vogel, 2006; Monsan et al., 2001; van Hijum et al., 2006).

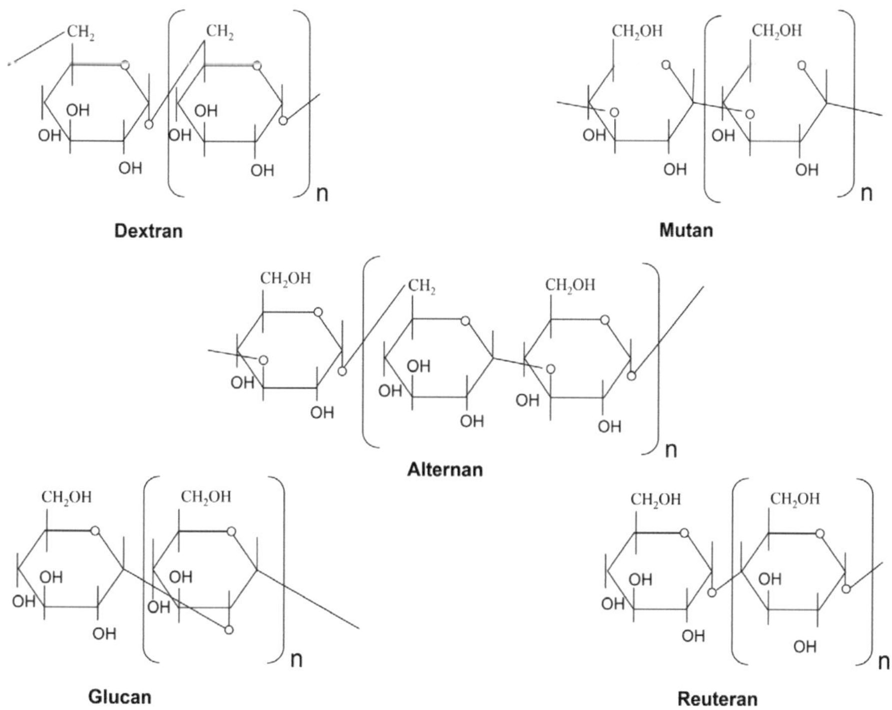

Fig. 1. Schematic representation of the repeating units of: dextran, mutan, alternan, glucan and reuteran (Korakli & Vogel, 2006)

(1,3) β-glucans are found in bacteria and eukaryotic organisms. These polysaccharides include the linear glucans and 6-substituted (1,3) β-glucans that have branch-on-branch or cyclic structures. Concerning prokaryotes, several bacteria including *Agrobacterium* and *Rhyzobium* species can produce these polymers. One such structure, curdlan, has been approved as a food additive by the Food and Drug Administration (FDA), and essentially is a linear (1,3) β-D-glucan which may have a few inter- or intra- chain (1,6) linkages (McIntosh et al., 2005).

β-glucan production is rarely found in LAB. It has only been reported to be synthesized and secreted by a small number of strains isolated from alcoholic beverages, namely: *Pediococcus parvulus* IOEB8801 and *Oenococcus oeni* IOEB0205 from wine and *P. parvulus* 2.6R, CUPV1, CUPV22, *Lb. diolivorans* G77 and *O. oeni* I4 from cider (Dueñas-Chasco et al., 1997, 1998; Garai-Ibabe et al., 2010; Ibarburu et al., 2007; Llauberes et al., 1990).

In all cases, these β-D-glucans have a common structure comprising a main chain of (1,3)-linked β-D-glucopyranosyl units along with more or less frequent side chains of β-D-glucopyranosyl units attached by (1,2) linkages (Figure 2).

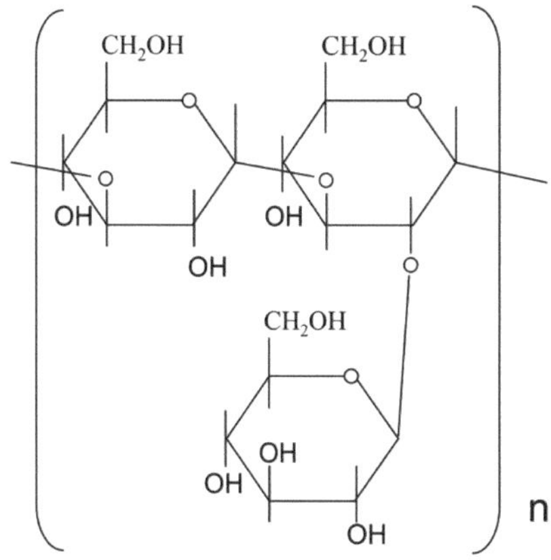

Fig. 2. Schematic representation of the repeating unit of β-(1,3-1,2)-D-glucan of *P. parvulus* 2.6 (Dueñas-Chasco et al., 1997)

In addition, a similar polymer constitutes the capsule of *S. pneumoniae* serotype 37 and the EPS secreted by *Propionibacterium freundenreichii* subsp. *shermanii* TL34 (Adeyeye et al., 1988; Nordmark et al., 2005).

3. Biosynthesis of EPS

3.1 Genetic determinants and mechanisms of production and secretion

The genes involved in the biogenesis of the HePS are usually organized in clusters that can be located either in the chromosome of the thermophilic LAB (e. g. *S. thermophilus* Sfi6) or in plasmids of mesophilic bacteria (e.g. *L. lactis* subsp. *cremoris* NIZO B40) (Laws et al., 2001). This structural organization is highly conserved among LAB and is very similar to that observed for the operons and clusters involved in the synthesis of: (i) O-antigen lipopolysaccharides in enterobacteria, (ii) capsules (CPS) of pathogens, such as *S. pneumoniae* or *Staphylococcus aureus* and (iii) the EPS from *Sinorhizobium meliloti* (García et al., 2000; Glucksmann et al., 1993; Lin et al., 1994).

The clusters from LAB have been reported for *S. thermophilus, Lb. helveticus, L. lactis, Lb. delbrueckii* subsp. *bulgaricus, Lb. rhamnosus* and the *Lb. casei* group (Ruas-Madiedo et al, 2008). The genes are oriented in a single direction and transcribed as a single mRNA. The genes are grouped into four regions within the cluster: The first contains genes whose products are regulatory proteins; the second includes genes encoding proteins involved in polymerization/chain length determination; the third contains genes encoding enzymes required for the biosynthesis of the HePS repeating units, and the genes of the last region encode proteins implicated in transport and polymerization (Figure 3) (Jolly & Stingele, 2001).

Additionally, these clusters may occasionally include genes involved in biosynthesis of nucleotide sugars from which the repeating units are constructed. Thus, in *Lb. rhamnosus*, they are associated to the EPS operon and they can be transcribed either from their own promoter or together with the EPS operon genes (Péant et al., 2005). There is a great variability in the genes involved in the synthesis of the repeating units. This region is responsible for the production of a specific EPS type; whereas the genes of polymerization and transport are more conserved.

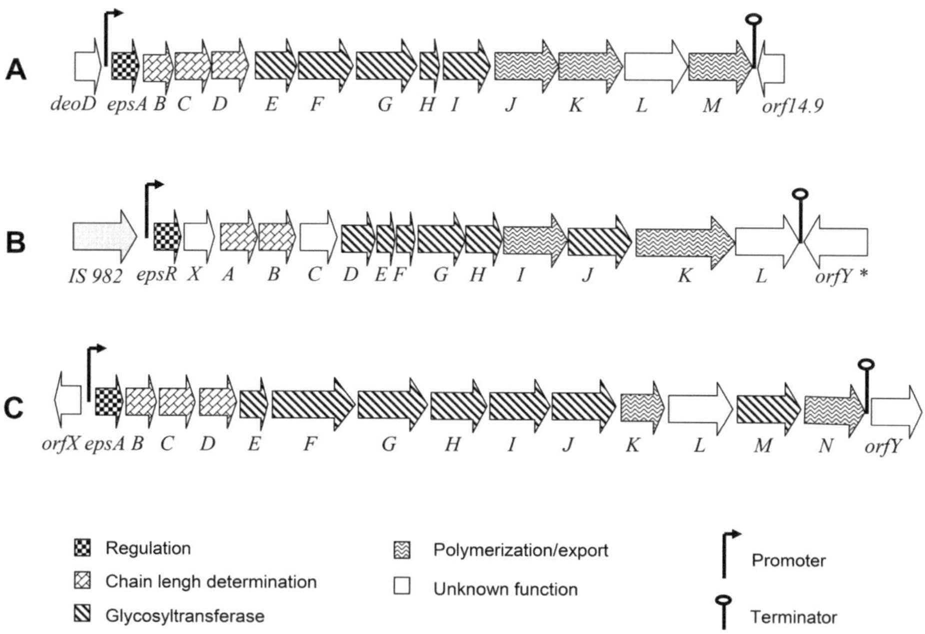

Fig. 3. Organization of the *eps* gene clusters of (A) *S. thermophilus* Sfi6, (B) *L. lactis* subsp. *cremoris* NIZO B40 and (C) *Lb. bulgaricus* Lfi5 (Lamothe et al., 2002; Laws et al., 2001). The proposed function of the different gene products is indicated

HePS are made by the polymerization of repeating unit precursors formed in the cytoplasm. They are assembled at the membrane by the sequential addition of nucleotide sugars (e.g., UDP-glucose, UDP-galactose and dTDP-rhamnose) or nucleoside diphosphate sugars to the growing repeated units through the action of the specific glycosyltransferases. This unit is most probably anchored to a lipid carrier molecule located in the plasma membrane and the first monosaccharide is linked by the action of a type of glycosyltransferase called priming-glycosyltransferase. It has been proposed that this lipid carrier might be an isoprenoid derivative such as undecaprenyl-phosphate (C55-P) by analogy with the synthesis of other EPS of gram-negative bacteria, as well as in the assembly of peptidoglycan, lipoteichoic acids and O-antigen lipopolysaccharides (Ruas-Madiedo et al., 2009). However some studies based in the resistance to bacitracin (a compound that blocks the transformation of the C55-PP to C55-P), suggested that HePS biosynthesis in *S. thermophilus* Sfi6 uses a lipid carrier different from undecaprenyl carrier (Stingele et al., 1999).

The mechanisms of polymerization, chain length determination and export still remain poorly understood. However, the similarity of gene products involved in these processes to those participating in the polymerization and export of O-antigens from *Escherichia coli* and the EPS of *S. meliloti*, suggests that probably LAB utilize similar mechanisms for polymerization and export of EPS. Thus, an enzyme of the flipase family could transfer the lipid-bound repeating units from the cytoplasmic face of the membrane to the external face.

Using the same line of argument, a polymerase could catalyze the linking of the repeating units and an enzyme could uncouple the lipid-bound polymer and control chain length (Laws et al., 2001).

The regulatory mechanisms involved in expression of the HePS genes remain unclear. However, a transcriptional regulatory role has been proposed for a group of gene products in different LAB genera, due to their similarity with a family of transcriptional regulators whose prototype is LytR, the regulator of the autolysin operon of *Bacillus subtilis* (Ruas-Madiedo et al., 2009).

Finally, a variety of nucleotide sugars is needed for the synthesis of a range of polysaccharides which is not specific to EPS biosynthesis. The production of these precursors occurs in the cytoplasm, mainly from glucose 1-P which is synthesized from glucose 6-P. These two forms of phosphorylated glucose are part of the central metabolism of the bacterium, which begins with the transport of sugar to the interior of the cell (de Vuyst et al., 2001).

Concerning to HoPS, most of them (α-glucans and fructans) are synthesized from sucrose through the action of extracellular enzymes commonly named glycansucrases. Enzymes synthesizing α-glucan polymers are limited to LAB, while enzymes synthesizing fructans, are present in other gram-positive and gram-negative bacteria. A large number of these glycansucrase encoding genes have been identified in the chromosomes of *Streptococcus*, *Leuconostoc* and *Lactobacillus* strains and they usually form part of a monocistronic transcriptional unit. Some of these genes are expressed constitutively whilst other are sucrose-inducible (Gänzle & Schwab, 2009; van Hijum et al., 2006).

Only one gene, named *gtf*, is required for the β-glucan biosynthesis in LAB. This gene is located in a 35 kb plasmid of *P. parvulus* 2.6 or in a 5.5 kb plasmid of *P. parvulus* IOEB8801, CUPV1, CUPV22, *Lb. diolivorans* G77 and *Lb. suebicus* CUPV221, while *gtf* is located in the chromosome of *Oenococcus oeni* I4 and IOEB0205 (Dols-Lafargue et al., 2008; Garai-Ibabe et al., 2010; Walling et al., 2001; Werning et al., 2006). All of these bacteria produced the same 2-substituted (1,3)-β-D-glucan and their *gtf* genes show high level of homology (at least 97%).

The *gtf* gene encodes a β-glycosyltransferase (GTF). It is a membrane protein whose topological prediction indicates that β-glucan or, at least its repeating unit precursors, are synthesized in the cytosol (Werning et al., 2006). In agreement, *in vitro* experiments indicate that the β-glucan polymer is synthesized by GTF directly from UDP-glucose (Werning et al., 2008) which, as mentioned above, is part of the central metabolism of the bacterium and is available within the cell. The translocation through the membrane is performed by a mechanism that is not yet known. The fact that heterologous expression of GTF in other LAB leads to the synthesis and secretion of β-glucan in the recipient bacteria, strongly suggests that this polymer does not require specific transporters to be released into the extracellular space (Stack et al., 2010; Werning et al., 2008). Supporting this hypothesis, the capsular 2-substituted (1,3)-β-glucan of *S. pneumoniae* serotype 37, is synthesized by a single β-glycosyltransferase called Tts (which shares a 33% identity with GTF), and its heterologous expression in other gram-positive bacteria, also result in the capsular EPS formation (Llull et al., 2001).

The presence of four potential transmembrane regions at the C-terminal region of GTF glycosyltransferase and two more at its N-terminus suggest the formation of a membrane pore by the enzyme to facilitate the extrusion of the polymer (Werning et al., 2006). In this regard, there is experimental evidence (Heldermon et al., 2001; Tlapak-Simons et al., 1998, 1999) suggesting that in the HAS glycosyltransferase from *S. pyogenes* only four transmembrane domains and two membrane-associated regions are sufficient to interact with the membrane phospholipids and to create a pore-like structure through which a nascent hyaluronan chain can be extruded to the exocellular environment. A similar mechanism could happen in the translocation of curdlan through the *Agrobacterium* membrane, for which it has been proposed that a pore could be formed by interaction of the CrdS synthase with phosphatidylethanolamine (Karnezis et al., 2003). Thus, the association of several GTF monomers could form a pore for extrusion of the EPS and, that might promote the processive catalysis of GTF carried out by this enzyme. However other alternative mechanisms might allow the translocation or secretion of β-glucan across LAB membranes. This includes the use of an ABC-like transporter, since this class of transporter can export various bacterial polysaccharides (Silver et al., 2001).

Finally, it should be noted that the presence of mobile genetic elements is a common feature in the genetic organization of the DNA region involved in HePS or HoPS synthesis (Bourgoin et al., 1999; Dabour & LaPointe, 2005; Peant et al., 2005; Tieking et al., 2005; van Hijum et al., 2004). It is well known that these elements allow horizontal transfer between different genera. In this regard, its presence could explain the instability of the HePS-producer phenotype of some strains (Ruas-Madiedo et al., 2009) as well as the loss of expression of some glycansucrases or the presence of chimeras in different HoPS-producing strains (Gänzle & Schwab, 2009). The *gtf* genes of LAB are flanked by genes which could be involved in functions of conjugation and recombination, respectively (Werning et al., 2006; Dols-Lafargue et al., 2008). Thus, horizontal transfer mediated by plasmids or transposition events, might explain the wide distribution and high degree of *gtf* gene preservation in β-glucan-producing strains belonging to different genera.

3.2 Enzymes for production

HePS and β-glucans are produced by glycosyltransferases which use nucleotide sugars as substrate. On the other hand, α-glucans and fructans are synthesized by glycansucrases which are able to use the energy of the glycosidic bond of sucrose to ligate glucose or fructose to the growing polysaccharide chain. In addition, these enzymes can synthesize hetero-oligosaccharides, when the acceptors are maltose and isomaltose (Monsan et al., 2001).

Enzymes synthesizing α-glucan polymers are called glucansucrases (GS) and those synthesizing fructans are named fructansucrases (FS). Unlike the glycosyltransferases (discussed below), GS and FS are transglycosidases evolutionarily, structurally, and mechanistically related to the glycosyl-hydrolases (GH). Therefore, according to the classification of the GH into families (which is based on the amino acid sequences) GS and FS can be respectively placed within the GH70 and the GH68 families (Henrissat & Bairoch, 1996). In addition, according to the Nomenclature Committee of the International Union of Biochemistry and Molecular Biology (UIBMB), GS enzymes are classified (according to the reaction they catalyze and the type of product) into: (i) dextransucrases (E.C.2.4.1.5) and (ii) alternansucrases (E.C.2.4.1.140). The mutansucrases and reuteransucrases are classified together with the dextransucrase enzymes. Also, FS enzymes can be classified, based on the different products synthesized, into: (i) inulosucrases (E.C.2.4.1.9) and (ii) levansucrases (E.C.2.4.1.10) (van Hijum et al., 2006).

Although GS and FS enzymes perform very similar reactions on the same substrate, they do not show great similarities in their amino acid sequence and strongly differ in their protein structures. Regarding the amino acid sequence, these enzymes are composed of 4 structural domains from the N to the C terminus: (i) a signal peptide involved in the secretion of the enzyme; (ii) a variable N-terminal domain of unknown function; (iii) a conserved catalytic domain comprising sucrose-binding and sucrose-cleaving domain and in FS also a calcium ion binding site and (iv) a C-terminal domain, that is composed of a series of tandem repeats which is thought to be involved in the control of product size as well as in α-glucan binding (GS), or in cell wall anchoring (FS) (Korakli & Vogel, 2006).

The mechanism of action of GS is still not fully understood. The key step in the transfer of D-glucosyl units is the formation of a covalent glucosyl-enzyme intermediate, in which an amino acid triad, composed of two aspartate and one glutamate residues, is involved. From this intermediate, the glucosyl unit is transferred to the acceptor (the polymer in growth) by a processive catalytic mechanism. The overall synthesis can be described as three steps: initiation, elongation and termination. The last corresponds to the dissociation of the α-glucan-enzyme complex. Regarding the elongation, this can proceed by two alternative mechanisms, one acts at the reducing end and the other at the non-reducing end of the growing α-glucan chain (Monchois et al., 1999; Monsan et al., 2001).

So far little is known about the mechanism of action of the FS enzymes. Fructan biosynthesis could be carried out by a multiple elongation mechanism, where the fructose residues are added to the growing fructan chain. The catalytic mechanism proposed for the transfructosylation reaction occurs in two steps, involving both a nucleophilic and an acidophilic site, through the formation of a covalently linked intermediary fructosyl-enzyme (Monsan et al., 2001; Sinnott, 1991).

The glycosyltransferases are ubiquitous enzymes in prokaryotes and eukaryotes. They are involved in the biosynthesis of oligosaccharides, polysaccharides and glyconjugates (e.g. lipopolysaccharides and glycoproteins). These enzymes are responsible for the biosynthesis of glycosidic bonds by the transfer of a sugar residue from activated donor molecules to specific acceptor molecules. Donor sugar substrates are mostly nucleotide sugars; however they can also be nucleoside monophosphate sugars, lipid phosphate sugars and sugar 1-phosphate. Frequently the acceptors are other sugars but they can be lipids, nucleic acids, antibiotics, etc. Additionally, two stereochemical outcomes are possible for reactions that result in the formation of a new glycosidic bond: the anomeric configuration of the product can be retained (α-glycosyltransferases) or inverted (β-glycosyltransferases) with respect to the donor substrate (Lairson et al., 2008).

The HePS biosynthesis in LAB involves several enzymes for production of repeating units: a *priming*-glycosyltransferase that transfers the first sugar from sugar 1-phosphate onto a phosphorylated carrier lipid and one or more α- or β-glycosyltransferases that sequentially add new sugars from nucleotide sugars to the growing repeating unit. Some already characterized examples of these enzymes are: EpsE, a phospho-galactosyltransferase from *S.*

thermophilus Sfi6 and EpsD a phospho-glucosyltransferase from *L. lactis* subsp. *cremoris* B40, which are priming-glycosyltransferases. Examples of other glycosyltransferases are: EpsG, an N-acetyl-glucosaminetransferase from *S. thermophilus* Sfi6, which transfers N-acetylglucosamine to a β-galactose precursor anchored to a carrier lipid or EpsG, that catalyzes the linkage of galactose to a cellobiose precursor anchored to a carrier lipid from *L. lactis* subsp. *cremoris* (Ruas-Madiedo et al., 2009).

As mentioned above, the 2-substituted (1,3)-β-D-glucan biosynthesis of *P. parvulus* 2.6 is carried out by the GTF glucosyltransferase. GTF was overproduced in *L. lactis* NZ9000 and purified as a membrane-associated enzyme (Werning et al., 2008). These membrane preparations used UDP-glucose as donor substrate to catalyze the biosynthesis of a high molecular weight polysaccharide that corresponds to the 2-substituted (1,3)-β-D-glucan (Werning et al., 2008 and unpublished results). The acceptor is so far unknown, but it could be the growing β-glucan polymer or any lipid molecule present in the cellular membranes.

Traditionally, glycosyltransferases have been classified on the basis of their donor, acceptor and product specificity according to the recommendations of the IUBMB. However, this system requires full characterization before an Enzyme Commission (EC) number can be assigned. To overcome this limitation, these enzymes have been classified into families on the basis of amino acid sequence similarities as in the case of GH. At present there are already 91 families (referred to as GTx), available at URL: http://www.cazy.org/fam/acc_GT.html. This classification is periodically reviewed and updated (Coutinho et al., 2003).

According to protein sequence similarity, enzymes that produce HePS can be grouped into various families. With regard to HoPS, the dextransucrases have been included in the family of glycosyl hydrolases GH70. In addition, GTF belongs to the GT-2 family that includes other glycosyltransferases such as cellulose synthases, β-1,3 glucan synthases, chitinsynthases, HAS and β-glucosyltransferases. All of these enzymes have in common that they use a nucleotide sugar as substrate for the synthesis of a polymer, with inversion of anomeric configuration of the donor substrate.

Two general three-dimensional (3D) folding models, called GT-A and GT-B have been observed for all the structures of nucleotide-sugar-dependent glycosyltransferases solved to date, and two mechanisms for retaining or inverting enzymes can be proposed within both classes. The GT-A fold may be considered as two tightly associated and abutting β/α/β Rossmann domains that tend to form a continuous central sheet of at least eight β-strands. The GT-B fold consists of two β/α/β Rossmann domains that face each other and are linked flexibly. These domains correspond to the donor substrate and acceptor binding sites (Breton et al., 2006; Lairson et al., 2008).

So far, no 3D-structure has been resolved for any glycosyltransferase involved in EPS synthesis from LAB, which would be essential for a better understanding of the mode of action of these enzymes. However, using the GT classification system proposed by Coutinho et al (2003) (in which families can be classified into clans on the basis of their folding and stereochemical outcome of the reaction that they catalyze) it is possible to predict that the glycosyltransferases can adopt one of two possible foldings. Thus, GTF from *P. parvulus* 2.6 belongs to the inverting-clan I of GT-A glycosyltransferases. A 3D-dimensional model based on the sequence of the putative active domain of GTF was built using as template the experimentally resolved structures of SpsA synthase from *Bacillus subtilis* (PDB code 1qgq) and of a putative glycosyltransferase from *Bacteroides fragilis* (PDB code 3bcv). The model proposed for GTF corresponds to a GT-A fold, which consists of an open twisted β-sheet surrounded by α-helices on both sides, where the N- and C-terminus domains are respectively the donor substrate and the acceptor binding sites (Figure 4).

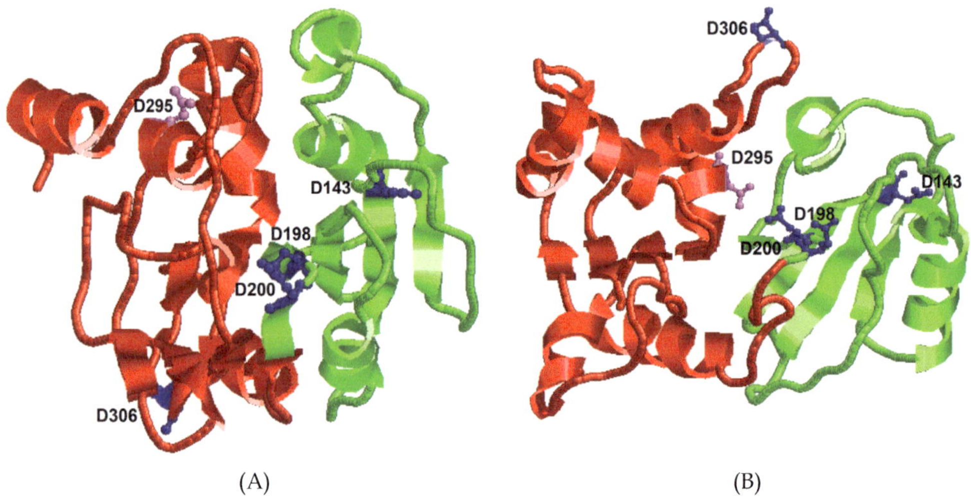

Fig. 4. Three-dimensional model for the putative active site of GTF constructed using resolved structures of SpsA from *B. subtilis* (A) and a putative glycosyltransferase from *Bc. fragilis* (B). The conserved N-terminal nucleotide-binding domain is shown in green and the C-terminal acceptor domain in red. The ball-and-stick representations show putative residues involved in the catalytic center: D143, D198 and D200 at the N-terminal domain and D306 at the C-terminal domain (in blue) as well as D295 (in violet), which is an alternative residue to D306, whose putative functions are discussed in the text

Based on structural data and on site directed mutagenesis experiments, it has been proposed that only four aspartic or carboxylate groups are required to form a single catalytic center of GT-A inverting enzymes (Charnock et al., 2001; Garinot-Schneider et al., 2000; Keenleyside et al., 2001; Tarbouriech et al., 2001). Three Asp located at the N-terminus domain are involved in nucleoside diphosphate (NDP) coordination. The first Asp residue is implicated in the recognition of uracil or thymine base. The other two, commonly referred to as DXD motif, bind to hydroxyl groups on the ribose moiety and the divalent metal ion (Mg^{2+} or Mn^{2+}), which in turn coordinate the phosphate groups from NDP, facilitating the release of the NDP from the donor substrate. A fourth residue, Asp or Glu, in the acceptor domain (C-terminus) acts at the catalytic site by activating the acceptor hydroxyl group, which would subsequently perform a nucleophilic attack on the C1 of the donor substrate (Lairson et al., 2008).

Candidate aspartate residues exist in GTF and it is possible to predict their location in the active site of this protein based on the 3D-structural model shown in Figure 4. Three aspartates (D143, D198 and D200) are coincident with those conserved and proposed in other glycosyltransferases from the GT-2 family, including glycosyltransferase SpsA of *B. subtilis*. The fourth aspartate (D306) is not a good candidate to be part of the active site according to the current models and one alternative could be that at D295, though this should be evaluated by substrate docking experiments and site directed mutagenesis.

4. Production, purification and analysis of EPS

4.1 Methods of production

Most LAB species show low yields of polysaccharide production which is the main reason for their lack of commercial exploitation. Generally (with a few exceptions) the yield of production is under 1 g L^{-1} for HoPS, when culture conditions are not optimized, and even less for the majority of the HePS. Van der Meulen et al. (2007) reported the EPS production by 10 LAB strains isolated from dairy and cereal products. Nine out of ten produced glucans in amounts from 0.8 to 1.7 g L^{-1}. The only HePS producer was *Lb. curvatus*, which synthesized the EPS to levels of 22 mg L^{-1}. Mozzi et al. (2006) reported that 31 HePS producers screened from 201 LAB strains (including thermophilic and mesophilic species) synthesized from 10 to 166 mg L^{-1}. Only seven of them produced > 100 mg L^{-1}.

Efforts have been made to improve yields of EPS production by LAB resulting in a variety of methods well detailed in the literature. All of them focus on parameters that have a strong influence on the production of HoPS and HePS. As a general rule the amount and composition of the EPS produced by LAB is strongly influenced by the culture and fermentation conditions such as pH, temperature and medium composition (Dueñas et al., 2003).

The production of α-glucans by LAB can be obtained in the presence of sucrose, after optimization of sucrose concentration in the growth medium and the time of incubation. The depletion of the sucrose source would cause the arrest of the enzymatic reaction of the dextransucrases. It has been reported that high producers of dextran are primarily *Leuconostoc*, but many other strains of LAB are able to produce this bacterial polysaccharide (Sarwat et al., 2008). The importance of improving its production is related to the industrial applications in the food, pharmaceutical and chemical industries as adjuvant, emulsifier, carrier or stabilizer (Goulas et al., 2004). *Lc. mesenteroides* CMG713 produces the highest concentration of dextran after 20 hours of incubation at 30°C in the presence of 15% sucrose at pH 7.0, with an EPS yield of 6 g L^{-1} (Sarwat et al., 2008). Recently, Capek et al. (2011) reported an exceptionally high production of this HoPS (50 g L^{-1}) by *Lc. garlicum* PR.

On the contrary, β-glucan production is very tedious, because of the low yield obtained. Therefore, it demands new strategies to improve synthesis apart from the optimization of growth parameters. For production of the 2-substituted (1,3) β-glucan heterologous gene expression has been tested. A plasmid, pNGTF, was constructed in order to express the *P. parvulus gtf* in *L. lactis* NZ9000. This plasmid allows inducible expression of the *gtf* gene from the nisA gene promoter by the addition of nisin to the growth medium (Werning et al., 2008). The EPS released to the medium by NZ9000[pNGTF] was quantified and purified. The expression of GTF glycosyltransferase by NZ9000[pNGTF] yielded levels of purified EPS of 300 mg L^{-1}, when the bacteria was grown in batch conditions (Werning et al., 2008). The structural characterization of the purified EPS confirmed that the recombinant strain synthesizes and secretes the same 2-substituted (1,3)-β-D-glucan as *P. parvulus* 2.6 (Werning et al., 2008). The synthesis of the EPS was still not very high, but it could probably be improved by optimizing growth conditions of the producing recombinant strain in continuous culture in a chemostat at controlled pH.

With regard to HePS, their production has been improved in the native isolates by optimizing growth conditions and media composition. As an example, Vijayendra and Babu (2008) optimized the EPS production by *Leuconostoc* sp. CFR-2181, a strain isolated from dahi, an Indian traditional fermented dairy product, in a simple, low cost semisynthetic medium. Maximum biomass and HePS production was observed when sucrose was used as carbon source and a yield of 30 g L^{-1} of the biopolymer was obtained, although the concentration was estimated by dry weight and purity of the HePS was not assayed. As an alternative, to improve production of HePS, the *eps* clusters of various LAB have been overexpressed, mainly in *L. lactis*, by construction of recombinant strains carrying the genes in multicopy plasmids (e.g. the HePS of *S. thermophilus* Sfi39). In addition, metabolic engineering of *L. lactis* has been used to redirect carbon distribution between glycolysis and nucleotide sugar biosynthesis, with the aim of increasing intracellular levels of UDP-glucose, UDP-galactose and UDP-rhamnose, the substrates of the glycosyltransferases encoded by plasmid pNZ4000 and involved in HePS biosynthesis in the NIZO B40 strain (Boels et al., 2002).

The importance that EPS has gained in the food industries has been responsible for the development of other strategies to improve the total amount produced. Some good examples are their *in situ* production in food matrices and their *in vitro* production by the use of immobilized enzymes.

LAB can produce a large variety EPS during elaboration of dairy products. Since the use of LAB is historically considered safe (GRAS microorganisms), production *in situ* of novel functional EPS means that toxicological testing will be reduced, or not required, and the products can be quickly brought to the market (De Vuyst et al., 1999).

Yogurt is a well-known dairy product derived from milk fermentation by cultures producing EPS (e.g. *Lb. delbrueckii ssp. bulgaricus* and *S. thermophilus* produce, respectively, 60-150 mg L^{-1} and 30-890 mg L^{-1} of HePS) (Marshall and Rawson, 1999). The use of EPS-producing starter cultures for yogurt elaboration is increasing, because these biopolymers improve water retention and texture and confer thickness without altering the organoleptic characteristics of the final product. Thus, there is no need to add stabilizers, many of which are prohibited in a wide range of countries. Although the role of pure EPS has not been studied, authors agree that the key points of improving the texture are the conformation of the EPS and their interactions with casein (Badel et al., 2011).

In the cheese making process strains such as *Lb. delbrueckii ssp. bulgaricus*, *Lb. helveticus* and *Lb. casei*, produce HePS. These polysaccharides, as occurs in yogurt, help to improve the rheological properties of the cheese. Their role in cheese elaboration depends on associations with other strains and also on the presence or absence of charges in the EPS produced (Badel et al., 2011).

Furthermore, there are other examples of *in situ* EPS production, such as kefir. This is a very important beverage in Eastern European countries. It is a fermented milk product produced by a population of different species of bacteria and yeasts. Several functional properties of kefir, such as is ability to modulate immune responses, to diminish allergic reactions and to inhibit tumour growth have been postulated for this beverage (Liu et al., 2002). LAB produce lactic acid and yeast synthesize ethanol, which is used by *Lb. kefiranofaciens* to produce kefiran, a polysaccharide resistant to enzymatic degradation. Kefiran is a natural biopolymer that could be used as a thickener in fermented products.

Finally, the use of enzymes for the *in vitro* synthesis of polysaccharides is showing promise. Immobilized enzymes are preferred as this allows the recovery and reuse of the enzyme, and may improve the properties of the enzyme such as stability, activity, specificity and selectivity (Mateo et al., 2007). The technique is used for production of isomaltooligosaccharides by immobilized dextransucrase. Isomaltooligosaccharides are oligosaccharides with prebiotic activity that can be produced either by acceptor reactions of dextransucrase or hydrolysis of dextran by dextranase. In the case of dextransucrase, it can also produce leucrose (a disaccharide used as sweetener). Other uses are also currently being developed e.g. immobilization of β-galactosidase to produce galactooligosaccharides. Concerning immobilization technology, the alginate encapsulation method has the best performance, rendering yields of up to 90% (Tanriseven & Dogan, 2002). Dextranase has been immobilized on various supports including glutaraldehyde-activated chitosan, porous glass, bentonite and the commercially available matrix, Eupergit C, with high yield (90%) (Aslan & Tanriseven, 2007). Dextransucrase and dextranase share the optimum pH (pH 5.4) which facilitates their combined use. However, few studies using co-immobilization of dextransucrase and dextranase are yet available (Erhardt et al., 2008; Olcer & Tanriseven, 2010).

4.2 Methods of purification and characterization of EPS

Purification is the physical separation of a chemical substance of interest from contaminating substances. In the case of EPS, purification from bacterial culture supernatants means elimination of producer microorganisms and their secreted metabolites as well as components of the growth media. Ruas-Madiedo et al. (2005) extensively reviewed this subject, thus we shall present here only an overview of the more usual procedures, with a more detailed description of methods related to the determination of EPS structure.

The first step of purification of EPS depends on the bacterial growth medium utilized for its production. In complex media or in food matrix, such as milk, the first requirement is the elimination of proteins. For their removal a precipitation with TCA as well as treatment with proteases are the most commonly used methods. Then, the supernatant as well as the supernatant of bacterial cultures grown in defined media are usually subjected to one or more cycles of precipitation with either ethanol or acetone. The biopolymers present in the supernatants, if they are soluble, are dissolved in water, and then dialysed to remove the low molecular weight contaminants, in general a membrane with a cut-off of MWCO 12-14.000 Da is used.

After lyophilisation of the samples, the EPS is often further purified using a chromatographic technique. The parameters involved in the choice of the appropriate chromatography are: charge, solubility and molecular weight of the EPS. Size-exclusion chromatography (SEC) is a chromatographic method in which molecules in solution are separated by their size, not by molecular weight. It is usually applied to large molecules or macromolecular complexes. Another example is ion-exchange chromatography, which allows the separation of ions and polar molecules based on their charge. In general the procedure involves passing a mixture dissolved in a mobile phase through a stationary phase. For high molecular weight polysaccharides, such as the 2-substituted (1,3)-β-glucan, SEC is the most suitable method, because the charge of the EPS is zero. In this case, dried EPS are dissolved in 0.3 M NaOH (to eliminate extra contaminants and to improve the solubility of the EPS) and centrifuged to eliminate insoluble material. The supernatant is loaded into a column of Sepharose CL-6B equilibrated with NaOH, which is also used as eluent. Fractions are collected, and monitored for carbohydrate content by the phenol-sulphuric method (Dubois et al., 1956).

Polysaccharides are polydisperse polymers, and consequently only an apparent average molecular weight (*M*w) can be determined. To this end, the average *M*w can be estimated after SEC fractionation. A calibration curve is performed by fractionation of standards (Dextran Blue, T70, T10, and vitamin B12) and used for the determination of the Mw. As an alternative, high-performance size-exclusion chromatography (HPSEC) equipped with multi-angle laser-light scattering (MALLS) and refractive index (RI) detectors can be used to determine (*M*w) and *z*-average radius of gyration (R_z) of the EPS.

To determine the monosaccharide composition of the EPS, the analysis of neutral sugars is performed by polysaccharide hydrolysis with 3M TFA. The resulting monosaccharides are converted into their corresponding alditol acetates by reduction with $NaBH_4$ and subsequent acetylation (Laine et al., 1972). Identification and quantification is performed by gas-liquid chromatography (GLC) using a HP5 fused silica column, with a temperature program and a flame ionization detector. With this technique the chromatogram shows only one peak per each monosaccharide, leading to an easy identification of the monosaccharide composition. However, if the polysaccharide contains uronic acid(s) it must be subjected to methanolysis after the hydrolysis. An *O*-methyl glycoside is formed, the acid function is transformed into an ester group and the sugar derivative can then be acetylated and analyzed by GLC.

To determine the type of bond between each residue present in the EPS molecule a methylation analysis is usually performed. The polysaccharides are methylated according to the method of Ciucanu and Kerek (1984). The partially methylated polysaccharides are hydrolyzed with 3M TFA and the products are reduced with $NaBD_4$, acetylated and analyzed by gas chromatography/mass spectroscopy (GC-MS) (Leal et al., 2008). Each peak of the chromatogram is identified by the retention time and mass spectra parameters. The quantification is associated to the peaks area.

To resolve the 3D-structure of an EPS molecule, both the ring size (pyranose/furanose) of the monosaccharide residues and the relative orientations of the adjacent monosaccharides have to be determined. Nuclear magnetic resonance (NMR) is the technique most often used to study the conformation of the polysaccharides and allows elucidation of the type of glycosidic linkages and the structure of the repeating units that constitute the EPS molecules (review by Duus et al., 2000). Before NMR analysis, the purified EPS is dissolved in D_2O so that exchangeable protons are replaced by deuterium (deuteration). This procedure is repeated several times and may involve intermediate lyophilisation steps. A ^1H NMR spectrum of the EPS gives information about the number of monosaccharides present in the repeating unit by counting the resonances in the anomeric region (4.4-5.5 ppm). The common hexoses are detected as well in ^{13}C NMR spectra (95-110 ppm). If there are resonances just downfield of 1 ppm in the ^{1}H NMR spectrum it is a sign of CH_3-groups of e.g. a fucose or a rhamnose residue. Resonances close to 2 ppm reveal *N*-acetyl and/or *O*-acetyl functionalities. From the splitting of the anomeric peaks in ^{1}H spectra ($J_{H1,H2}$) the anomeric configuration can be established; a *J*-coupling of ~4 Hz indicates the α-configuration and a value of ~8 Hz indicates the β-form for common monosaccharides like D-glucopyranose and D-galactopyranose. The corresponding values of $J_{C1,H1}$ are ~170-175 Hz for the α-form and ~160-165 Hz for the β-form obtained from a coupled ^{13}C NMR analysis. Since most polysaccharide NMR spectra show peak overlap in the ring region

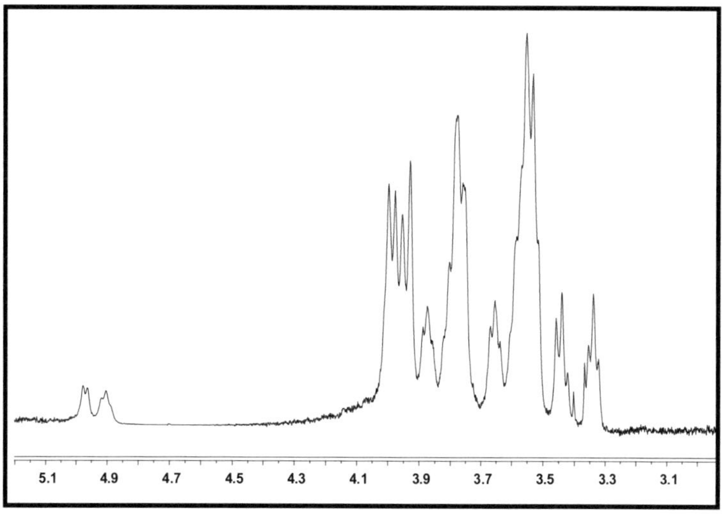

(A)

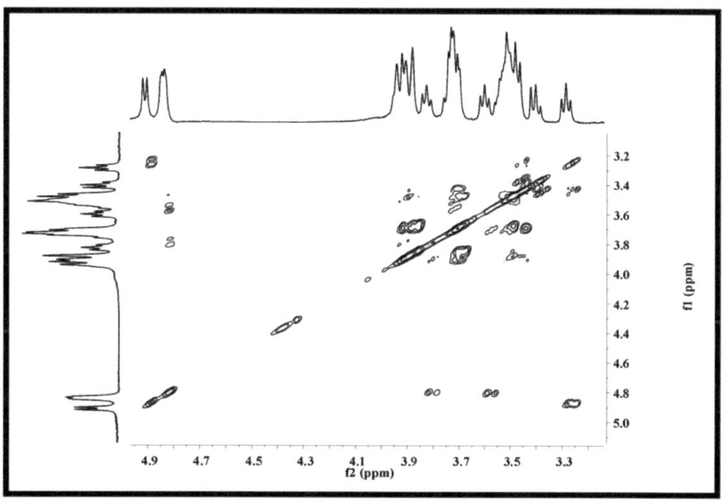

(B)

Fig. 5. NMR analysis of the purified the 2-substituted (1,3)-β-D-glucan. ^{1}H-NMR (A) and ^{1}H,^{1}H-COSY (B) spectra are depicted

(δ_H 3.1-4.4), 2D-NMR techniques are often applied. The proton chemical shifts are linked to their respective carbon by a $^1H,^{13}C$-HSQC NMR analysis or, when the resonances in the ^{13}C dimension overlap too much, a $^{13}C,^1H$-HETCOR NMR analysis is performed. To distinguish methylene protons (CH_2) from methyl (CH_3) and methine (CH), a multiplicity-edited $^1H,^{13}C$-HSQC analysis can be used. The protons in each spin system can be assigned using $^1H,^1H$-TOCSY and/or $^1H,^1H$-DQF-COSY analysis, both techniques allow the magnetization to travel over bonds with the help of *J*-couplings and thereby connecting the protons. EPS contain protons and carbons, but sometimes also nitrogen and even phosphorus. Their respective chemicals shifts can be assigned and correlated to ^1H using e.g. $^1H,^{15}N$-HSQC and 1H,31PTOCSY analysis. The exchangeable amide protons, measured in a H_2O-D_2O mixture, have *J*-couplings to the ring protons and can be assigned correctly in the ring by a $^1H,^1H$-TOCSY with water suppression. As an example, the uni- and bi-dimensional NMR analysis of the 2-substituted (1,3)-β-D-glucan is depicted in Figure 5. The 1D-NMR spectrum (Fig. 5A) revealed several peaks in the region between 3.2 and 4.1 ppm and 2 peaks in the anomeric region characteristics of this type of polysaccharide. 2D-COSY analysis (Fig. 5B) showed couplings between anomeric protons and C2 protons ($H1_C/H2_C$, $H1_B/H2_B$ y $H1_A/H2_A$) characteristic of the 2-substituted (1,3)-β-D-glucan.

Finally, to determine the supramolecular structure and conformation of an EPS, atomic force microscopy (AFM) is currently used. It has been successfully applied to visualize a range of polysaccharides including curdlan (Ikeda & Shishido, 2005), and oat *β*-glucan (Wu et al., 2006). The (1,3)-β-D-glucan helixes dissociate into random coils when the strength of the bonds keeping the helix together are decreased below a critical limit. The helix-coil transition is referred to as denaturation (Sletmoen & Stokke, 2008). Denaturation of (1,3)-β-D-glucan triplexes occurs in alkaline solutions or in dimethyl sulfoxide. In alkaline solutions due to the ionization of the hydroxyl groups and the subsequent electrostatic repulsion between chains, a previous dissociation of the aggregates takes places and then, as the alkalinity increases, the helix structure is believed to denature to yield individual disordered single chains (Sletmoen & Stokke, 2008).

5. Biotechnological applications of EPS from LAB

5.1 Current applications of the EPS in the food industry

High molecular weight polysaccharides are used as additives in the manufacture of a wide variety of food products, because they act as thickeners, stabilisers, viscosifiers, emulsifiers or gelling agents. Most of these polysaccharides are derived from plants (e.g. pectin, cellulose) and seaweeds (e.g. alginate, carrageenan) (Kleerebezem et al., 1999). In terms of applications in the food industry, microbial extracellular polysaccharides including HePS such as xanthan from the phytopathogenic bacterium *Xanthomanas campestris* and gellan from *Sphingomonas paucimobilis*, are also alternative sources of biothickners approved by the FDA for use as foods additives (Laws et al., 2001). Although these are prepared in reliable quantities, their physical properties might not suit all applications, given that there is also a demand for novel materials that can improve rheological characteristics and health promoting properties. On the other hand, the use of bacterial polysaccharides as food additives requires their production by non-pathogenic bacteria. In this sense, LAB have QPS (qualified presumption of safety) status and EPS produced by these bacteria can be considered as food-grade additives (Ruas-Madiedo et al., 2008).

LAB are routinely used in food preparations, not only due to their metabolic activities, but also due to their preservative effects such as: (i) acidification or production of hydrogen peroxide and (ii) the production of bacteriocins (e.g. nisin), which restrict microbial contamination (Kuipers et al., 1998; Wood, 1997). In addition, EPS production by LAB has received considerable attention, since they provide thickening properties and contribute to improve the texture and mouth-feel of the resulting fermented milks or other dairy products. Moreover, certain EPS produced by LAB, have beneficial effects on human health such as cholesterol-lowering, immunomodulation and prebiotic effect, features that are discussed later. It is therefore considered an advantage to use these polymers as food additives rather than the gram-negative polymers, for which no health promoting abilities have been proposed (Ruas-Madiedo et al., 2008).

EPS may act both as a texturizer, improving the rheology (viscosity and elasticity) of a final product, and as physical stabilizers by binding hydration water and interacting with other milk constituents (ions and proteins) thus limiting syneresis. These physical and rheological properties depend on features such as chemical composition, molecular size, charge, presence of side chains, rigidity of the molecules and 3D-structures of the EPS polymers. In addition to physical characteristics, the interactions between EPS and various components in food products contribute to the development of the final product. Nevertheless, many studies have shown that rheological properties of fermented milk products do not correlated well with the amount of EPS content (de Vuyst et al., 2001; Duboc & Mollet, 2001; Folkenberg et al., 2006).

Despite the above, EPS from LAB have not yet been exploited industrially as food additives and one of the main drawbacks to use these polymers for such purpose is the low production level compared with xanthan (de Vuyst & Degeest, 1999). Furthermore, low cost culture media and easy isolation procedures, both rendering high yields are essential for the application of EPS as food grade additives. For this reason, production *in situ* by LAB can be an alternative to the use of biopolymers from plants or non-GRAS bacteria. In particular, HePS producing LAB are used in the dairy industry, mostly belonging to the genera *Streptococcus*, *Lactobacillus* and *Lactococcus* to improve the texture and organoleptic properties of the product. Some examples of these are the production of fermented milks such as viili and langmjolk in Nordic countries as well as the production of kefir, yogurt and low fat cheese type mozzarella and Cheddar cheese (de Vuyst et al., 2001; Ricciardi & Clementi, 2000; Ruas-Madiedo et al., 2008). Another application in development for the bakery industry is the *in situ* production of glucans or fructans by the use of *Lactobacillus* or *Weissella* strains in sour dough manufacture (Tieking et al., 2003).

The use of cereal-base substrates is considered as a promising alternative to fermented dairy products due to their high nutritional value and the presence of both soluble and insoluble dietary fiber (Angelov et al., 2005; Martensson et al., 2005). In this sense, regarding the development of new functional foods, and their particular ability to produce 2-substituted (1,3)- β-glucans, *P. parvulus* 2.6 and *Lb. diolivorans* G77 have been studied as starter cultures in the preparation of oat-based fermented foods. It has been found that these bacteria can grow and produce the EPS in the oat-base substrate, improving the viscosity and texture of the fermented product (Martensson et al., 2003). In addition, analysis of the rheological properties of the β-glucan synthesized by *P. parvulus* 2.6 showed that it has potential utility as a biothickener (Velasco et al., 2009). Also, it has been reported that differences in the viscosity of two cultures of different strains of *Pediococus parvulus* were not attributable to differences in the primary structure or molecular mass of the β-glucan produced (Garai-Ibabe et al., 2010). Other factors such as EPS conformation or interactions between EPS and growth media microstructure could also affect the rheological features. Thus, presumably in the near future, 2-substituted (1,3)- β-glucan-producing LAB will be used for elaboration of non-dairy fermented food. Moreover, the β-glucan could be used as a food additive, due to its gelling properties.

5.2 Potential applications of EPS for production of functional food

5.2.1 Prebiotics

The concept of prebiotic was originally defined as "non-digestible food ingredient that beneficially affects the host by selectively stimulating the growth and/or activity of one or a limited number of bacteria in the colon, and thus improves host health" (Gibson & Roberfroid, 1995). Nine years later, this definition was revised by Gibson et al. (2004) and redefined the concept of prebiotic as "a selectively fermented ingredient that allows specific changes, both in the composition and/or activity in the gastrointestinal microbiota that confers benefits upon host wellbeing and health".

The non-digestible oligosaccharides (NDO) are the prototypes of prebiotic saccharides. The oligosaccharides are compounds with lower molecular weight due to a lower degree of polymerization (DP). Although the IUB-IUPAC defined oligosaccharides as saccharides composed of 3 to 10 monosaccharide units, other sources define them as compounds with 3 to 20 monosaccharide units. Since, there is not a standard definition, the use of "short-chain carbohydrates" as a term to include oligosaccharides and smaller polysaccharides seems to be more appropriate. The NDO are oligosaccharides with monosaccharide units, having a configuration that makes their osidic bonds non-digestible by the hydrolytic activity of the human digestive enzymes (Roberfroid & Slavin, 2000). They have a low calorific value, non-cariogenicity, are associated with a lower risk of infections and diarrhoea, promote the growth of beneficial bacteria in the colon and an improvement of the immune system response (Mussatto & Mancilha, 2007). The ability of the gut microbiota to ferment oligosaccharides depends on a variety of factors including the degree of polymerization, type of sugar, the glycosidic linkage and the degree of branching, as well as the synergy between bacteria during fermentation, the relationship between substrate bacteria and fermentation products, the nature of the fermentations and the saccharolytic capacity of the bacteria (Voragen, 1998).

The production of oligosaccharides in food started to be investigated in Japan, between 1970-1975, and since then a number of these biopolymers have been identified (Table 1). The USA and Europe have recently become leaders in fructan, fructo-oligosaccharide (FOS) and inulin production. The reason is linked to their low cost production as well as the reproducibility of prebiotic effects in humans. Galacto-oligosaccharides (GOS) are also commercialized in these countries but not yet as widely used as fructans. (Rastall & Maitin, 2002).

The oligosaccharides have been widely used in foods, beverages and confectionery due to their properties as hygroscopicity, stabilization of active substances (involved in e.g. flavour and colour), water activity, sweetness and bitterness. They can be obtained by three different ways: (i) extraction with hot water from roots (e.g. inulin) or seeds (e.g. soybean oligosaccharides), (ii) enzymatic synthesis from one or a mixture of disaccharides using osyl-transferases (e.g. fructooligosaccharides) or (iii) partial enzymatic hydrolysis of oligosaccharides (e.g. oligofructose) or polysaccharides (e.g. xylooligosaccharides) (Roberfroid & Slavin, 2000).

Compound	Molecular structure*	Raw material	Process
Cyclodextrins	$(Glu)_n$	Starch	transglycosylation
Fructooligosaccharides	$(Fru)_n$-Glu	Sucrose	transglycosylation
Galactooligosaccharides	$(Gal)_n$-Glu	Lactose	transglycosylation
Gentiooligosaccharides	$(Glu)_n$	Starch	transglycosylation /hydrolysis
Glycosylsucrose	$(Glu)_n$-Fru	Sucrose	transglycosylation
Isomaltooligosaccharides	$(Glu)_n$	Starch	transglycosylation /hydrolysis
Isomaltulose (palatinose)	$(Glu\text{-}Fru)_n$	Sucrose	transglycosylation
Lactosucrose	Gal-Glu-Fru	Lactose/Sucrose	transglycosylation
Lactulose	Gal-Fru	Lactose	isomerization
Maltooligosaccharides	$(Glu)_n$	Starch	hydrolysis
Raffinose	Gal-Glu-Fru	Sucrose	extraction
Stachyose	Gal-Gal-Glu-Fru	Sucrose	extraction
Soybean oligosaccharides	$(Gal)_n$-Glu-Fru	Starch	extraction
Xylooligosaccharides	$(Xyl)_n$	Xylan	hydrolysis

Table 1. Non-digestible oligosaccharides. * Glu, glucose; Fru, fructose; Gal, galactose; Xyl, xylose

Polysaccharides are often the main source of bioactive oligosaccharides and therefore new sources of them are continuously investigated. In this context, LAB have become a promising target due to its GRAS/QPS status. Currently, bioactive commercialized oligosaccharides are extracted from plants but not yet from LAB. However, the high diversity of LAB and their EPS offer new possibilities for detection and production of bioactive oligosaccharides. To obtain oligosaccharides, the post-synthetic engineering strategies consist in enzymatic or chemical actions involving two types of enzymes, glycosyl-hydrolases (EC 3.2.1.y) and polysaccharide lyases (EC 4.2.2.y). A strategy to make the action of the enzyme specific is to grow the EPS-producing bacteria on plates with their own polysaccharide as a carbon source so that in order to survive, they themselves secrete

EPS degrading enzymes. After detection and purification, a specific enzyme for catalysis of the polysaccharide is obtained (Badel et al., 2011).

5.2.2 Immunomodulators

An immunomodulator is a substance which has an effect on the immune system. This system can be regulated in different ways by the use of immunosuppressors or immunostimulants to inhibit or to induce the immune response. In particular, their use, included as additives in food, could be useful to combat infections, to prevent digestive tract cancers or to treat sicknesses due to immunodeficiency, such as inflammatory bowel diseases (Crohn's disease and ulcerative colitis). One strategy to modulate the immune system is the modulation of cytokine expression through the use of herbal medicines. The immunomodulators alter the activity of immune function through the dynamic regulation of informational molecules such as cytokines (Spelman et al., 2006).

The mechanism involved in the immunomodulation can be explained by the interaction of the immunomodulators to their receptor in the membrane of an immune system cell. This interaction activates an internal cascade of phosphotranfer of proteins mediated by kinases and related to a specific pathway. As a consequence, a change of binding affinity of transcriptional regulators for their operators takes place, which results in activation or repression of gene expression (Figure 6).

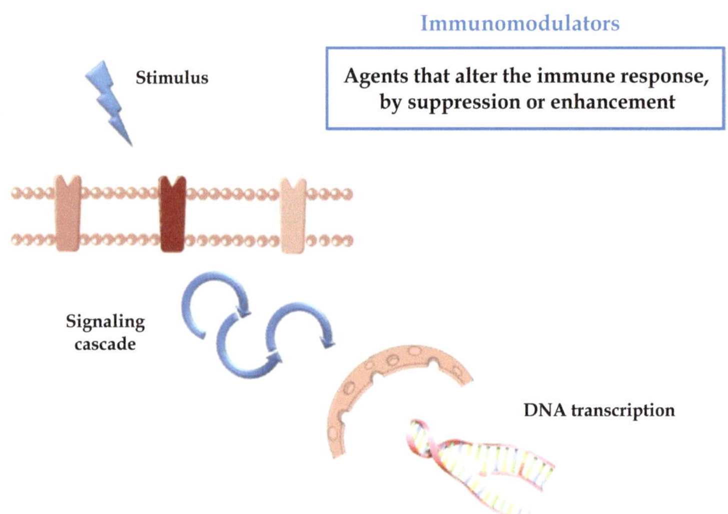

Fig. 6. Stimulation and/or inhibition mediated by immunomodulators

Neither HoPS nor HePS produced by LAB have been used to elaborate functional food, nor directly tested as immunomodulators. However, most of the LAB used as probiotics (according to FAO/WHO, "live microorganisms which when administered in adequate amounts confer a health benefit on the host") for elaboration of functional food, have immunomodulatory properties (Foligne et al., 2007), and produce EPS. However, their ability to immunomodulate is strain specific and can not yet be directly connected to the EPS. In addition, current knowledge (see below) indicates that the nature and structure of some of these biopolymers synthesized by LAB have the potential to be used as immunomodulating food additives.

β-glucans are known as "biological response modifiers" (Wasser, 2011), due to their ability to activate the immune system. β-glucans are integral cell wall components of a variety of fungi, plants and bacteria. In the early 40's Pillemer and Ecker (1941) described the effect of a crude yeast cell wall, called Zymosan, and described that this extract was able to activate the non-specific innate immunity. Only in the 80's Czop and Austen (1985) described the action of Zymosan that involved its binding to a β-1-3-glucan specific receptor, latter on identified and named Dectin-1 (Brown et al., 2001), found in the cell membrane of macrophages, which activates an internal cascade of events.

The immunomodulating function of the β-glucans is related to their structure; in fact different polysaccharides are able to interact with specific receptors, due to their 3D-structure. It has been recently demonstrated that β-glucans with a linear backbone containing 1-3 linkage (e. g. Zymosan) have the ability to activate several receptors: Dectin-1, complement receptor 3 (CR3), scavenger receptors class A and class B, and Toll-like receptors (TLRs) 2 and 6 (Chlubnova et al., 2011). The interaction with these receptors triggers a cascade of internal effects, including production of cytokines (TNF-α, IL-6 and IL-10).

Also branched 6-substituted (1,3)- β-glucans isolated from mushroom, *Candida albicans* and *Pneumocystis carinii*, show high affinity for Dectin-1 (Palma et al., 2006) and for TLR-2 and TLR-4 receptors in elicitation of immune response (Chlubnova et al., 2011). Mushroom, especially Basidiomycetes, are a source of β-glucans with a high biological activity. They mostly have a 1-3 linkage in the main chain, and sometimes an additional 1-6 branch point. This branching point increase their antitumoral and immunomodulating effects (Barreto-Bergter & Gorin, 1983). It is well established that the structural composition offers a higher capacity for carrying biological information, because they have a greater potential for structural variability, specially related to the triple-helical tertiary conformation (Yanaki et al., 1983). A good example of a preventive effect is given by a Japanese study on their popular edible and medicinal mushroom *Hypsizygus marmoreus* (Ikekawa, 2001). In this study mice were divided into two groups: untreated and treated with a diet containing 5% of the dried fruiting body of *H. marmoreus*. All mice were injected with a strong carcinogen, methyl-cholanthrene, and carcinogenesis was investigated. The results obtained allowed to the authors to conclude that the mechanism of prevention and inhibition of carcinogenesis was due to immunopotentiation (Ikekawa, 2001). Therefore, (1,3) β-glucan produced by LAB and their producing strains have potential as immunomodulators.

In this line, it has been shown that 2-substituted (1,3) β-glucan producer LAB belonging to the *Lactobacillus* and *Pediococcus* genera are able to immunomodulate macrophages *in vitro* (Fernández de Palencia et al., 2009; Garai-Ibabe et al., 2010). Moreover, treatment of the macrophages with the purified biopolymer resulted in an increase of secretion of the anti-inflammatory IL-10 cytokine (Fernández de Palencia et al., unpublished results). In addition, four β-glucan-producing LAB strains have been tested for their survival under gastrointestinal stress (Fernández de Palencia et al., 2009; Garai-Ibabe et al., 2010) using an *in vitro* model that simulates the human gut conditions (Fernández de Palencia et al., 2008). Among them, *P. parvulus* 2.6 and *Lb. suebicus* CUPV221 showed significant resistance to digestive tract gut conditions. Furthermore, the presence of the EPS conferred to the producing strains increased capability to adhere to Caco-2 human epithelial intestinal cells (Garai-Ibabe et al., 2010; Fernández de Palencia et al., 2009). Thus, the use of the 2-substituted (1,3) β-glucan as an additive, or produced *in situ,* in fermented food or in the gut has potential as an immunomodulator to alleviate inflammatory bowel diseases. In addition, human consumption of oat-based food prepared with *P. parvulus* 2.6 resulted in a decrease of serum cholesterol levels, boosting the effect previously demonstrated for (1,3)-β-D-glucans in oat (Martensson et al., 2005). Finally, the production of yogurt and various beverages with 2-substituted β-D-glucan-producer LAB indicate advantageous techno-functional properties of these strains (Elizaquível et al., 2011). Therefore, LAB producing this EPS have potential as probiotic strains useful for the manufacture of functional foods.

Dextrans have been also investigated as immunomodulators. Previous studies on dextran-70 justified its beneficial effect in the prevention of acute respiratory distress syndrome after trauma and sepsis as well as pancreatitis (Modig, 1988). Recently, it was reported that dextran-70 reduced the leukocyte-endothelium interaction. In a clinical trial forty patients who were undergoing coronary bypass surgery were divided into 2 groups of 20. In group A a dextran-70 infusion was administrated at a concentration of 7.5 ml kg^{-1} before the surgery, and 12.5 ml kg^{-1} after the cardiopulmonary bypass. Group B was the control and received a gelatin infusion at the same concentration. Several parameters were measured including determination of IL 8, IL 10 and troponin-I levels. The conclusion was that this α-glucan was able to reduce the systemic inflammatory response and the release of the cardiac troponin-I after cardiac operation (Gombocz et al., 2007).

Another α-glucan from the edible mushroom *Tricholoma matsutake* has been investigated, and reported to have excellent biological activities; exerting modulating effects on the immune competence of mice and rats. In this study, a sodium hydroxide extract of the mushroom was defatted followed by fractionation with a combination of ion exchange chromatography and gel filtration to identify the active component. A single-peak fraction (MPG-1) was obtained after reverse-phase chromatography. MPG-1 was a glycoprotein with molecular mass of 360 kDa, and contained about 90% glucose. NMR and methylation analysis revealed that the α-1,4-linkage was the predominant glucan linkage with α-1,6- and α-1,2-branching in the minority. It was demonstrated that the mycelium preparation is effective in improving immunological functions in stressed individuals. In an *in vitro* model the compound formed a complex with the active form of TGF-β1. These results indicate that the mycelium contains a novel α-glucan-protein complex with immunomodulatory effect (Hoshi et al., 2005).

Therefore, the high production of α-glucans by LAB and the immunomodulatory properties of these bio-molecules as described above, predict that in the near future studies will be performed to evaluate the beneficial properties of these EPS, with the aim to use them as food additives.

Finally, the low yield of HePS produced by LAB and their complex biosynthetic pathway, suggests that in the short term they are not very good candidates as food additives, although it is expected that the immunomodulatory properties of the producing strains will be further investigated.

6. Conclusion

Currently it is clear that diversification of functional foods, which have been scientifically validated as having beneficial properties, will increase in the near future, and the EPS synthesized by LAB could have a place in the market as an ingredient of this type of food. To this end, EPS can be synthesized *in situ* by their producing strains or can be used in isolated form as a food additive. Their chances of reaching the Market place will be improved by the discovery and utilization of new EPS-producing strains isolated from sources other than dairy products. The discovery and characterization of new EPS-producers isolated from food (e.g. processed meat products) and beverages (e.g. wine and cider) will increase the variety of EPS and the use of their producing strains for the elaboration of novel solid and liquid functional food. For example, *Lc. mesenteroides Lb. plantarum* and *Lb. sakei* strains have been isolated from Spanish sausages. They produce highly homogeneous, α-glucan HoPS synthesised by a dextransucrase, and are able to immunomodulate macrophages (Nácher-Vázquez et al., unpublished results). Microorganisms that are native to the human gut and produce EPS (or could be engineered to produce EPS) would also be of great interest as their chances of survival in the gut environment would be much higher than other microbes. It has been shown that microorganisms that produce 2-substituted-(1,3)-β-glucan are able to adhere to human gut epithelial cells. Such organisms would presumable be able to colonise the gut and compete effectively with pathogens, at the same time as producing a beneficial immunomodulation.

However, the use of EPS producing LAB strains will require a thorough scientific evaluation both *in vitro* and *in vivo*. It has to be stated that currently most of the general claims for components of functional food (though this does not apply to oat β-glucans and their blood LDL-cholesterol lowering properties) have not been approved for use after evaluation by the European Food Safety Authority (EFSA, http://www.efsa.europa.eu). The main reason for rejection of probiotic bacteria has been the lack of enough characterization of the bacteria (determination of the nucleotide sequence of their entire genome is advisable), and/or insufficient scientific evidence to correlate the use of the bacteria with health benefits (*Lb. delbrueckii* subsp. *bulgaricus* AY/CSL and *S. thermophilus* 9Y/CSL and beneficial modulation of the gut microbiota). Therefore, each particular strain has to be subjected to evaluation, although its EPS has been already experimentally validated. The evaluation should be performed, first *in vitro*, then in animal models and finally in human trails. In addition, due to the rules of the EFSA the use of genetically modified organisms (GMO) is restricted (although they are not totally forbidden). Moreover, the opinion of consumers in Europe and USA regarding the use of GMO in food is not favourable. Consequently, well-characterized (preferably GRAS) EPS-producers from natural ecological environments are the best candidates for use in functional food.

However, if the EPS are going to be used as food additives, after purification, then there is no restriction concerning the use of a GMO producing strain. The use of GMO able to produce high levels of EPS or newly designed biopolymers is still very limited, and the production levels of most EPS are not very high. Therefore, provided that enzymes, and hence the genes, involved in their biosynthetic pathways are known, the future improvement of EPS production, will be by DNA recombinant technology and metabolic engineering to generate GMO EPS-producing LAB, that will be used for production of the biopolymers in large-scale fermenters. Moreover, genetic engineering could be used to alter substrate specificity of the EPS biosynthetic enzymes to generate new polysaccharides and oligosaccharides with improved prebiotic properties. In the case of the glycosyltransferases, which synthesize HoPS this strategy could result in the synthesis of not only new HoPS, but also new HePS. Finally, the requirement of only one protein for the synthesis opens the window for production of new biopolymers by immobilized enzymes.

Overall, there should be a rapid expansion in the development of novel LAB probiotic organisms and their prebiotic EPS products. However, their ultimate success in the market place will require a rigorous scientific evaluation.

7. Acknowledgments

We thank Dr Stephen Elson for critical reading of the manuscript. This work was supported by the Spanish Ministry of Science and Innovation (grant AGL2009-12998-C03-01). Sara Notararigo and Montserrat Nácher are recipients of predoctoral fellowships from Consejo Superior de Investigaciones Científicas.

8. References

Adeyeye, A., Jansson, P.E., & Lindberg, B. (1988). Structural studies of the capsular polysaccharide from *Streptococcus pneumoniae* type 37. *Carbohydrate Research,* Vol.180, No.2, (September 1988), pp. 295-299

Angelov, A., Gotcheva, V., Hristozova, T., & Gargova, S. (2005). Application of pure mixed probiotic lactic acid bacteria and yeast cultures for oat fermentation. *Journal of the Science of Food and Agriculture,* Vol.85, No.12, (September 2005), pp. 2134-2141

Aslan, Y., & Tanriseven, A. (2007). Immobilization of *Penicillium lilacinum* dextranase to produce isomaltooligosacharides from dextran. *Biochemical Engineering Journal,* Vol.34, No.1, (April 2007), pp. 8-12

Badel, S., Bernardi, T., & Michaud, P. (2011). New perspectives for Lactobacilli exopolysaccharides. *Biotechnology Advances,* Vol.29, No.1, (January-February 2011), pp. 54-66

Barreto-Bergter, E., & Gorin, P.A. J., (1983). Structural chemistry of polysaccharides from fungi and lichens In: *Advances in Carbohydrate Chemistry and Biochemistry,* R.S. Tipson and H. Derek, pp.67-103, Academic Press, ISBN: 0065-2318

Boels, I.C., Kleerebezem, M., & de Vos, W.M. (2002). Engineering of carbon distribution between glycolysis and sugar nucleotide biosynthesis in *Lactococcus lactis. Applied Environmental Microbiology,* Vol.69, No.2 (February 2002), pp. 1129-1135

Bourgoin, F., Pluvinet, A., Gintz, B., Decaris, B., & Guédon, G. (1999). Are horizontal transfers involved in the evolution of the *Streptococcus thermophilus* exopolysaccharide synthesis loci?. *Gene,* Vol.233, No.1-2, (June 1999), pp. 151-161

Breton, C., Snajdrová, L., Jeanneau, C., Koca, J., & Imberty, A. (2006). Structures and mechanisms of glycosyltransferases. *Glicobiology,* Vol.16, No.2, (February 2006), pp. 29R-37R

Brock, T.D. (2008). Biology of Microorganisms. 12th ed. M. Madigan, J. Martinko, J. Parker. Pearson/Prentice-Hall. ISBN 0321536150

Brown, G.D., & Gordon, S. (2001). A new receptor for β-glucans. *Nature,* Vol.413, (September 2001), pp. 36-37

Capek, P., Hlavonová, E. Matulová, M., Mislovicová, D. Ruzicka, J., Koutný, M., & Keprdová, L. (2011). Isolation and characterization of an extracellular glucan produced by *Leuconostoc garlicum* PR. *Carbohydrate Polymers,* Vol.83, No.1, (January 2011), pp. 88-93

Charnock, S.J., Henrissat B., & Davies, G.J. (2001). Three-dimensional structures of UDP-sugar glycosyltransferases illuminate the biosynthesis of plant polysaccharides. *Plant Physiology,* Vol.125, No.2, (February 2003), pp. 527-531

Chlubnová, I., Sylla, B., Nugier-Chauvin, C., Daniellou, R., Legentil, L., Kralova, B., & Ferrières, V. (2011). Natural glycans and glycoconjugates as immunomodulating agents. *Natural Product Reports,* Vol.28, No. 5, (March 2011), pp. 937-952

Ciucanu, I., & Kerek, F. (1984). Rapid and simultaneous methylation of fatty and hydroxy fatty acids for gas-liquid chromatographic analysis. *Journal of Chromatography A,* Vol. 284, (1984), pp.179-185

Coutinho, P.M., Deleury E., Davies G.J., & Henrissat B. (2003). An Evolving Hierarchical Family Classification for Glycosyltransferases. *Journal of Molecular.Biology,* Vol.328, No.2, (April 2003), pp. 307-317

Czop, J., & Austen K. (1985). A β-glucan inhibitable receptor on human monocytes: its identity with the phagocytic receptor for particulate activators of the alternative complement pathway. *The Journal of Immunology,* Vol.134, No.4, (April 1985), pp. 2588-2593

Dabour, N., & LaPointe, G. (2005). Identification and molecular characterization of the chromosomal exopolysaccharide biosynthesis gene cluster from *Lactococcus lactis subsp. cremoris* SMQ-461. *Applied .and Environmental Microbiology,* Vol.71, No.11, (November 2005), pp. 7414-7425

de Vuyst, L., & Degeest, B. (1999). Heteropolysaccharides from lactic acid bacteria. *FEMS Microbiology Reviews,* Vol.23, No.2, (April 1999), pp. 153-17

de Vuyst, L., De Vin, F., Vaningelmem, F., & Degeest B. (2001). Recent developments in the biosynthesis and applications of heteropolysaccharides from lactic acid bacteria. *International Dairy Journal,* Vol.11, No.9, (October 2001), pp. 687-707

Dols-Lafargue, M., Lee, H.Y., Le Marrec, C., Heyraud, A., Chambat, G., & Lonvaud-Funel, A. (2008). Caracterization of *gtf,* a glucosyltransferase gene in the genome of *Pediococcus parvulus* and *Oenococcus oeni,* two bacterial species commonly found in wine. *Applied and Environmental Microbiology,* Vol.74, No.13, (July 2008), pp. 4079-4090

DuBois, M., Gilles, K.A., Hamilton, J.K., Rebers, P.A., & Smith, F. (1956). Colorimetric Method for Determination of Sugars and Related Substances. *Analytical Chemistry,* Vol.28, No.3, (March 1956) , pp. 350-356

Duus, J.Ø., Gotfredsen, C.H., & Bock, H. (2000). Carbohydrate structural determination by NMR spectroscopy: modern methods and limitations. *Chemical Reviews,* Vol.100, No.12, (November 2000), pp. 4589–4614

Dueñas, M., Munduate, A., Perea, A., & Irastorza, A. (2003). Exopolysaccharide production by Pediococcus damnosus 2.6 in a semidefined medium under different growth conditions. *International Journal of Food Microbiology,* Vol.87, No.1-2, (October 2003), pp. 113-120

Dueñas-Chasco, M.T., Rodríguez-Carvajal, M.A., Mateo, P.T., Franco-Rodríguez, G., Espartero, J., Irastorza-Iribas, A., & Gil-Serrano, A.M. (1997). Structural analysis of the exopolysaccharide produced by *Pediococcus damnosus* 2.6. *Carbohydrate Research,* Vol.303, No.4, (October 1997), pp. 453-458

Dueñas-Chasco, M.T., Rodríguez-Carvajal, M.A., Tejero-Mateo, P., Espartero, J.L., Irastorza-Iribas, A., & Gil-Serrano, A.M., (1998). Structural analysis of the exopolysaccharides produced by *Lactobacillus* spp. G-77. *Carbohydrate Research,* Vol.307, No.1-2, (February 1998), pp. 125-133

Duboc, P., & Mollet, B. (2001). Applications of exopolysaccharides in the dairy industry. *International Dairy Journal,* Vol.11, No.9, (October 2001), pp. 759-768

Elizaquível, P., Sánchez, G., Salvador, A., Fiszman, S., Dueñas, M.T., López, P., Fernández de Palencia, P., & Aznar, R. (2011). Evaluation of yogurt and various beverages as carriers of lactic acid bacteria producing 2-branched (1,3)-β-d-glucan. *Journal of Dairy Science,* Vol.94, No. 7 (July 2011), No. pp. 3271-3278

Erhardt, F.A., Kügler, J., Chakravarthula, R.R., & Jördening, H.J. (2008). Co-Immobilization of dextransucrase and dextranase for the facilitated synthesis of isomalto-oligosaccharides: preparation, characterization and modeling. *Biotechnology and Bioengineering,* Vol.100, No.4, (July 2008), pp. 673-683

Fernández de Palencia, P., López, P., Corbí, A., Peláez, C., & Requena, T. (2008). Probiotic strains: survival under simulated gastrointestinal conditions, in vitro adhesion to Caco-2 cells and effect on cytokine secretion. *European Food Research and Technology,* Vol. 227, No.5, (March 2008), pp. 1475-1484

Fernández de Palencia, P., Werning, M.L., Sierra-Filardi, E., Dueñas, M.T., Irastorza, A., Corbí, A., & López, P. (2009). Probiotic properties of the 2-substituted (1,3)-β-D-glucan producing bacterium *Pediococcus parvulus* 2.6. *Applied and Environmental Microbioly.* Vol.75, No.14, (July 2009), pp. 4887-4891

Foligne, B., Nutten, S., Grangette, C., Dennin, V., Goudercourt, D., Poiret, S., Dewulf, J., Brassart, D., Mercenier, A., & Pot, B. (2007). Correlation between in vitro and in vivo immunomodulatory properties of lactic acid bacteria. *World Journal of Gastroenterology,* Vol.14, No.2, (January 2007), pp. 236-243

Folkenberg, D.M., Dejmek, P., Skriver, A., Guldager, H.S., & Ipsen, R. (2006). Sensory and rheological screening of exopolysaccharide producing strains of bacterial yoghurt cultures *International Dairy Journal,*Vol.16, No.2, (February 2006), pp. 111-118

Garai-Ibabe, G., Areizaga, J., Aznar, R., Elizaquivel, P. Prieto, A., Irastorza, A., & Dueñas, M.T.(2010). Screening and selection of 2-branched (1,3)-β-D-glucan producing lactic acid bacteria and exopolysaccharide characterization. *Journal of Agriculture Food Chemistry,* Vol. 58, No. 10, (December 2010), pp. 6149-6156

Garai-Ibabe, G., Werning, M.L., López, P., Corbí, A.L., & Fernández de Palencia, P. (2010) Naturally occurring 2-substituted (1,3)-β-D-glucan producing *Lactobacillus suebicus* and *Pediococcus parvulus* strains with potential utility in the food industry. *Bioresource Technology,* Vol.101, No.23, (December 2010), pp. 9254-9263

García, E., Llull, D., Muñoz, R., Mollerach M., & López R. (2000) Current trends in capsular polysaccharide biosynthesis of *Streptococcus pneumoniae. Research in Microbiology,* Vol.151, No.6, (July 2000), pp. 429-435

Garinot-Schneider, C., Lellouch, A.C., & Geremia, R.A. (2000) Identification of essential amino acid residues in the *Sinorhizobium meliloti* glucosyltransferase ExoM. *Journal of Biological Chemistry,* Vol.183, No.1, (October 2000), pp. 31407-31413

Gänzle, M., & Schwab, C. (2009). Ecology of exopolysaccharide formation by lactic acid bacteria: sucrose utilisation, stress tolerance, and biofilm formation, In: *Bacterial Polysaccharides: Current Innovations and Future Trends,* Matthias Ullrich, pp. (263-278) Caister Academic Press, ISBN 978-1-904455-45-5, Norfolk, UK.

Gibson, G.R., & Roberfroid, M. (1995). Dietary modulation of the human colonic microbiota: introducing the concept of prebiotics. *The Journal of Nutrition,* Vol.125, No.6, (June 1995), pp. 1401-1412

Gibson, GR., Probert, H.M., Van Loo, J., Rastall, R.A., & Roberfroid, M. (2004). Dietary modulation of the human colonic microbiota: updating the concept of prebiotics. *Nutrition Research Reviews,* Vol.17, No.2, (December 2004), pp. 259-275

Glucksmann, M.A., Reuber, T.L., & Walker, G.C. (1993). Genes needed for the modification, polymerization, export, and processing of succinoglycan by *Rhizobium meliloti*: a model for succinoglycan biosynthesis. *Journal of Bacteriology,* Vol.175, No.21, (November 1993), pp. 7045-7055

Gombocz, K., Beledi, A., Alotti, N., Kecskes, G., Gabor, V., Bogar, L., Koszegi, T., & Garai, J. (2007). Influence of dextran-70 on systemic inflammatory response and myocardial ischaemia-reperfusion following cardiac operations. *Critical Care,* Vol.11, No.4, (Augus 2007), pp. R87

Goulas, A. K., Cooper, J.M, Grandison, A.S., & Rastall, R.A. (2004). Synthesis of isomaltooligosaccharides and oligodextrans in a recycle membrane bioreactor by the combined use of dextransucrase and dextranase. *Biotechnology and Bioengineering,* Vol.88, No.6, (December 2004), pp. 778-787

Heldermon, C.; DeAngelis, P.L., & Weigel, P.H. (2001). Topological organization of the hyaluronan synthase from *Streptococcus pyogenes. Journal of Biological Chemistry,* Vol.276, No.3, (January 2001), pp. 2037-2046

Henrissat, B., & Bairoch, A. (1996). Updating the sequence-based classification of glycosyl hydrolases. *Biochemistry Journal,* Vol. 316, No.2, (June 1996), pp. 695-696

Hoshi, H., Yagi, Y., Iijima, H., Matsunaga, K., Ishihara, Y., & Yasuhara, T. (2005). Isolation and characterization of a novel immunomodulatory alpha-glucan-protein complex from the mycelium of Tricholoma matsutake in basidiomycetes. *Journal of Agricultural and Food Chemistry,* Vol.53, No.23, (November 2005), pp. 8948-8956

Ibarburu, I., Soria-Diaz, M.E., Rodriguez-Carvajal, M.A., Velasco, S.E., Tejero-Mateo, P., Gil-Serrano, A. M., Irastorza, A., & Dueñas, M.T. (2007). Growth and exopolysaccharide (EPS) production by *Oenococcus oeni* I4 and structural characterization of their EPSs. *Journal of Applied Microbiology,* Vol.103, No.2, (August 2007), pp. 477-486

Ikeda, S., & Shisido, Y. (2005). Atomic force studies on heat-induced gelation of curdlan. *Journal of Agricultural Chemistry,* Vol.53, No.3 (January 2005), pp. 786-791

Ikekawa, T. (2001). Beneficial Effects of Edible and Medicinal Mushrooms. *Health Care,* Vol.3, No.4, (December 2001), pp. 8

Jolly, L., & Stingele, F. (2001). Molecular organization and functionality of exopolysaccharide gene clusters in Lactic acid bacteria. *International Dairy Journal,* Vol.11, No.9, (October 2001), pp. 733-745

Karnezis, T., Epa, V.C., Stone, B.A., & Stanisich, V.A. (2003). Topological characterization of an inner membrane (1->3)-β-D-glucan (curdlan) synthase from *Agrobacterium* sp. strain ATCC31749. *Glycobiology,* Vol.13, No.10, (October 2003), pp. 693-706

Keenleyside, W.J., Clarke, A.J., & Whitfield, C. (2001). Identification of Residues Involved in Catalytic Activity of the Inverting Glycosyl Transferase WbbE from *Salmonella enterica* serovar borreze. *Journal of Bacteriology,* Vol.183, No.1, (January 2001), pp. 77-85

Kimmel, S.A., Roberts, R.F., & Ziegler, G.R. (1998). Optimization of exopolysaccharide production by *Lactobacillus delbrueckii subsp. bulgaricus* RR grown in a semidefined medium. *Applied and Environmental Microbiology,* Vol.64, No.2, (February 1998), pp. 659-664

Kleerebezem, M., van Kranenburg, R., Tuinier, R., Boels, I.C., Zoon, P., Looijesteijn, E., Hugenholtz, J., & de Vos, W.M. (1999). Exopolysaccharides produced by *Lactococcus lactis*: from genetic engineering to improved rheological properties? *Antonie van Leeuwenhoek,* Vol.76, No.1-4, (July-November 1999), pp. 357-365

Korakli, M., & Vogel, R.F. (2006). Structure/function relationship of homopolysaccharide producing glycansucrasesand therapeutic potential of their synthesized glycans. *Applied Microbiology and Biotechnology,* Vol.71, No.6, (April 1986), pp. 790-803

Kuipers, O.P., de Ruyter, P.G.G.A., Kleerebezem, M., & de Vos, W.M. (1998). Quorum sensing-controlled gene expression in lactic acid bacteria. *Journal of Biotechnology,* Vol.64, No.1, (September 1998), pp. 15-21

Laine, R.A., Esselman, W.J., & Sweeley, C.C. (1972). Gas-liquid chromatography of carbohydrates. In: *Methods in Enzymology.* V. Ginsburg, pp. 159-167, Academic Press, ISBN: 0076-6879

Lairson, L.L., Henrissat, B., Davies, G.J., & Withers, S.G. (2008). Glycosyltransferases: structures, functions, and mechanisms. *Annual Review of Biochemistry,* Vol.77, (July 2008), pp. 521-555

Lamothe, G.T., Jolly, L., Mollet, B., & Stingele, F. (2002). Genetic and biochemical characterization of exopolysaccharide biosynthesis by *Lactobacillus delbruekii* subsp. *Bulgaricus. Archives of Microbiology*, Vol. 178, No.3, (September 2002), pp. 218-228.

Laws, A, Gu, Y., & Marshall, V. (2001) Biosynthesis, characterisation, and design of bacterial exopolysaccharides from lactic acid bacteria. *Biotechnology Advances,* Vol.19, No.8, (December 2001), pp. 597-62

Leal, D., B. Matsuhiro, B., Rossi, M., & Caruso, F. (2008). FT-IR spectra of alginic acid block fractions in three species of brown seaweeds. *Carbohydrate Research*, Vol.343, No.2, (February 2008), pp. 308-316

Lin, W.S.; Cunneen, T., & Lee, C.Y. (1994). Sequence analysis and molecular characterization of genes required for the biosynthesis of type 1 capsular polysaccharide in *Staphylococcus aureus. Journal of Bacteriology,* Vol.176, No.22, (November 1994), pp. 7005-7016

Liu, J.R., Wang, S.Y., Lin, Y.Y., & Lin, C.W. (2002). Antitumor activity of milk-kefir and soymilk-kefir in tumor-bearing mice. *Nutrition and Cancer,* Vol.44, No.2, (November 2009), pp. 182-187

Llaubères, R. M., Richard B., Lonvaud, A., Dubourdieu, D., & Fournet, B. (1990). Structure of an exocellular β-D-glucan from *Pediococcus sp*, a wine lactic bacteria. *Carbohydrate Research,*Vol.203, No.1, (August 1990), pp. 103-107

Llull, D., Garcia, E., & Lopez, R. (2001). Tts, a processive β-glucosyltransferase of *Streptococcus pneumoniae*, directs the synthesis of the branched type 37 capsular polysaccharide in pneumococcus and other gram-positive species. *Journal of Biological Chemistry,* Vol.276, No.24, (June 2001), pp. 21053-21061

Macedo, M.G., Lacroix, C., & Champagne, C.P. (2002). Combined effects of temperature and medium composition on exopolysaccharide production by lactobacillus rhamnosus rw-9595m in a whey permeate based medium. *Biotechnological Progress,* Vol.18, No.2, (March-April 2002), pp. 167-173

Mateo, C., Palomo, M., Fernandez-Lorente, G., Guisan, J.M., & Fernandez-Lafuente, R. (2007). Improvement of enzyme activity, stability and selectivity via immobilization techniques. *Enzyme and Microbial Technology,* Vol.40, No.6, (May 2007), pp. 1451-1463

Marshall, V.M., & Rawson, H.L. (1999). Effects of exopolysaccharide-producing strains of thermofilic lactic acid bacteria on the texture of stirred yoghurt. *International Journal of Food Science and Technology,* Vol.34, No.2, (April 1999), pp. 137–143

Martensson, O., Dueñas, M., Irastorza, A., Öste, R., & Holst, O. (2003). Comparison of growth characteristics and exopolysaccharide formation of two lactic acid bacteria strains, *Pediococcus damnosus* 2.6 and *Lactobacillus brevis* G-77, in an oat-based, nondairy medium. *Lebensm.-Wiss.Technol, Vol.*36, No.3, (May 2003), pp. 353-357

Martensson, O., Biörklund, M., Lambo, M.A., Dueñas-Chasco, M.T., Irastorza, A., Holst, O., Norin, E., Walling, G., Öste, R., & Önning, G. (2005). Fermented ropy, oat-based products reduce cholesterol levels and stimulate the bifidobacteria flora in humans. *Nutrition Research,* Vol.25, No.5, (May 2005), pp. 429-442

McIntosh, M., Stone, B.A., & Stanisich, V.A. (2005). Curdlan and other bacterial (1,3)- β-D-glucans. *Applied Microbiology and Biotechnology,* Vol.68, No.2, (August 2005), pp. 163-173

Modig, J. (1988). Comparison of effects of dextran-70 and ringer's acetate on pulmonary function, hemodynamics, and survival in experimental septic shock. *Critical Care Medicine,* Vol.16, No.3, (March 1988), pp. 266-271

Monchois, V.; Willemot, R.M., & Monsan, P. (1999). Glucansucrases: Mechanism of action and structure-function relationships. *FEMS Microbiology Reviews,* Vol.23, No.2, (April 1999), pp. 131-151

Monsan, P., Bozonnet, S., Albenne, C., Joucla, G., Willemot R.M., & Remaud-Simeon, M. (2001). Homopolysaccharides from lactic acid bacteria. *International Dairy Journal,* Vol.11, No.9, (October 2001), pp. 675-685

Montersino, S., Prieto, A., Muñoz, R., & de las Rivas, B. (2008). Evaluation of exopolysaccharide production by *Leuconostoc mesenteroides* strains isolated from wine. *Journal of Food Science,* Vol.73, No.4, (May 2008), pp. 196-199

Mozzi, F., Vaningelgem, F., Hebert, E.M., Van der Meulen, R., Foulquie Moreno, M.R., Font de Valdez, G., & De Vuyst, L. (2006). Diversity of heteropolysaccharide-producing lactic acid bacterium strains and their biopolymers. *Applied and Environmental Microbiology,* Vol.72, No.6, (June 2006), pp. 4431-4435

Mussatto, S.I., & Mancilha, I.M. (2007). Non-digestible oligosaccharides: A review. *Carbohydrate Polymers,* Vol.68, No.3, (April 2007), pp. 587-597

Nordmark, E.L., Yang Z., Huttunen, E., & Widmalm G., (2005). Structural studies of the exopolysaccharide produced by *Propionibacterium freudenreichii ssp. shermanii* JS. *Biomacromolecules,* Vol.6, No.1, (November 2004), pp. 521-523

Ölcer, Z., & Tanriseven, A. (2010). Co-immobilization of dextransucrase and dextranase in alginate. *Process in Biochemistry,* Vol.45, No.10, (October 2010), pp. 1645-1651

Palma, A.S., Feizi, T., Zhang, Y., Stoll. M.S., Lawson, A.M., Díaz-Rodríguez, E., Campanero-Rhodes, M.A., Costa, J., Gordon, S., Brown, G.D., & Chai, W. (2006). Ligands for the β-glucan receptor, Dectin-1, assigned using "designer" microarrays of oligosaccharide probes (neoglycolipids) generated from glucan polysaccharides. *Journal of Biological Chemistry*, Vol.281, No.9, (March 2006), pp. 5771-5779

Péant, B., LaPointe, G., Gilbert, C., Atlan, D., Ward, P., & Roy, D. (2005). Comparative analysis of the exopolysaccharide biosynthesis gene clusters from four strains of *Lactobacillus rhamnosus. Microbiology*, Vol.151, No.6, (June 2005), pp. 1839-1851

Pillemer, L., & Ecker, E.E. (1941). Anticomplementary factor in fresh yeast. *Journal of Biological Chemistry, Vol.*137, No.1, (January 1941), pp. 139-142

Rastall, R.A., & Maitin, V. (2002). Prebiotics and synbiotics: towards the next the generation. *Current Opinion in Biotechnology*, Vol.13, No.5, (October 2002), pp. 490-496

Ricciardi, A., & Clement, F. (2000). Exopolysaccharides from lactic acid bacteria: structure, production and technological applications. In: *Italian Journal of Food Science*, Vol.12, No.1, (June 2000), pp. 23-45, ISSN 1120-1770

Roberfroid, M., & Slavin, J. (2000). Nondigestible oligosaccharides. *Critical Reviews in Food Science and Nutrition*, Vol.40, No.6, (November 2000), pp. 461-480

Ruas-Madiedo, P., & de los Reyes-Gavilan, C.G. (2005). Invited review: methods for the screening, isolation, and characterization of exopolysaccharides produced by lactic acid bacteria. *Journal of Dairy Science*, Vol.88, No.3 (March 2005), pp. 843–856

Ruas-Madiedo, P., Abraham, A., Mozzi, F., & de los Reyes-Gavilán, C.G. (2008). Functionality of exopolysaccharides produced by lactic acid bacteria, In: *Molecular aspects of lactic acid bacteria for traditional and new applications*. B. Mayo, P. López, and G. Pérez-Martín (Ed.), 137-166, Research Signpost, ISSN 978-81-308-0250-3, Kerala, India

Ruas-Madiedo, P., Gueimonde, M., Arigoni, F., de los Reyes-Gavilan, C.G., & Margolles, A. (2009) Bile Affects the Synthesis of Exopolysaccharides by *Bifidobacterium animalis. Applied and Environmental Microbiology*, Vol.75, No.4, (February 2009), pp. 1204-1207

Ruas-Madiedo, P., Salazar, N., & de los Reyes-Gavilan, C.G. (2009) Biosyntesis and chemical composition of exopolysaccharides produced by lactic acid bacteria, In: *Bacterial Polysaccharides: Current Innovations and Future Trends,* Matthias Ullrich (Ed.), pp. (279-310), Caister Academic Press, ISBN 978-1-904455-45-5, Norfolk, UK

Sarwat, F., Ul Qader, S.A., Aman, A., & Ahmed, N. (2008). Production & characterization of a unique dextran from an indigenous *Leuconostoc mesenteroides* cmg713. *International Journal of Biological Sciences*, Vol.4, No.6, (July 2008), pp. 379-386.

Silver, R.P., Prior, K., Nsahlai, C., & Wright, L.F. (2001). ABC transporters and the export of capsular polysaccharides from Gram-negative bacteria. *Research in Microbiology*, Vol.152, No.3-4, (April-May 2001), pp. 357-364

Sinnott, M.L. (1991). Catalytic mechanisms of enzymic glycosyl transfer. *Chemical Reviews*, Vol.90, No.7, (November 1990), pp. 1170-1202

Spelman, K., Burns, J., Nichols, D., Winters, N., Ottersberg, S., & Tenborg, M. (2006). Modulation of cytokine expression by traditional medicines: A review of herbal immunomodulators. *Alternative Medicine Review*, Vol.11, No.2, (June 2006) pp. 128-150

Sletmoen, M., & Stokke, B. T. (2008). Higher order structure of (1,3)-β-D-glucans and its influence on their biological activities and complexation abilities. *Biopolymers*, Vol.89, No.4, (January 2008), pp. 310-321

Stack, H.M., Kearney, N., Stanton, C., Fitzgerald, G.F., & Ross, R.P. (2010). Association of β-glucan endogenous production with increased stress tolerance of intestinal Lactobacilli. *Applied and Environmental Microbiology*, Vol.76, No.2, (January 2010), pp. 500-507

Stingele, F., Newell, J.W., & Neeser, J.R. (1999). Unraveling the function of glycosyltransferases in *Streptococcus thermophiles* Sfi6. *Journal of. Bacteriology*, Vol.181, No.20, (October 1999), pp. 6354-6360

Tanriseven, A., & Dogan S., (2002). Production of isomalto-oligosaccharides using dextransucrase immobilized in alginate fibres. *Process Biochemistry*, Vol. 37, No.10, (October 2001), pp. 1111–1115

Tarbouriech, N., Charnock, S.J., & Davies, G.J. (2001). Three-dimensional structures of the Mn and Mg dTDP complexes of the family GT-2 glycosyltransferase SpsA: a comparison with related NDP-sugar glycosyltransferases. *Journal of Molecular Biology*, Vol.314, No.4, (December 2001), pp. 655-661

Tieking, M., Korakli, M., Ehrmann, M.A., Gänzle, M.G., & Vogel, R.F. (2003). In situ production of exopolysaccharides during sourdough fermentation by cereal and intestinal isolates of lactic acid bacteria. *Applied and Environmental Microbiology*, Vol.69, No.2, (February 2003), pp. 945-952

Tieking, M., Ehrmann, M.A., Vogel, R.F., & Gänzle, M.G. (2005). Molecular and functional characterization of a levansucrase from the sourdough isolate *Lactobacillus sanfranciscensis* TMW 1.392. *Applied Microbiology and Biotechnology*, Vol.66, No.6, (March 2005), pp. 655-663

Tlapak-Simmons, V. L., Baggenstoss, B. A., Clyne, T., & Weigel, P. H. (1999). Purification and lipid dependence of the recombinant hyaluronan synthases from *Streptococcus pyogenes* and *Streptococcus equisimilis*. *Journal of Biological Chemistry*, Vol.274, No.7, (February 1999), pp. 4239-4245

Tlapak-Simmons, V.L., Kempner, E.S., Baggenstoss, B.A., & Weigel, P.H. (1998). The active streptococcal hyaluronan synthases (HASs) Contain a Single HAS Monomer and multiple cardiolipin molecules. *Journal of Biological Chemistry*, Vol.273, No.40, (October 1998), pp. 26100-26109

Van der Meulen, R., Grosu-Tudor, S., Mozzi, F., Vaningelgem, F., Zamfir, M., Font de Valdez, G., & De Vuyst, L. (2007). Screening of lactic acid bacteria isolates from dairyand cereal products for exopolysaccharide production and genes involved. *International Journal of Food Microbiology*, Vol.118, No.3, (July 2007), pp. 250-258

van Hijum, S.A.F.T., Kralj, S., Ozimek, L.K., Dijkhuizen, L., & van Geel-Schutten, I.G.H. (2006). Structure-function Relationships of glucansucrase and fructansucrase enzymes from lactic acid bacteria. *Microbiology and Molecular Biology Reviews*, Vol.70, No.1, (March 2006), pp. 157-176

van Hijum, S.A.F.T., Szalowska, E., van der Maarel, M.J.E.C., & Dijkhuizen, L. (2004). Biochemical and molecular characterization of a levansucrase from *Lactobacillus reuteri*. *Microbiology*, Vol.150, No.3, (March 2004), pp. 621-630

Velasco, S. E., Areizaga, J., Irastorza, A., Dueñas, M.T., Santamaria, A., & Muñoz, M.E. (2009). Chemical and rheological properties of the β-glucan produced by

Pediococcus parvulus 2.6. *Journal of Agricultural and Food Chemistry*, Vol.57, No.5, (March 2009), pp. 1827-1834

Vijayendra, S.V., & Babu R.S. (2008). "Optimization of a new heteropolysaccharide production by a native isolate of *Leuconostoc* sp. Cfr-2181. *Letters in Applied Microbiology*, Vol.46, No.6, (Jun 2008) pp. 643-648

Voragen, A.G.J. (1998). Technological aspects of functional food-related carbohydrates. *Trends in Food Science and Technology*, Vol.9, No. 8-9, (August 1998) pp. 328-335

Wasser, S.P. (2011). Current findings, future trends, and unsolved problems in studies of medicinal mushrooms. *Applied Microbiology and Biotechnology*, Vol.89, No.5, (March 2011) pp. 1323-1332

Walling, E., Gindreau, M., & Lonvaud-Funel, A. (2001). La biosynthèse d'exopolysaccharide par des souches de *Pediococcus damnosus* isolées du vin: mise au point d'outils moléculaires de détection. *Lait*, Vol.81, No.1-2, (January-April 2001), pp. 289-300

Werning, M.L., Ibarburu, I., Dueñas, M.T., Irastorza, A., Navas, J., & López, P. (2006). *Pediococcus parvulus gtf* gene encoding the GTF glycosyltransferase and its application for Specific PCR detection of β-D-glucan-producing bacteria in foods and beverages. *Journal of Food Protection*, Vol.69, No.1, (January 2006), pp. 161-169

Werning, M.L., Corrales, M.A., Prieto, A., Fernández de Palencia, P., Navas, J., & López, P. (2008). Heterologous expression of a position 2-substituted (1-->3)-β-glucan in *Lactococcus lactis*. *Applied Environmental Microbiology*, Vol.74, No.16, (August 2008) pp. 5259-5262

Wood, B.J.B. (1997). *Microbiology of fermented food*, Blackie Academic & Professional, ISBN 0-7514-0216-8, London, United Kingdom

Wu, J., Zhang, Y., Wang, L., Xie, B., Wang, H., & Deng, S. (2006). Visualization of single and aggregated hulless oat (*Avena nuda* L.) (1→3),(1→4)-β-D-glucan molecules by atomic force microscopy and confocal scanning laser microscopy. *Journal of Agriculture Food Chemistry*, Vol.54, No.3, (January 2005), pp. 925-934

Yanaki, T., Ito, W., Tabata, K., Kojima, T., Norisuye, T., Takano, N., & Fujita, H. (1983). Correlation between the antitumor activity of a polysaccharide schizophyllan and its triple-helical conformation in dilute aqueous solution. *Biophysical Chemistry*, Vol.17, No.4, (June 1983), pp. 337-342

6

Guar Foaming Albumin – A Foam Stabilizer

Ami Shimoyama and Yukio Doi
Department of Food and Nutrition, Kyoto Women's University,
Higashiyama-ku, Kyoto,
Japan

1. Introduction

Various surface active agents are used as food additives in food processing when a decrease in surface tension is required e.g., in production and stabilization of all kinds of dispersions, which include emulsion, foams, aerosols and suspensions. Emulsions and foams are of particular interest in food processing and the basic principles involved in their formation and maintenance are very similar: foaming agents and emulsifiers, due to their amphipathic nature, form interface films and thus prevent the disperse phases from flowing together (1).

Protein stabilized foam is important to the structure and texture of many food products, including various cakes, confections, meringues, etc. (2). To produce stable foams the following abilities of the protein responsible for foam formation become important: 1) the ability to adsorb rapidly at the air-water interface, 2) the ability to denature promptly at the interface for maintaining the appropriate balance between hydrophobicity and hydrophilicity, and 3) the ability to interact mutually among the proteins that unfold at the interface and form a strong cohesive, viscoelastic film that can withstand thermal and mechanical agitation (1).

The molecular characteristics of a protein surface and the conformational flexibility determine the mode of adsorption at the interface (3). Flexible disordered proteins like beta-casein can undergo rapid conformational changes at the interface, being excellent foaming proteins. On the other hand, rigid structured globular proteins such as lysozyme and soy protein cannot undergo extensive conformational changes at the interface, being less foaming proteins. The mechanical strength of a protein film at the interface depends on cohesive intermolecular interactions and the stiffness is due to small bubble size and high viscosity. In particular, the formation of sufficiently strong protein film is needed for baked foam products to endure rupture and collapse due to heat expansion of air. Thus the gelling properties in addition to foaming properties are also required for this purpose (1).

Proteins from egg white and milk are widely used for many processed foods and foam-type products. Egg white contains various globular proteins with enough flexibility to make strong cohesive interactions, being a prime foam stabilizer for a variety of baked foam products. In fact, egg white appears to be only protein suited to producing baked foam products; other protein-stabilized foams such as gelatin (used for marshmallow) and whipped whey isolate cannot endure the thermal agitation upon heating and are melt away when baked. In spite of its excellent stabilizing ability to form heat resistant bubbles, egg white has a serious drawback, strong allergenicity. Egg white contains ovomucoid and ovalbumin, which make it the major food allergen (4).

Recently we isolated an albumin fraction with high foaming ability and foam stabilizing ability from guar meal, and designated guar foaming albumin (GFA) (5). The foaming activity of GFA was 10 times higher than that of egg white at low protein concentrations. GFA mainly composed of a simple protein with the molecular mass of 13 kDa. As a plant protein, GFA has a rather high nutritional value, and would be best suited to allergic patients against animal proteins. These features of GFA make it a promising candidate as a foaming agent in lieu of egg white. Guar meal is a byproduct during extraction of a guar gum, a galactomannnan gum, widely used as stabilizer in various processed food (6). Although proteins isolated from guar meal are characterized in some extent (7-10), GFA was first identified as a protein responsible for their foaming property.

With the particular attention to the high foaming ability and foam stabilizing ability of GFA, in this chapter we would like to investigate its foaming functionality, especially from the perspective of its application for baked food products. We focused on possible application of GFA to substitute egg white, a major food allergen.

2. Materials and methods

All chemicals were of analytical grade and were used as supplied. Commercial guar meals imported from Pakistan were provided by Taiyokagaku Co. Ltd. (Yokka-ich, Japan)

Materials. GFA was prepared according the method we reported elsewhere (5). The GFA solution was dialyzed against 0.1 M phosphate buffer (pH 6.8) for assessing foaming properties. To examine the effect of added sucrose and NaCl, sample is dialyzed against 5 mM phosphate buffer (pH 6.8), and diluted with using the same buffer if necessary. For a control experiment, thin albumen (EW), obtained from fresh egg white strained through fine gauze, was diluted with an appropriate buffer.

Foaming Studies. The sample (8 g each) diluted to 5 ~ 70 mg/ml of protein concentration was placed in a bowl (14.5 cm in diameter and 9.5 cm in depth), and whipped with a two-blade hand mixer (model HF-230, Hitachi Appliance, Inc.) for 2 min at the lowest speed setting of 1, followed for 2.5 min at speed 5 and finally for 0.5 min at speed 1. To determine the effect of sucrose and NaCl, samples were whipped for 2 min at speed 1 followed for 7.5 min at speed 5 and for 0.5 min at a speed setting of 1. The concentration of sucrose was adjusted by adding an appropriate amount of saturated sucrose solution to sample solutions in order to achieve a mild and complete mixing. The addition of sucrose and NaCl was carried out before whipping. To measure the specific volume, the foam prepared was transferred to a glass container (2.7 cm in diameter and 2.0 cm in depth) by using a plastic spatula and the overflowed foam was removed by sliding the spatula along the edge of the container without pressing foam. The precise volume of the container was previously obtained by measuring the weight of the container filled with water. The specific volume of foam was obtained by dividing the foam volume by the foam weight.

For baking, the foam in a bowl were transferred to a plastic tube (2.5 cm in diameter and 2.0 cm in height) which was placed on cooking paper by using a spatula without squeezing foam. After removing the tube off by carefully lifting it up (no foam left attached to the tube), the shaped foam samples were baked in an electric oven (model EMO-VA4, Sanyo Appliance, Inc.) for 20 min at 100ºC. To measure the volume of a baked foam sample, a plastic tube big enough to cover the whole sample without touching it was placed surrounding the sample. Then a clump of rapeseed was carefully introduced into the tube without damaging the sample just to fulfill the tube, and the weight of the rapeseed was measured. The volume of the baked foam was calculated by subtracting the displaced rapeseed volume from the total tube volume. Control experiments were carried out in the same manner using EW as samples. All determinations in the foam volumes were performed in triplicate samples of at least two independent experiments.

Bubble size measurement. A portion of the foam sample, usually around 0.5 cm^3, was carefully placed on a microscope slide and observed through an inverted microscope equipped with a digital camera (Olympus, C5060). The protein concentrations of samples were 5 and 40 mg/ml in 0.1 M phosphate buffer (pH 6.8). All pictures were taken within 3 min after the formation of foam. The image analysis to measure the size distribution of foam was carried out by using ImageJ (version 1.43b, NIH) program and statistical analysis by Excel program (Microsoft).

Enzyme-linked immunosorbent assay. An enzyme-linked immunosorbent sassy kit to determine the peanut protein content was purchased from Morinaga (Tokyo, Japan). The kit is manufactured according to the guideline for detecting food allergens, approved by Ministry of Health, Labour and Welfare in Japan. The antibody used in the kit was elicited by using a mixture of peanut proteins containing Ara h2, which is one of major peanut allergenic proteins.

Other Analytical Procedures. Protein concentrations were determined according to the Bradford method (11) with bovine serum albumin as a standard.

3. Results and discussion

Protein concentration dependence. When the concentration dependence of the foam volumes were examined, the specific foam volumes of GFA increased as the protein concentration increased to 20 mg/ml and plateaued thereafter (Figure 1 A). The specific volumes of GFA were 1.2-1.9 times higher than those of egg white when compared at the same protein concentrations. It should be mentioned that in the previous study (5) we showed 10 times higher foaming ability of GFA than that of EW where the foamability was assessed at much lower protein concentrations, less than 0.1 mg/ml. In the present study the foaming ability as well as stability were assessed at much higher protein concentrations. Interestingly, the foam volumes of the 1:1 mixture of GFA:EW were close to those of EW at lower than 20 mg/ml, but they became close to those of GFA as the protein concentration increased. At high protein concentrations (> 40 mg/ml), the foam volumes of the 1:1 mixture much exceeded those of GFA, indicating a synergistic effect of GFA on the foaming activity at high protein concentrations. It should be noted that the protein concentration here represents that of the total protein, i.e., the protein concentration of GFA and EW is 30 mg/ml each for the 1:1 mixture sample at 60 mg/ml.

It is known that basic proteins such as lysozyme and clupeine improve the foaming properties of acidic proteins such as bovine serum albumin and β-lactoglobulin due to enhanced electrostatic interactions at the bubble surface (12). In these cases, the difference in

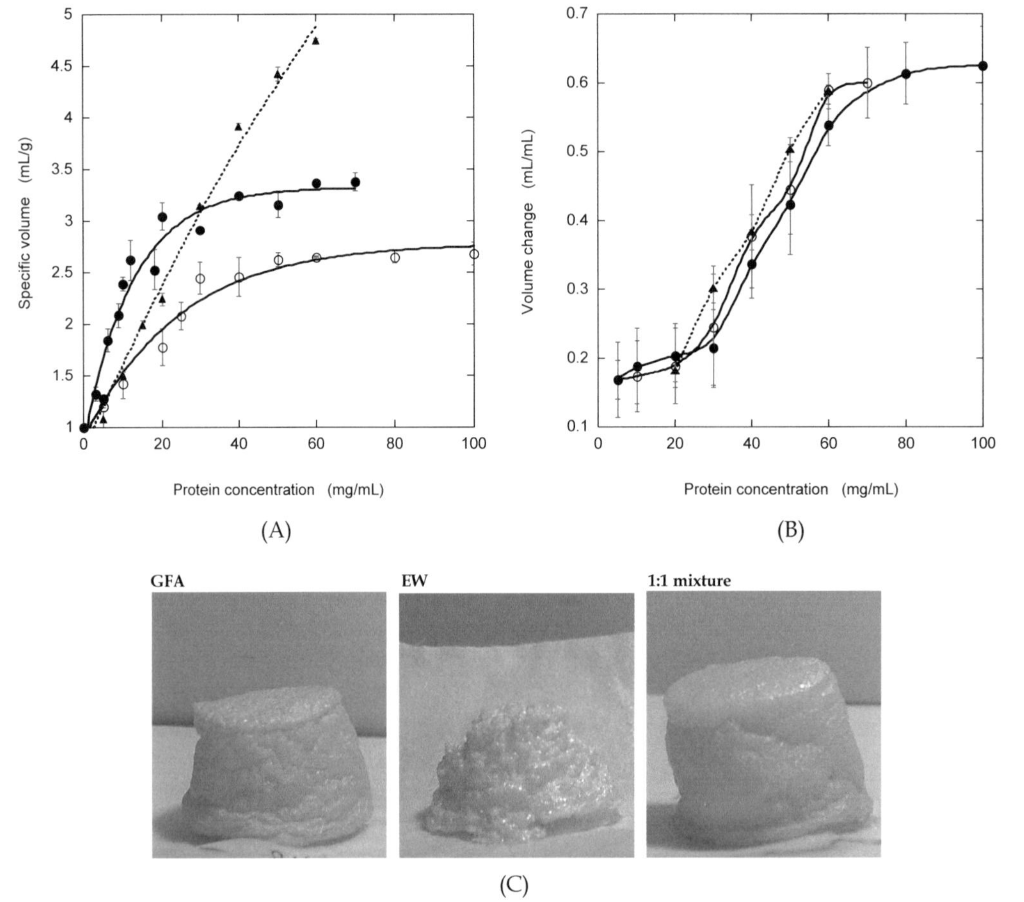

Fig. 1. (A) Changes in the specific volume at various protein concentrations of GFA, EW and the 1:1 mixed sample. Samples were dialyzed against 0.1 M phosphate buffer (pH 6.8), and GFA, EW and the 1:1 mixed sample were whipped by a hand mixer for 5 min. The specific foam volumes calculated were plotted against protein concentrations. GFA (●), EW (○) and the 1:1 mixed sample (▲). Bars indicate the standard deviation. (B) Change in the foam volumes before and after baking of GFA, EW and the 1:1 mixed sample. The samples prepared in Figure 1A were baked for 20 min at 100 °C using an electric oven. The volumes of baked foams were always smaller than the foams before baking. GFA (●), EW (○) and the 1:1 mixed sample (▲). Bars indicate the standard deviation. (C) Appearance of the baked foam produced by GFA, EW and the 1:1 mixture. The protein concentration used was 50 mg/ml each

isoelectric points (pI) between the acidic and basic proteins must be sufficiently large so that at intermediate pHs, interactions are strong enough to yield good foaming properties. GFA is prepared by the acid precipitation at its pI of 4 and most proteins of egg white have their pIs at the acidic region. Therefore, the synergistic effect observed here may not be explained merely by a simple electrostatic interaction between oppositely charged proteins. It is also interesting to mention that mixtures of two dissimilar proteins exhibit thermodynamic incompatibility upon mixing resulting in phase separation at the air-water interface (13).

The phase separation occurred at high protein concentrations (10-20% w/v) then brings about the instability of the film since the high interfacial energy between the phase-separated regions may act as zones of instability. Considering the increased foaming ability observed in the 1:1 mixtures at high protein concentrations, GFA must be a protein that is compatible to most of proteins present in EW. Any intermolecular disulfide interactions between proteins, if any, existed in the 1:1 mixture appears not to be involved in the synergistic effect observed here since a similar augmented foam formation was also observed in the presence of beta-mercaptoethanol (data not shown). Although exact nature

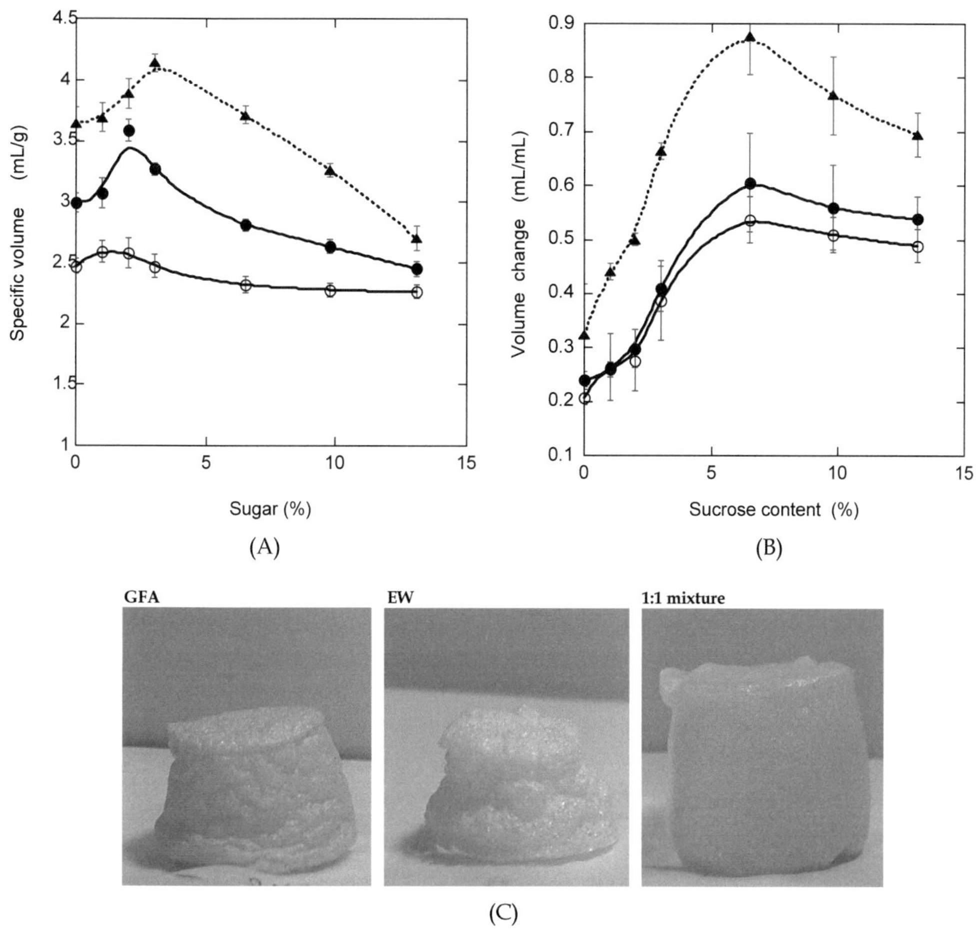

Fig. 2. (A) The effect of sucrose on the specific volumes of GFA, EW and the 1:1 mixed sample. The protein concentration used was 32 mg/ml each. Samples were prepared as described in Figure 1 A, except sucrose was added to each solution before foaming by a hand mixer for 10 min. GFA (●), EW (○) and the 1:1 mixed sample (▲). Bars indicate the standard deviation. (B) The effect of sucrose on the volume changes before and after baking. The protein concentration used was 32 mg/ml each. Samples were baked as described in Figure 1 B. GFA (●), EW (○) and the 1:1 mixed sample (▲). Bars indicate the standard deviation. (C) Appearance of the baked foam produced in the presence of 5% sugar by GFA, EW and the 1:1 mixture. The protein concentration used was 50 mg/ml each

of the synergistic effect remains to be investigated, some specific, weak interaction operative only at high protein concentrations might be involved.

The ability of proteins to stabilize foams is mainly determined by the rheological properties of protein adsorption layers: viscoelastic behaviour, irreversible structures and rheological changes under compression and expansion (14). An extreme expansion of foams may occur upon heating, for example, during baking meringue. Since GFA was shown to possess better foam stability than EW (5), changes in foam volume upon baking were examined. In all samples of GFA, EW and the 1:1 mixture, the foam volumes after baking reduced to 15% to 65% of the original volumes depending on the protein concentrations and the protein concentration dependence of the volume reduction were similar among the samples (Figure 1 B). Therefore, the stability of the GFA adsorbed film upon baking is comparable to that of the EW film. It should be noted, however, that the actual foam volume after baking in the present experiment was maximum in the 1:1 mixture at 60 mg/ml (Figure 1 C). The fact that the strength of the GFA adsorbed film remained even after baking is in contrast to that of β-casein, for example, whose foams are easily ruptured upon baking.

Effect of sucrose addition. The addition of sucrose up to 13% to EW did not significantly affect its foaming ability (Figure 2 A). The addition of 2-3% sucrose to GFA slightly enhanced the foam specific volume but decreased at higher sucrose concentrations. In the 1:1 mixture, the positive effect of sucrose was observed between 2-6%, but at higher concentrations sucrose affected adversely. In fact sucrose appears to damage the foaming ability of GFA more severely than that of EW as the foam volume of GFA reduced considerably in the presence of 45% sucrose in contrast to EW, which did not show an appreciable reduction (data not shown).

In general, addition of sugars to protein solutions often impairs foamability due to enhanced stability of protein structure, but improves foam stability due to increased viscosity (2). When the foams prepared with various sucrose concentrations were baked, stabilizing effect of sucrose is pronounced (Figure 2 B, C). Both GFA and EW foams were similarly stabilized and the maximum stability was observed at around 5% sucrose. The stabilizing effect of sucrose in the 1:1 mixture was similar but more conspicuous, suggesting that the synergistic interaction between GFA and EW was effectively augmented by the presence of small amount of sucrose. The increase in viscosity by the addition of sucrose may not contribute significantly to the observed stability.

Effect of NaCl addition. The addition of NaCl to EW decreased the foam specific volume only slightly (Figure 3 A). In contrast, the foaming ability of GFA as well as the 1:1 mixture were slightly enhanced by the NaCl addition. The synergistic interaction between GFA and EW appears not to be influenced by charge neutralization upon addition of salt ions. Upon baking, however, the addition of NaCl notably influenced the foam stability (Figure 3 B, C). EW reduced its foam volume upon heating as increased the NaCl concentration. On the contrary, in both GFA and the 1:1 mixture, the foam volume reduction upon heating was restrained by the addition of NaCl. Especially, the foams produced by the 1:1 mixture were greatly stabilized by NaCl, as was the case by sucrose, implying that both electrostatic interactions and hydrogen bondings between proteins in the film may contribute to exhibiting the synergistic effect.

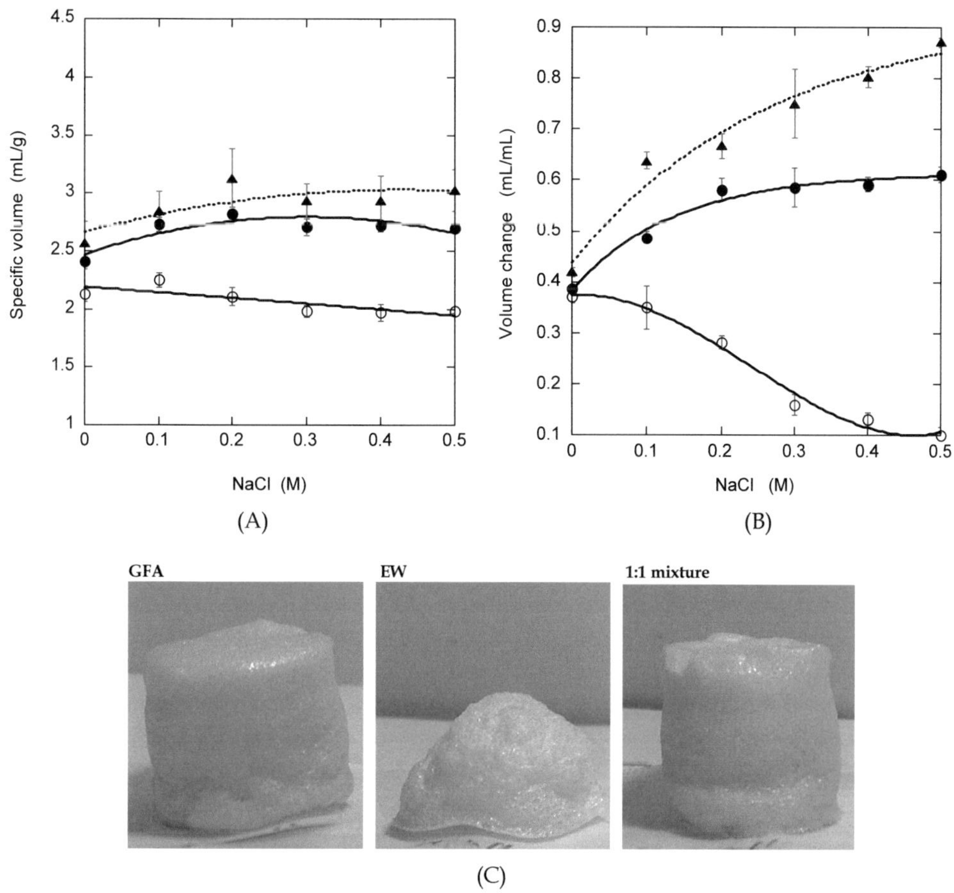

Fig. 3. (A) The effect of NaCl on the specific volumes of GFA, EW and the 1:1 mixed sample. The protein concentration used was 25 mg/ml each. Samples were prepared as described in Figure 1 A, except NaCl was added to each solution before foaming and samples were dialyzed against 5 mM phosphate buffer (pH 6.8). GFA (●), EW (○) and the 1:1 mixed sample (▲). Bars indicate the standard deviation. (B) The effect of NaCl on volume changes before and after baking. The protein concentration used was 25 mg/ml each. Samples were baked as described in Figure 1 B. GFA (●), EW (○) and the 1:1 mixed sample (▲). Bars indicate the standard deviation. (C) Appearance of the baked foam produced in the presence of 0.3 M NaCl by GFA, EW and the 1:1 mixture. The protein concentration used was 50 mg/ml each

Bubble size distribution. Since the foam produced by GFA seemed much smoother in appearance than that by EW (5), the size distribution of the GFA bubbles was examined through microscopic observation (Figure 4). GFA can produce smaller and more uniform bubbles than EW; the average bubble size of GFA is half of that of EW with a much narrower size distribution (Table 1). The average bubble size obtained with the 1:1 mixture was between those with GFA and EW. A similar difference in size distribution was also observed at lower protein concentration (5 mg/ml); the average bubble areas with GFA and EW were 0.025 and 0.052 mm^2, respectively.

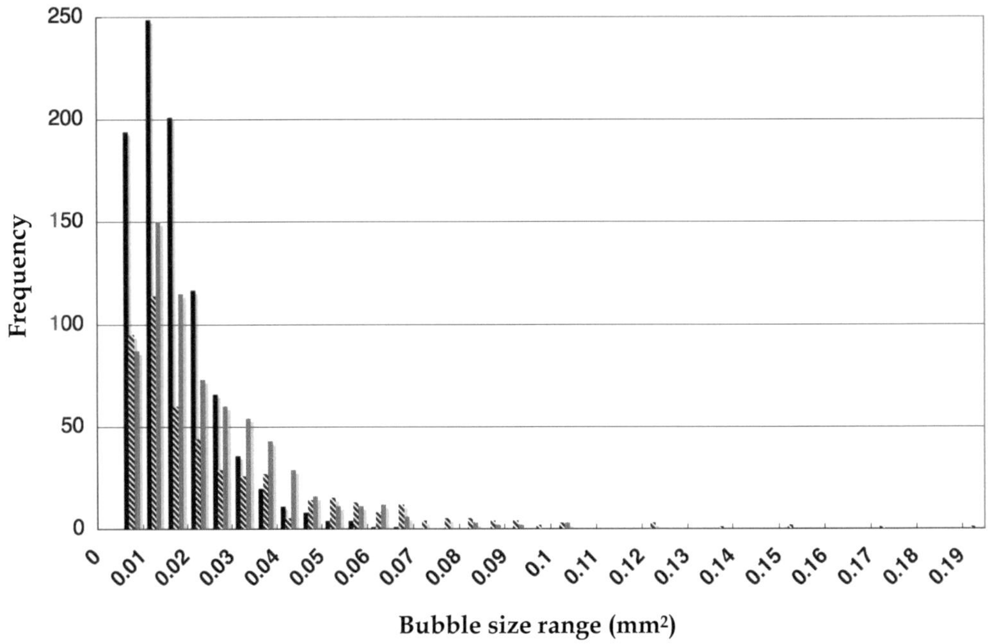

Fig. 4. Bubble size distribution of foams made by GFA (solid), EW (oblique) and the 1:1 mixture (shaded). The protein concentration in all samples was 40 mg/ml

	GFA	EW	1:1 mixture
Bubble area, Mean x 10^{-3} (mm^2) ± SD	12.5 ± 9.31	23.3 ± 26.4	19.4 ± 16.2
Bubble area, Median x 10^{-3} (mm^2)	10.3	13.1	14.2
Number of Bubbles (mm^{-2})	43.4	23.7	32.3

Table 1. Summary of the bubble size distribution in foams obtained with GFA, EW and the 1:1 mixture

According to the Laplace principle, the internal pressure of a bubble is greater than the external pressure, and the pressure difference is inversely proportional to radius of the foam bubble (1). Therefore, smaller foam bubbles can withstand heavier load than larger bubbles; the smaller the bubble sizes, the stiffer and stronger the foams become. The observed small bubble sizes of GFA substantiate the stability of GFA foams. However, the fact that the average foam size of the 1:1 mixture is larger than that of GFA did not appear to be in line with the synergistic effect observed in the mixture (Figure 1 A). The mechanism for the synergistic effect should be further investigated in many respects.

Immunological analysis. Previously no apparent immunoreactivity of GFA was observed against the antisera obtained from the allergic patients to egg, wheat and soybean (5). Since GFA was obtained from guar (*Cyamopsis tetragonolobus*) which belongs to the pea family (*Fabaceae*), the possible immunological relation to peanut proteins (*Arachis* *hypogaea*) was investigated by using an enzyme-linked immunosorbent assay kit, which is manufactured according to the guideline for detecting food allergens approved by Ministry of Health, Labour and Welfare in Japan. In Table 2 were listed the amount of protein in various nuts detected by the antibody against peanut proteins. The content of the protein in guar beans reactive to the anti-peanut antibody were less than 10^{-5} fold compared with that found in peanut. In the protein isolate of GFA, the amount of protein detected by the present method was 10^{-5} μg/g, which could be negligible when GFA was used in processed food as foam enhancer. It should be pointed out that an allergic warning against peanuts such as "contains peanuts" should be labelled properly on the surface of processed food products according to Japanese Food Sanitation Act if food contains the peanut protein more than 10 ppm.

Sources	Reactive Protein (μg/g)*
Peanut (*Arachis hypogaea*)	62880
Guar (*Cyamopsis tetragonolobus*)	5.9
Makadamia nut (*Macadamia integrifolia*)	0.9
Soybean (*Glycine max*)	0.5
Common bean (*Phaseolus vulgaris*)	0.5
Azuki bean (*Vigna angularis*)	0.3
Pistachio (*Pistacia vera*)	0.2
GFA	10.0

*The amount of the protein reactive to the anti-peanut antibody is calculated on the basis of mass of nut (bean), except for GFA which is based on the total protein content measured by the Bradford method (11).

Table 2. The amount of protein detected by ELISA using the antibody against peanut proteins

Abbreviations Used: EW, egg white (thin albumen); GFA, guar forming albumin; pI, isoelectric point

4. References

[1] Damodaran, S. Amino acids, peptides, and protein. In *Fennema's Food Chemistry*, 4th ed.; Damodaran, S.; Parkin, K.L.; Fennema, O. R., Eds.; CRC Press: New York, 2007; p.277-281.

[2] Campbell, G.M.; Mougeot, E. Creation and characterisation of aerated food products. *Trends Food Sci. Technol.* 1999, 10, 283-296.

[3] Dickinson, E. Adsorbed protein layers at fluid interfaces: interactions, structure and surface rheology. *Colloid Surface B: Biointerfaces.* 1999, 15, 161-176.

[4] Davis, P.J. Historical and cultural background to plant food allergies. *In Plant Food Allergens*; Mills, E.N.C.; Shewry, P.R., Eds; Blackwell: Oxford, 2004; p1-15.

[5] Shimoyama, A.; Kido, S.; Kinekawa, Y.; Doi, Y. Guar foaming albumin: a low molecular mass protein with high foaming activity and foam stability isolated from guar meal. *J. Agric. Food Chem.* 2008, 56, 9200–9205.

[6] Undersander, D. J.; Putnam, D. H.; Kaminski, A. R.; Kelling, K. A.; Doll, J. D.; Oplinger, E. S.; Gunsolus, J. L. Guar. In *Alternative Field Crops Manual*. University of Wisconsin-Extension;
http://www.hort.purdue.edu/newcrop/afcm/guar.html, accessed July, 2008.
[7] Lee, J. T.; Connor-Appleton, S.; Haq, A. U.; Bailey, C. A.; Cartwright, A. L. Quantitative measurement of negligible trypsin inhibitor activity and nutrient analysis of guar meal fractions. *J. Agric. Food Chem.* 2004, 52, 6492–6495.
[8] Nath, J. P.; Subramanian, N.; Narasinga-Rao, M. S. Isolation and characterization of the major fraction of guar proteins. *J. Agric. Food Chem.* 1980, 28, 844-847.
[9] Nath, J. P.; Subramanian, N.; Narasinga-Rao, M. S. Effect of detoxification treatments on the proteins of guar meal. *J. Agric. Food Chem.* 1981, 29, 529-532.
[10] M. Biochemical and technological studies on the production of isolated guar protein. *Nahrung/Food* 2001, 45, 21-24.
[11] Bradford, M. M. A rapid and sensitive method for the quantitation of microgram quantities of protein utilizing the principle of protein-dye binding. *Anal. Biochem.* 1976, 72, 248-245.
[12] Poole, S.; West S.I.; Walters C.L. Protein-protein Interactions: Their importance in the foaming heterogeneous protein systems. *J. Sci. Food Agric.* 1984, 35, 701-711.
[13] Razumovsky, L.; Damodaran, S. Incompatibility of mixing of proteins in adsorbed binary protein films at the air-water interface. *J. Agric. Food Chem.* 2001, 49, 3080-3086.
[14] Yampolskaya, G.; Platikanov, D. Proteins at fluid interfaces: Adsorption layers and thin liquid films. *Adv. Colloid Interface Sci.* 2006, 128-130, 159-183.

Utilization of *Aspergillus niger* Phytase Preparation for Hydrolysis of Phytate in Foods

Akiko Matsuo[1] and Kenji Sato[2]
[1]*Department of Nutrition, College of Nutrition, Koshien University, Momijigaoka, Takarazuka, Hyogo*
[2]*Division of Applied Life Sciences, Graduate School of Life and Environmental Sciences, Kyoto Prefectural University, Shimogamo, Kyoto*
Japan

1. Introduction

The majority of phosphorous in plant seed, especially cereal grains and legumes, is myo-inositol hexaphosphate (phytate or IP_6) (Rosa et al., 1999). Structure of IP_6 is shown in Figure 1. Figure 2 shows distribution of IP_6 in some plant foods. Unrefined cereals and soybean products contain high levels of IP_6. Rice bran contains IP_6 more than 6,000 mg/100 g of dry matter. IP_6 has a strong capability to chelate multivalent metal ions, particularly zinc, calcium, and iron ions, which results in the formation of highly insoluble salts (Nolan et al., 1987; Hotz et al., 2001). The IP_6 can be hydrolyzed by enzyme (phytase) and converted to lower myo-inositol phosphates; from inositol pentaphosphate (IP_5) to inositol monophosphate (IP_1) and myo-inositol. It has been demonstrated that the metal complexes with IP_1–IP_4 are more soluble than those with IP_5 and IP_6 (Sandberg et al., 1989). Animal experiments have demonstrated that the ingestion of the meal that contains more than 1% IP_6 decreases intestinal absorption of metal ions and induces the metal ion deficiency (Hirabayashi et al., 1998; Grases et al., 2001). In some communities, unrefined cereals are still main ingredients for the diet. In addition, the unrefined grains of rice, wheat, rye, etc., are richer in various minerals, dietary fiber, vitamins, and other bioactive components than refined ones. Therefore, these unrefined cereals are used as food ingredients for bread,

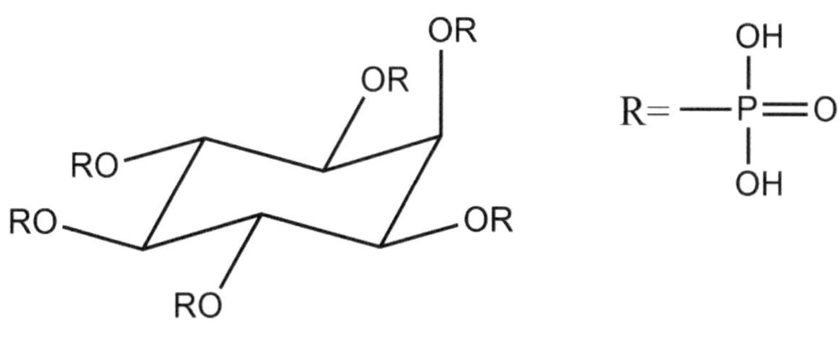

Fig. 1. Structure of phytic acid (IP_6)

breakfast cereals and so on due to their high health-promoting activities (Fukui et al., 1997; Haros et al., 2001; Lopez et al., 2001; Porres et al., 2001). In addition to the unrefined cereals, legumes, especially soybean have long history as food ingredient. Recently, soy flour, soy protein isolate, its protease digest, etc. have been formulated and used as protein source for infant formulae, sports drinks, enteral nutrients and also diets for animal, poultry, and fish. However, as mentioned above, these products are also rich in IP_6. Then, nutritional disturbances in the absorption of iron and zinc may occur by ingestion of IP_6–rich foods (Shaw et al., 1995; Sandberg et al., 1996; Minihane et al., 2002).

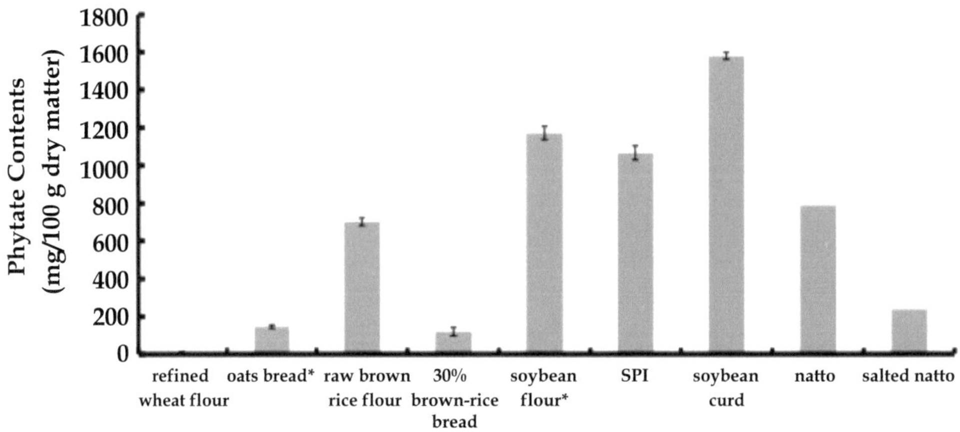

* Data from Rosa et al. 1999. Natto; steamed soybean fermented with *Baccillus subtillis*. Salted natto; steamed soybean fermented with *Aspergillus oryzae* and further fermented with wild yeast and lactic acid bacteria in the presence of salt (more than 5%).

Fig. 2. Phytate contents of some cereals and soybean products

It has been demonstrated that the IP_6 in foods can be degraded by phytase, which would improve mineral absorption in humans (Sandberg et al., 1996; 2002). On the basis of these facts, some phytase preparations have been formulated and used to reduce IP_6 and IP_5 levels of some plant-based foods (Haros et al., 2001; Saito et al., 2001; Porres et al., 2001: Matsuo et al., 2005; 2010) and also animal and fish diets (Pallauf & Rimbach, 1997; Sajjadi & Carter, 2003).

In this chapter, the recent application of fungal phytase preparation as food additive and its problem are reviewed. In addition, recent studies for control of the contamination of other enzymes in the phytase preparation are focused.

2. Source of phytase

It has been demonstrated that fermentation using lactic acid bacteria and fungus can reduce IP_6 levels in soybean flour and bran-enriched bread (Hirabayashi et al., 1995; Harose et al., 2001; Andlid et al., 2004; Leenhardt et al., 2005; Matsuo et al., 2005; Reale et al., 2004; 2007; Palacious et al., 2007; Li et al., 2008; Jorquera et al., 2008; Sanz-Penella et al., 2009). These facts indicate microorganisms involved in the fermentation process can produce phytase.

Indeed, some bacteria, yeasts, and fungi have been demonstrated to produce phytases. As mentioned above, the enzymatic hydrolysis of the IP_6 in unrefined cereals and legumes improves mineral absorption in humans (Sandberg et al., 1996; 2002). These facts can lead an idea that microbial phytase can be used as food additive to reduce IP_6 level in the plant-based foods. So far to now, phytase preparations have been prepared from *Asperigillus niger*,

recombinant *Aspergillus oryzae, Pichia pastoris* etc. in an industrial scale and used as food additive (Greiner & Konietzny, 2006). However, these microbial also produce protease, amylase and etc. As demonstrated in the following sections, a food additive-grade *Asperigillus niger* phytase preparation has significant protease and amylase activities (Matsuo et al., 2010). In some cases, occurrences of protease and amylase in the food additive-grade phytase preparation exerts adverse effects on texture and appearance of the final products.

In the following sections, application of a food additive-grade *Aspergillus niger* phytase preparation on some plant-based foods are introduced. Effect of the phytase treatment not only on IP_6 content but also on the other properties will be discussed.

3. Application for improvement of nutritional value of animal, poultry, and fish diets

Plant seeds and grains have been used for animal, poultry, and fish diets. As shown in Figure 2, the unfermented soy and cereal products contain high levels of IP_6. The occurrence of IP_6 in diet may interfere mineral absorption (Sandberg, 2002). In some cases, it may induce serious nutritional problem. Indeed, feeding IP_6-containing soy protein isolate retarded growth of fish larvae. To solve these problems, *Aspergillus niger* and *Pichia pastoris* phytase preparation was used to decrease IP_6 in the plant-based diet. This treatment significantly improve feeding efficacy. Now, formulated *Aspergillus niger* and *Pichia pastoris* phytases for improvement of feeding efficacy for fish, chicken, animals are commercially available (Pallauf & Rimbach, 1997; Sajjadi & Carter, 2003).

4. Application of phytase to isolate conglycinin and glycinin from soy protein isolate

Phytase treatment of soy protein isolate can be carried out in aqueous solution. During the *Asperigillus niger* phytase treatment, protein precipitation occurred (Saito et al., 2001). As well known, major soluble proteins in soybean are glycinin and β-conglycinin, which are classified as globulin. Both proteins are precipitated in acidic condition at approximately pH 4.5. However, predicted isoelectric point of glycinin based on the protein sequence is neutral pH. The glycinin forms complex with IP_6 and then it has apparent acidic isoelectric point, which makes glycinin precipitate in acidic condition. The phytase treatment degrades the glycinin-bound IP_6 and shifts isoelectric point to neutral pH. Then selective precipitation of glycinin occurs at neutral pH after the phytase treatment of soy protein isolate. This technique is successfully applied to sepration of glycinin and β-conglycinin for food ingredient. It has been demonstrated that β-conglycinin can reduce blood neutral lipid level of animal and human by ingestion (Fukui et al., 2004; Moriyama et al., 2004; Kohno et al., 2006). Now, food-grade β-conglycinin fraction is prepared on the basis of phytase treatment and following selective precipitation technique. The β-conglycinin fraction is approved to present health claim on lipid metabolism in "Food for the Specific Health Use" in Japan.

5. Application for unrefined cereal flour-containing bread making.

Whole grains of wheat, rice, rye, etc., are richer than refined grains in various minerals, dietary fiber, and vitamins. Besides minerals, vitamins, and dietary fiber, the unrefined grains are rich in bioactive components that have health-promoting activities; for example, γ-aminobutyric acid (GABA), γ-oryzanol, polyphenols, and feruric acid. These components show antihypertensive (GABA), antioxidant, and anti-hypercholesterolemic (γ-oryzanol and feruric acid) activities (Sugano et al., 1997; Xu et al., 2001; Hayakawa et al., 2002). However, the consumption of the unrefined whole grains as the steamed and fried forms, are not very prevalent due to the poor swelling and textural properties of the unrefined whole grains. Then, the unrefined whole grains are usually processed to flour and used as bread ingredient, which has been gaining popularity worldwide (Fukui et al., 1997; Haros et al., 2001; Lopez et al., 2001; Porres et al., 2001). As shown in Figure 3, IP_6 in the dough mix containing 30% brown rice flour decreases during processing by the action of yeast phytase. However, the relatively higher levels of IP_6 remains in the bread containing brown rice flour. As shown in Figure 3, addition of food additive-grade *Aspergillus niger* phytase preparation significantly decreased IP_6 level of brown rice-added bread. *Aspergillus niger* phytase has

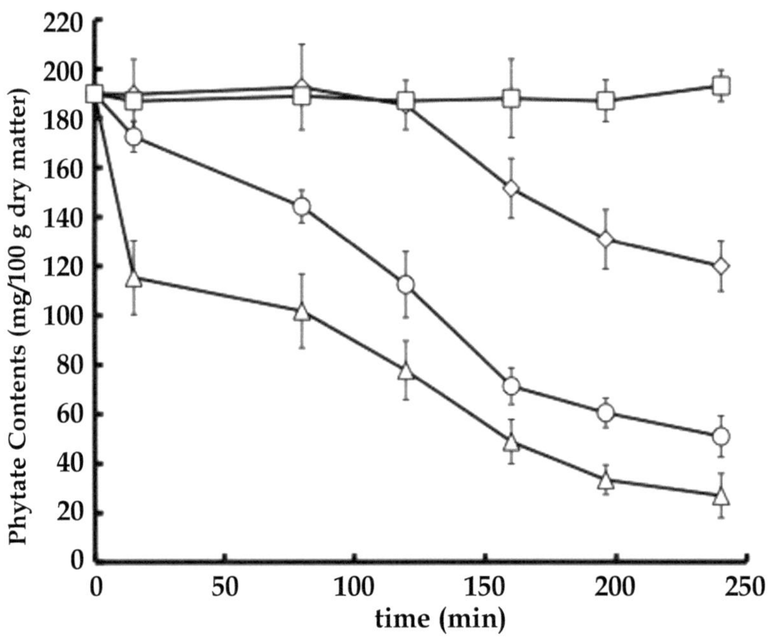

Fig. 3. Changes in phytate contents during bread making process. 0-195 min; mixing and fermentation, 240-min; baking. (□); 30% brown rice flour-containing dough without addition of yeast and phytase, (◊); added with yeast only, (○); added with yeast and 600 U of the *Asparagillus niger* phytase preparation, (△); added with yeast and 3000 U of the phytase preparation. Values are means ±SD, n=3. From Matsuo et al., 2005

been also used to decrease IP_6 level in the whole wheat fours (Haros et al., 2001). At early attempt, the flour was pretreated with phytase in water suspension system. It has been, however, demonstrated that direct addition of the phytase preparation into dough mix can reduce IP_6 content, which is the easier approach to decrease IP_6 level in the unrefined grain flour-containing bread (Haros et al., 2001; Matsuo et al., 2005; 2010). However, addition of high dose of the phytase preparation (3000 U) induced collapse of whole bread crust and deteriorates texture and taste of the final product as shown in Figurc 4 E (Matsuo et al., 2005).

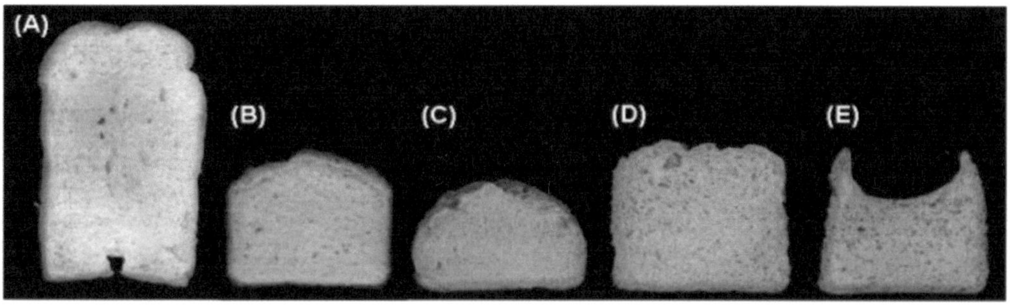

Fig. 4. Effect of addition of brown rice flour and phytase on the loaf volume. (A); wheat bread, (B); 30% brown rice flour bread, (C); 50% brown rice flour bread, (D); 30% brown rice bread flour bread added with 600 U of the phytase preparation; (E) 30% brown rice flour bread added with 3000 U of the phytase preparation. From Matsuo et al., 2005

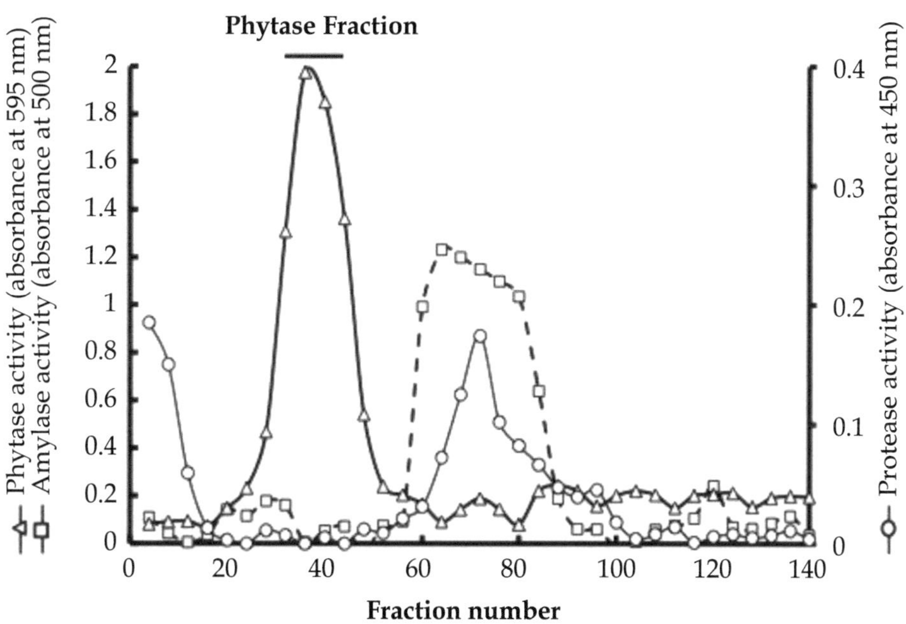

Fig. 5. Separation of phytase from amylase and protease in the crude phytase preparation by anion-exchange chromatography. Fraction indicated by the bar was collected and used as the purified phytase preparation

Two grams of the crude phytase preparation was dissolved in 50 mL of 0.02 M sodium acetate buffer, pH 6.0 and loaded on a column (15 cm × 10 mm i.d.) packed with TSK gel Super Q-Toyopearl 650S (Tosoh Co., Tokyo, Japan) that had been pre-equilibrated with the same buffer. The absorbed proteins were eluted using 200 mL of a linear gradient of 0-0.5 M NaCl in the same buffer at 1.5 mL/min. Fractions were collected every 1 min. From Matsuo et al., 2010.

Figure 5 shows that the food additive-grade phytase preparation contains significant activities of protease and amylase, which can be separated from the phytase activity by anion-exchange column chromatography (Matsuo et al., 2010). Addition of the purified phytase fraction free from protease and amylase activities into the dough mix decreased IP_6 level without affecting bread volume (Figure 6). These facts imply that amylase and/or protease are responsible for the collapse of the bread crust by addition of high dose of the commercial phytase preparation. However, it has been difficult to isolate phytase, protease, and amylase activities in enough amounts for the additive purpose by conventional chromatography technique due to high cost of preparative LC system. Then it is difficult to control protease and amylase activities in the phytase preparation, which has limited the use of the fungal phytase preparation as food additive. To solve this problem, a large-scale isolation method for food additive-grade enzyme is necessary. This method should be inexpensive, biocompatible, easy to scale-up. The following section introduces recent advance in purification of phytase by new technique.

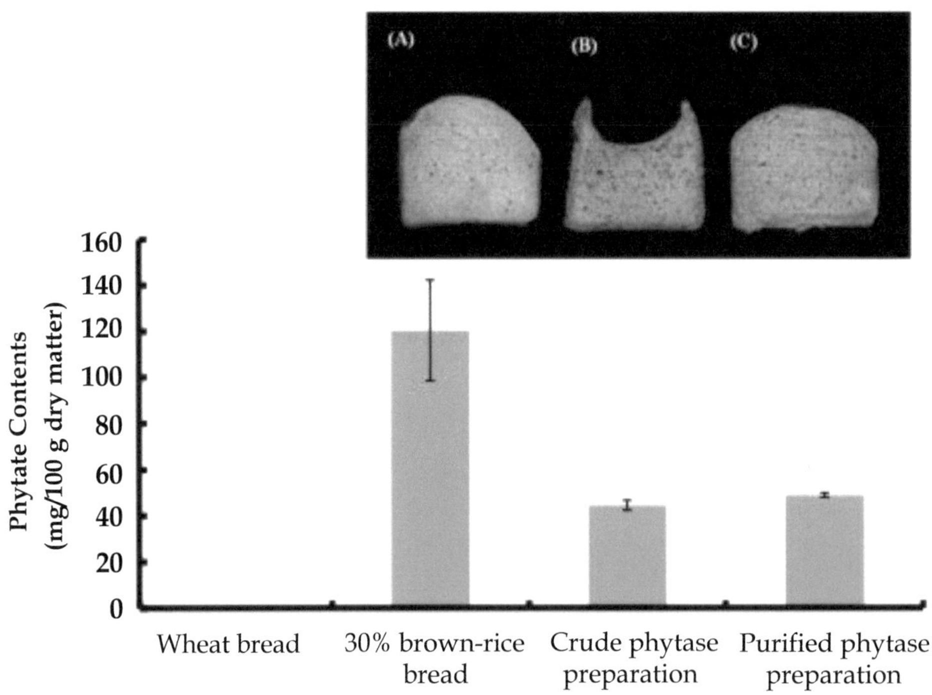

Fig. 6. Effect of crude and purified phytase (3,000 U) on phytate contents and swelling property of 30% brown rice bread. A; 30% brown rice bread, B; added with the crude phytase, C; added with the purified phytase. From Matsuo et al., 2010

6. Control of protease and amylase activities in phytase preparation by autofocusing

As mentioned in the previous section, it has been demanded to develop a method for large-scale fractionation of enzymes in crude extract and fermentation broth in order to control contamination of other enzymes in phytase preparation. For fractionation of peptides in the crude digests of food proteins, a preparative ampholyte-free isoelectric focusing has been developed (Hashimoto et al., 2005), which has a potential for purification of enzymes. This technique depends on amphoteric nature of the sample compounds. The crude enzymatic digest of food protein, crude extract, fermentation broth, etc. contain numerous compounds with different isoelectric points. Then these compounds can act as ampholine for isoelectric focusing. This phenomenon is referred to autofocusing (Yata et al., 1996; Akahosi et al., 2000; Hashimoto et al., 2005). This technique does not require harmful reagents and solvents just requires water and agarose gel in sample compartments and diluted phosphoric acid and sodium hydroxide in the electrode compartments. All of them can be used for food processing. Assembly of the apparatus is illustrated in Figure 7. A plastic plate with window is covered with nylon screen and the screen is wetted with hot agarose solution. After standing for few minutes, thin agarose gel layer is formed on the screen. The plates with agarose gel layer are inserted into the slots of the tank as shown in Figure 7. Then the tank can be separated into 12 compartments. Both ends of the compartments are used as electrode compartments. Others are used as sample compartments (10 compartments).

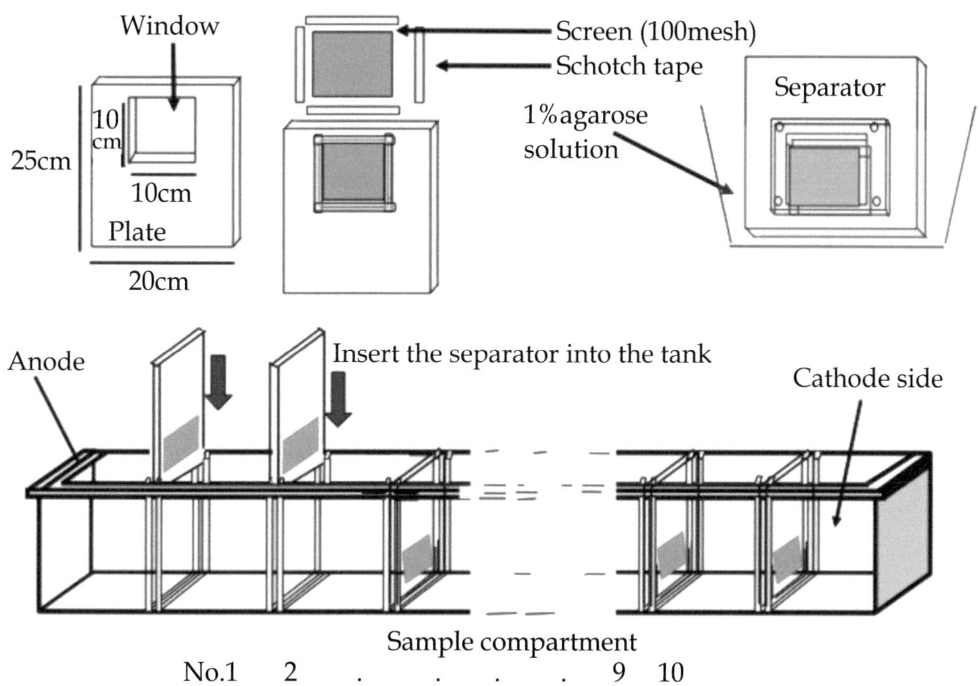

Fig. 7. Schematic drawing for assembly of Autofocusing apparatus. The tank is divided into 12 compartments separated with thin agarose gel layer. Volume of compartment can be changed from 100 mL to 5 L. From Hashimoto et al., 2005

Water-solution of sample is applied into the sample compartments. 0.1 N phosphoric acid and sodium hydroxide are loaded into cathode and anode, respectively. By loading direct electric current into the electrode compartments at constant voltage mode at 500 – 1000 V, the compounds in sample start to migrate to their own isoelectric points. This technique has been successfully used for peptide fractionation (Hashimoto et al., 2005; Higaki-Sato et al., 2006; Park et al., 2008; Murota et al., 2010; Elbarbary et al., 2010).

The food additive-grade phtase preparation also contain many compounds. Then autofocusing was used to purify phytase in the commercial preparation. As shown in Figure 8, pH gradient approximately from 2 to 11 was formed by autofocusing of food additive-grade *Aspergillus niger* phytase preparation. The protease activity was migrated to acidic fractions (Fr. 1-6). The amylase and phytase activities were widely distributed. Fr. 7 has the

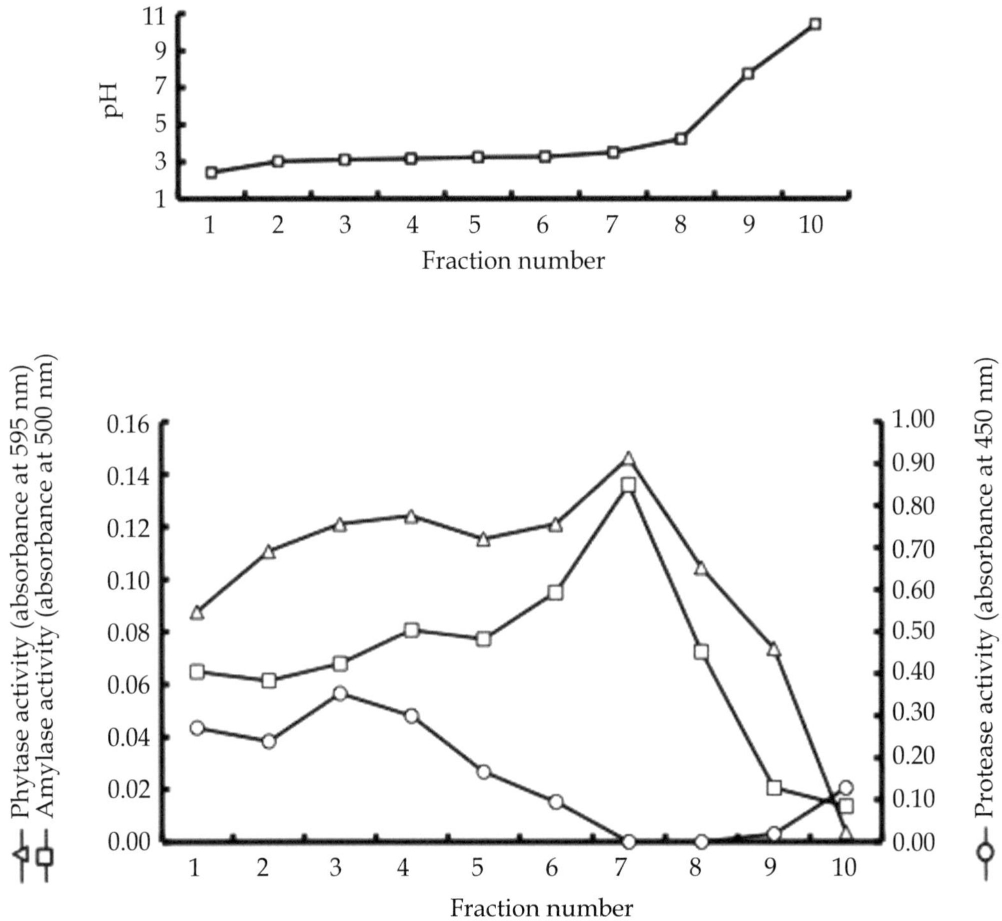

Fig. 8. Fractionation of phytase, amylase, and protease in a food additive-grade *Aspergillus niger* phytase preparation. pH gradient (upper) was formed by autofocusing for 24 hours. Fraction No. 7 and 9 were free from protease activity and used as high amylase and low amylase fractions, respectively

highest phytase and amylase activities. In the Fr. 9, the amylase activity dropped to 15% of the maximum activity, whereas relatively high level of the phytase activity (more than half of the activity in Fr. 7) was recovered in the Fr. 9. Then, Fr. 7 and 9 were collected and used as high amylase and low amylase fractions, respectively. These fractions were free from protease activity. The crude phyase preparation, which contains significant protease activity and two protease free autofocusing fractions were added to the 30% brown rice flour-added bread dough mix. Phytate content decreased by addition of all fractions. As shown in Figure 9, addition of the low amylase fraction did not significantly affect the loaf volume and did not induce collapse of crust up to 3000 U of phytase activity. Addition of the high amylase fraction showed positive effect on loaf volume up to 514 U of amylase activity. However, addition of amylase, more than 1029 U, decrease of loaf volume and partial collapse of crust (arrow) were observed. Addition of the crude preparation at 791 U of protease activity induced partial collapse of crust even amylase activity was less than 500 U. Addition of protease more than 1582 U resulted in extensive collapse of whole crust and deteriorated texture properties (mash-like texture).

The crude phytase preparation

F	0 U	1500 U	3000 U	6000 U	12000 U
A	0 U	435 U	875 U	1740 U	3480 U
P	0 U	791 U	1582 U	3162 U	6328 U

Fraction No. 7

F	0 U	536 U	1500 U	3000 U	6000 U
A	0 U	193 U	514 U	1029 U	2058 U
P	0 U	0 U	0 U	0 U	0 U

Fraction No. 9

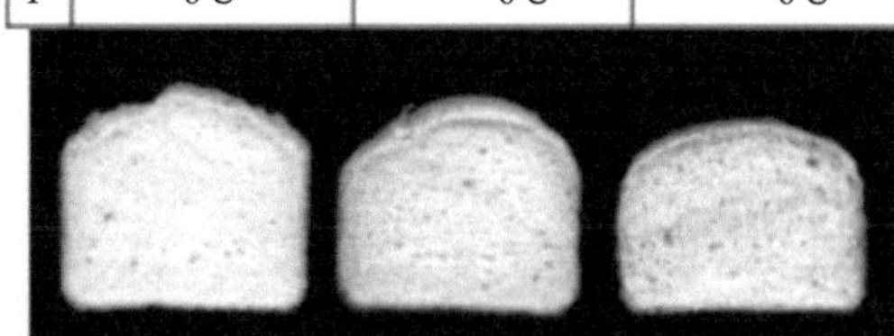

F	0 U	1161 U	3000 U
A	0 U	153 U	396 U
P	0 U	0 U	0 U

Fig. 9. Effects of addition of crude phytase preparation and high amylase (Fr. No. 7) and low amylase (Fr. No. 9) fractions on appearance and swelling properties of the 30% brown rice flour-containing bread. Phytase (F), amylase (A), and protease (P) activities are indicated in the tables

On the basis of these results, it can be concluded that IP_6 content in the unrefined grain flour-containing bread can be reduced by direct addition of the fungal phytase. However, presence of high protease activity in the phytase preparation can deteriorate swelling and textural properties and collapse the crust. On the other hand, amylase can improve the swelling property at relatively low activity (<500 U), while it deteriorates the swelling property at high activity (> 1000 U). These phenomena can be explained as follows. Protease in the preparation degrades gluten and deteriorates gluten network. Consequently, the dough can not retain the CO_2 gas generated by fermentation, which decreases loaf volume and finally collapses whole crust. The amylase in the preparation may induce oligosaccharides from starch and consequently enhance fermentation by yeast at the low amylase activity. On the other hand, the high amylase activity can deteriorate starch gel network after cooking process. Indeed, fungal amylase has been used in the baking industry to accelerate fermentation. However, addition of excessive amylase and protease has been demonstrated to make the dough slack (Fred et al., 1975). These facts clearly indicate that the control of protease and amylase activities is crucial for the application of the fungal phytase preparation to bread making to reduce IP_6 and obtain good quality of the product.

7. Conclusion and future prospects

It has been demonstrated that fungal phytase can be used to reduce IP_6 level of the soybean based products and also unrefined grain flour-containing products, especially bread. However, fungal phytase may contain protease and amylase, which may affect quality of the final products. In some cases, the product may be unacceptable due to excessive degradation of protein, which induces unfavorable effects on texture and appearance. Then it is crucial to control the contaminated enzyme activity in the phytase preparation. In the present chapter, ion exchange column chromatography and autofocusing can at least partially control contamination of protease and amylase activities. In regardless of the high resolution, column chromatography requires relatively expensive apparatus and high running cost. In addition, it may require salts, buffer, solvents etc, which should be removed from the final product. It also increases preparation cost. On the other hand, autofocusing does not require organic solvents and salts for fractionation and has inherited potential for scale-up. Recently, continuous type autofocusing apparatus has been developed (Hashimoto et al., 2006). Then this technique would be useful to control protease and amylase activities in the phytase preparation. By using this technique, a food additive phytase optimized for bread making and other applications could be produced.

IP_6 also suppresses divalent ions-induced Fenton's reaction; producing hydroxyl radical from hydrogen peroxide. Then IP_6 has been used to suppress oxidation of food (Graf and Eaton, 1990). However, little is known for the suppressive effect of IP_1-IP_5 against Fenton's reaction. If degradation products of IP_6 by phytase also suppress Fenton's reaction without reducing mineral bio-availability, phytase can be used for production of anti-oxidant food ingredient with excellent nutritional value. In addition, functional food based on suppression of Fenton's reaction without disturbing nutritional could be also produced by the limited digestion of IP_6 by phytase. For these purposes, further studies on control of hydrolysis of $IP_{2\text{-}6}$ by phytase are necessary.

8. References

Akahoshi, A.; Sato, K.; Nawa, N.; Nakamura, Y. & Ohtsuki, K. (2000). Novel Approach for Large-Scale, Biocompatible, and Low-Cost Fractionation of Peptides in Proteolytic Digest of Food Protein Based on the Amphoteric Nature of Peptides. *J. Agric. Food Chem.*, 48, 1955-1959.

Andlid, T. A.; Veide, J. & Sandberg, A. S. (2004). Metabolism of Extracellular Inositol Hexaphosphate (Phytate) by *Saccharomyces cerevisiae. Int. J. Food Microbiol.*, 97, 157-169.

Elbarbary, H. A.; Abdou, A. M.; Park, E. Y.; Nakamura, Y.; Mohamed, H. A. & Sato, K. (2010). Novel Antibacterial Lactoferrin Peptides Generated by Rennet Digestion and Autofocusing Technique. *Int. Dairy J.*, 20, 646-651.

Fred, F. B. (1975). Enzyme Uses in the Milling and Baking Industries. In *Enzymes in Food Processing,* 2nd Ed. (Reed, G., ed.) pp. 301−330, Academic Press, New York, NY.

Fukui, K.; Kuwata, G. & Imai, M. (1997). Effects of Phytate Removal from Soybean Protein on Calcium, Magnesium and Zinc Absorption of Rats. *J. Jpn. Soc. Nutr. Food Sci.,* 50, 273−278.

Fukui, K.; Kojima, M.; Tachibana, N.; Kohno, M.; Takamatsu, K.; Hirotsuka, M. & Kito, M. (2004). Effects of Soybean β-Conglycinin on Hepatic Lipid Metabolism and Fecal Lipid Excretion in Normal Adult Rats. *Biosci. Biotech. Biochem.*, 68, 1153-1155.

Graf, E. & Eaton, J. W. (1990). Antioxidant Functions of Phytic Acids. *Free Radical Biol. Med.,* 8, 61-69.

Grases, F.; Simonet, B. M.; Prieto, R. M. & March, J. C. (2001). Variation of $InsP_4$, $InsP_5$ and $InsP_6$ Levels in Tissues and Biological Fluids Depending on Dietary Phytate. *J. Nutr. Biochem.*, 12, 595−601.

Greiner, R. & Konietzny, U. (2006). Phytase for Food Application. *Food Technol. Biotechnol.,* 44, 125-140.

Haros, M.; Rosell, C. M. & Benedito, C. (2001). Use of Fungal Phytase to Improve Breadmaking Performance of Whole Wheat Bread. *J. Agric. Food Chem.*, 49, 5450−5454.

Hashimoto, K.; Sato, K.; Nakamura, Y. & Ohtsuki, K. (2005). Development of a Large-Scale (50 L) Apparatus for Ampholyte-Free Isoelectric Focusing (Autofocusing) of Peptides in Enzymatic Hydrolysates of Food Proteins. *J. Agric. Food Chem.*, 53, 3801-3806.

Hashimoto, K.; Sato, K.; Nakamura, Y. & Ohtsuki, K. (2006). Development of Continuous Type Apparatus for Ampholyte-Free Isoelectric Focusing (Autofocusing) of Peptides in Protein Hydrolysates. *J. Agric. Food Chem.*, 54, 650-655.

Hayakawa, K.; Kimura, M. & Kamata, K. (2002). Mechanism Underlying γ-Aminobutyric Acid-Induced Antihypertensive Effect in Spontaneously Hypertensive Rats. *Eur. J. Pharmacol.*, 438, 107−113.

Higaki-Sato, N.; Sato, K.; Inoue, N.; Nawa, Y.; Kido, Y.; Nakabou, Y.; Hashimoto, K.; Nakamura, Y. &

Ohtsuki, K. (2006). Occurrence of the Free and Peptide Forms of Pyroglutamic Acid in Plasma from the Portal Blood of Rats that Had Ingested a Wheat Gluten

Hydrolysate Containing Pyroglutamyl Peptides. *J. Agric. Food Chem.,* 54, 6984-6988.

Hirabayashi, M.; Matsui, T. & Yano, H. (1998). Fermentation of Soybean Flour with *Aspergillus usamii* Improves Availabilities of Zinc and Iron in Rats. *J. Nutr. Sci. Vitaminol.,* 44, 877–886.

Hotz, C. & Gibson, R. S. (2001). Assessment of Home-Based Processing Methods to Reduce the Phytate Content and Phytate/Zinc Molar Ratio of White Maize (Zeamays). *J. Agric. Food Chem.,* 49, 692–698.

Jorquera, M.; Martinez, O.; Maruyama, F.; Marschner, P. & Mora, M. D. L. (2008). Current and Future Biotechnological Application of Bacteria Phytases and Phytase-Producing Bacteria. *Microbes Environ.,* 23, 182-191.

Kohno, M.; Hirotsuka, M.; Kito, M. & Matsuzawa, Y. (2006). Decrease in Serum Triglycerol and Visceral Fat Mediated by Dietary Soybean β-Conglycinin. *J. Atherosclerosis Thrombosis,* 13, 247-255.

Leenhardt, F.; Levrat-Verny, M. A.; Chanliaud, E. & Rémésy, C. (2005). Moderate Decrease of pH by Sourdough Fermentation is Sufficient to Reduce Phytate Content of Whole Wheat Flour through Endogenous Phytase Activity. *J. Agric. Food Chem.,* 53, 98-102.

Li, X.; Chi, Z.; Liu, Z.; Yan, K. & Li, H. (2008). Phytase Production by Marine Yeast *Kodamea ohmeri* BG3. *Appl. Biochem. Biotechnol.,* 149, 183-193.

Lopez, H. W.; Krespine, V.; Guy, C.; Messager, A.; Demigne, C. & Remesy, C. (2001). Prolonged Fermentation of Whole Wheat Sourdough Reduces Phytate Level and Increases Soluble Magnesium. *J. Agric. Food Chem., 49,* 2657–2662.

Matsuo, A.; Sato, K.; Nakamura, Y. & Ohtsuki, K. (2005). Hydrolysis of Phytic Acid in Brown-Rice Bread by *Aspergillus niger* Phytase. *J. Jpn. Soc. Nutr. Food. Sci.,* 59, 267–272.

Matsuo, A.; Sato, K.; Park, E. Y.; Nakamura, Y. & Ohtsuki, K. (2010). Hydrolysis of Phytate in Brown Rice-Added Bread by Addition of Crude and Purified *Aspergillus niger* Phytase Preparations during Bread Making. *J. Food Biochemistry,* 34, 195–205.

Minihane, A. & Rimbach, G. (2002). Iron Absorption and the Iron Binding and Anti-Oxidant Properties of Phytic Acid. *Inter. J. Food. Sci. Technol., 37,* 741–748.

Moriyama, T.; Kishimoto, K.; Nagai, K.; Urade, R.; Ogawa, T.; Utsumi, S.; Maruyama, N. & Maebuchi, M. (2004). Soybean β-Conglycinin Diet Suppresses Serum Triglyceride Levels in Normal and Genetically Obese Mice by Induction of β-Oxidation, Downregulation of Fatty Acid Synthase, and Inhibition of Triglyceride Absorption. *Biosc. Biotech. Biochem.,* 68, 352-359.

Murota, I.; Tamai, T.; Baba, T.; Sato, R.; Hashimoto, K.; Park, E. Y.; Nakamura, Y. & Sato, K. (2010). Uric Acid Lowering Effect by Ingestion of Proteolytic Digest of Shark Cartilage and Its Basic Fraction. *J. Food Biochemistry,* 34, 182–194.

Nolan, K. B. & Duffin, P. A. (1987). Effects of Phytate on Mineral Bioavailability. *In Vitro* Studies on Mg^{2+}, Ca^{2+}, Fe^{3+}, Cu^{2+} and Zn^{2+} (also Cd^{2+}) Solubilities in the Presence of Phytate. *J. Sci. Food Agric.,* 40, 79–85.

Palacios, M.C.; Haros, M.; Rosell, C. M. & Sanz, Y. (2008). Selection of Phytate-Degrading Human *Bifidobacteria* and Application in Whole Wheat Dough Fermentation. *Food Microbiol.*, 25, 69-76.

Pallauf, J. & Rimbach, G. (1997). Nutritional Significance of Phytic Acid and Phytase. *Arch. Tieremahr.*, 50, 301-319.

Park, E. Y.; Morimae, M.; Matsumura, Y.; Nakamura, Y. & Sato, K. (2008). Antioxidant Activity of Some Protein Hydrolysates and Their Fractions with Different Isoelectric Points. *J. Agric. Food Chem.*, 56, 9246-9251.

Porres, J. M.; Etcheverry, P.; Miller, D. D. & Lei, X. G. (2001). Phytase and Citric Acid Supplementation in Whole-Wheat Bread Improves Phytate-Phosphorus Release and Iron Dialyzability. *J. Food Sci.*, 66, 614–619.

Reale, A.; Mannina, L.; Tremonte, P.; Sobolev, A. P.; Succi, M.; Sorrentino, E. & Coppola, R. (2004). Phytate Degradation by Lactic Acid Bacteria and Yeasts during the Wholemeal Dough Fermentation: a 31P NMR Study. *J. Agric. Food Chem.*, 52, 6300-6305.

Reale, A.; Konietzny, U.; Coppola, R.; Sorrentino, E. & Greiner, R. (2007). The Importance of Lactic Acid Bacteria for Phytate Degradation during Cereal Dough Fermentation. *J. Agric. Food Chem.*, 55, 2993-2997.

Rosa, M. G.; Eduardo, G. H. & Belen, G. V. (1999). Phytic Acid Content in Milled Cereal Products and Breads. *Food Res. Int.*, 32, 217-221.

Saito, T.; Kohno, M.; Tsumura, K.; Kugimiya, W. & Kito, M. (2001). Novel Method Using Phytase for Separating Soybean Beta-Conglycinin and Glycinin. *Biosci. Biotechnol. Biochem.*, 65, 884–887.

Sajjadi, M. & Carter, C. (2003). Dietary Phytase Supplementation and the Utilization of Phosphorus by Atlantic Salmon (*Salmo salar* L.) Fed a Canola-Based Diet. 2003. *Asia Pac. J. Clin. Nutri.*, 12, S32.

Sandberg, A. S.; Carlsson, N. G. & Svanberg, U. (1989). Effects of Inositol Tri-, Tetra-, Penta-, Hexaphosphates on *In Vitro* Estimation of Iron Availability. *J. Food Sci.*, 54, 159–186.

Sandberg, A.; Hulythen, L. & Turk, M. (1996). Dietary *Aspergillus niger* Phytase Increases Iron Absorption in Humans. *J. Nutr.*, 126, 476–480.

Sandberg, A. S. (2002). Bioavailability of Minerals in Legumes. *Br. J. Nutri.*, 88, S281-285.

Sanz-Penella, J. M.; Tamayo-Ramos, J. A.; Sanz, Y. & Haros, N. (2009). Phytate Reduction in Bran-Enriched Bread by Phytase-Producing Bifidobacteria. *J. Agric. Food Chem.*, 57, 10239-10244.

Shaw, N.; Chin, C. & Pan, W. (1995). A Vegetarian Diet Rich in Soybean Products Compromises Iron Status in Young Students. *J. Nutr.*, 125, 212–219.

Sugano, M. & Tsuji, E. (1997). Rice Bran Oil and Cholesterol Metabolism. *J. Nutr.*, 127, 521S-524S.

Xu, Z.; Hua, N. & Godber, J. S. (2001). Antioxidant Activity of Tocopherols, Tocotrienols, and γ-Oryzanol Components from Rice Bran against Cholesterol Oxidation Accelerated by 2,2'-Azobis (2-Methylpropionamidine)Dihydrochloride. *J. Agric. Food Chem.*, 49, 2077–2081.

Yata, M.; Sato, K.; Ohtsuki, K. & Kawabata, M. (1996). Fractionation of Peptides in Protease Digests of Proteins by Preparative Isoelectric Focusing in the Absence of Added Ampholyte: A Biocompatible and Low-Cost Approach Referred to as Autofocusing. *J. Agric. Food Chem.*, 44 , 76-79.

8

Biotechnological Production of Xylitol from Agro-Industrial Wastes

José Manuel Domínguez, José Manuel Salgado,
Noelia Rodríguez and Sandra Cortés
Vigo University
Spain

1. Introduction

Xylitol is a polyalcohol of five carbon atoms which is widely used in the food and chemical industry. Its interest is due to the sweetening power similar to that of sucrose. However, it has been shown that the use of xylitol as a sweetener is better due to its anticariogenecity, tooth rehardening and remineralization properties. It is suitable as sugar substitute for diabetics, and it limits the tendency to obesity when is continuously supplied in diet (Salgado et al., 2010a). Table 1 shows the properties of the food additives that are used as sweeteners and are permitted in the European Union (EU).

Xylitol can be found naturally in various fruits and vegetables such as strawberries, raspberries, yellow plum, lettuce and cauliflower (Prakasham et al., 2009). The extraction of xylitol from these products is low, for this reason is not a good source. It is produced mainly by chemical processes. This process is the hydrogenation of the five-carbon sugar D-xylose in the presence of nickel catalyst at elevated temperature and pressure (Prakasham et al., 2009). The chemical process has some drawbacks such as high cost of purification processes. To avoid aggressive stages of the chemical processes, biotechnological processes were studied for xylitol production. This alternative production is bioconversion of D-xylose to xylitol by microorganisms. To reduce costs and environmental problems renewable biomass from agro-industrial waste can be used as source of D-xylose.

The food, agricultural and forestry industries produce large volumes of wastes annually worldwide, causing a serious disposal problem (Rodríguez-Couto, 2008). In the EU waste hierarchy and legislation, prevention and minimization of waste is given the highest priority (Staniskis & Stasiskiene, 2005). The List of Wastes (formerly the European Waste Catalogue), is a catalogue of all waste types generated in the EU. This list of wastes is periodically reviewed on the basis of new knowledge and, in particular, of research results, and if necessary revised in accordance with Article 18 of Waste Framework Directive (75/442/EEC). The list of wastes has been published in the Spanish Official Gazette of 19th February 2002 by Order MAM/304/2002 in conformity with the Commission Decision 2000/532/EC of 3 May 2000, replacing Decision 94/3/EC establishing a list of wastes pursuant to Article 1(a) of Council Directive 75/442/EEC on waste and Council Decision 94/904/EC establishing a list of hazardous waste pursuant to Article 1(4) of Council Directive 91/689/EEC on hazardous waste. The List of Waste has been amended by Commission Decisions 2001/118/EU,

2001/119/EU and 2001/573/EU. The different types of waste in the List are fully defined by a six-digit code, with two digits each for chapter, sub-chapter and waste type. The List is used to categorize items and substances when they become waste, but does not itself define items and substances as waste.

Food additives	Code UE	Sweetness (% SP)	Calorie content (kcal · g^{-1})	Glycemic index	ADI (mg · kg^{-1} body weigh)	Approved In EU
Sorbitol	E 420	60	3.5	<10	Acceptable	Directive 94/35/EC
Mannitol	E 421	50	2.4	0	Acceptable	Directive 94/35/EC
Acesulfame K	E 950	200	0	0	0 - 15	Directive 94/35/EC
Aspartame	E 951	160 - 220	4	0	0 - 40	Directive 94/35/EC
Cyclamic acids and its Na and Ca salts	E 952	30	0	<10	0 - 7	Directive 94/35/EC
Isomalt	E 953	40	2.4	<10	Acceptable	Directive 94/35/EC
Saccharin and its Na, K and Ca salts	E 954	300	0	0	0 - 15	Directive 94/35/EC
Sucralose	E 955	500	0	0	0 - 5	Directive 2003/115/EC
Thaumatin	E 957	1400 - 2200	4	0	Acceptable	Directive 94/35/EC
Neohesperidine DC	E 959	1500	4	0	0 - 5	Directive 94/35/EC
Neotame	E 961	8000 - 13000	4	0	0 - 18	Directive 2009/163/EU
Salt of aspartame-acesulfame 1	E 962		4	0	0 - 40	Directive 2003/115/EC
Maltitol	E 965	90	2.4	36	Acceptable	Directive 94/35/EC
Lactitol	E 966	30 - 40	2.4	<10	Acceptable	Directive 94/35/EC
Xylitol	E 967	100	2.4	<10	Acceptable	Directive 94/35/EC
Erythritol	E 968	60 - 70	0.2	0	Acceptable	Directive 2006/52/EC

SP: Sweetener power respect to sucrose (100)
ADI: Acceptable daily intake established by the Scientific Committee on Food (SCF)

Table 1. Sweeteners permitted for food use in the European Union

The wastes studied in the current work, listed in Table 2, are included in paragraph 2 and 20 of the European List of Waste "Wastes from agricultural, horticultural, hunting, fishing and aquacultural primary production, food preparation and processing" and "Municipal wastes and similar commercial, industrial and institutional wastes including separately collected fractions", respectively. Their reutilization, as well as other wastes, is of great interest since, due to legislation and environmental reasons, the industry is increasingly being forced to find an alternative use for its residual matter, at the same time that the use of these wastes considerably reduces the production costs (Rodríguez-Couto, 2008).

Lignocelluloses in nature derive from wood, grass, agricultural residues, forestry wastes and municipal solid wastes (Pérez et al., 2002). They constitute a renewable resource from which many useful biological and chemical products can be derived. Accumulation of lignocellulose in large quantities in places where agricultural residues present a disposal problem results not only in deterioration of the environment but also in loss of potentially valuable material that can be used in paper manufacture, biomass fuel production, composting, human and animal feed among others (Sánchez, 2009). These lignocellulosic materials are renewable sources of energy where approximately 90 % of the dry weight of most plant materials is stored in the form of cellulose, hemicelluloses, lignin, and pectin (Kumar et al., 2009). Some authors have reported their fractionation to obtain a variety of marketable chemicals from the polymeric fractions of the raw materials (Moldes et al., 2007), including the xylitol production by *Debaryomyces hansenii*.

2. Wastes from agricultural, horticultural, hunting, fishing and aquacultural primary production, food preparation and processing			
Code	Origin	Type of waste	Characterized waste
02 01 03	Primary production wastes	Plant tissue waste	Leaf fruit Vine leaf
02 03 04	Wastes from fruit, vegetables, cereals, edible oils, cocoa, coffee and tobacco preparation and processing; tobacco processing; conserve production	Materials unsuitable for consumption or processing	Pistachio shells Chesnut shells Nut shells Hazelnut shells
02 06 01	Wastes from the baking and confectionery industry	Materials unsuitable for consumption or processing	Wheat bran leaves
02 07 02		Waste from spirits distillation	Distilled bagasse
02 07 04		Materials unsuitable for consumption or processing	White and red grape stems
20. Municipal wastes and similar commercial, industrial and institutional wastes including separately collected fractions.			
Code	Origen	Type of waste	Characterized waste
20 02 01	Garden and park wastes (including cemetery waste)	Compostable waste	Grass

Table 2. Classification of the characterized wastes according to the European List of Waste

This chapter deals with the study of different agroindustrial wastes through their characterization, fractionation by acid pre-hydrolysis and further analysis of the hemicellulosic fraction. The pre-hydrolysis of pistachio hulls was optimized; and finally, the liquors obtained under the optimal conditions, after supplementation, where assayed to carry out the xylose to xylitol bioconversion by *D. hansenii*.

2. Materials and methods

2.1 Quantitative acid hydrolysis

Aliquots from the homogenized lot were submitted to moisture determination and to quantitative hydrolysis in a two-stage acid treatment (the first stage with 72 wt % sulfuric acid at 30 °C for 1 h, the second stage after dilution of the media to 4 wt % sulfuric acid at 121 °C for 1 h) (Bustos et al., 2004). The solid residue after hydrolysis was considered as Klason lignin meanwhile hydrolyzates were analyzed by HPLC as described bellow.

2.2 Acid hydrolysis

Hydrolyzates were obtained in autoclave at 130 °C with 3 % H_2SO_4 solutions during 30 min, using a liquid/solid ratio of 8 g g^{-1} (Portilla-Rivera et al., 2007). The liquid phase from the acid hydrolysis was neutralized with $CaCO_3$ to a final pH of 5.8-6.0, and the $CaSO_4$ precipitated was separated from the supernatant by filtration.

2.3 Charcoal adsorption

Powdered charcoal (Probus, Madrid, Spain) was activated with hot water and dried at room temperature. Charcoal detoxification of hydrolyzates was carried out by contacting hydrolyzates and charcoal (mass ratio: 10 g g^{-1}) at room temperature under stirring for one hour (Rivas et al., 2002). The liquid phase was recovered by filtration and used for making culture media.

2.4 Experimental design and statistical analysis of the pistachio shells prehydrolysis

Pistachio shells were treated with solutions containing 1-3 % H_2SO_4 during 15–45 min. at 130 °C, according to a statistical experimental design (Design Expert version 5.0, Stat-Ease Inc., Minneapolis, USA), to optimize the prehydrolysis stage, and evaluated by the Response Surface Methodology using Statistica version 5.0 (Statsoft, USA) software considering the SS residual to evaluate the significance of the effects and the model. The influence of two operational variables (concentration of catalyser and reaction time) was tested on three levels in a 3**(2-0) full factorial design. All experiments were carried out in duplicate in randomized run order. Liquors were separated from the solid fraction by vacuum filtration through common laboratory paper filters and analyzed the concentration of xylose, glucose, arabinose, acetic acid, formic acid and furfural, which were the dependent variables.

This design allowed the estimation of the significance of the parameters and their interaction using Student's *t*-test. The interrelationship between dependent and operational variables was established by Eq.1, a model including linear, interaction and quadratic terms:

$$y = b_0 + b_1 x_1 + b_{11} x_1^2 + b_2 x_2 + b_{22} x_2^2 + b_{12} x_1 x_2 \tag{1}$$

where y represents the dependent variables, b denotes the regression coefficients (calculated from experimental data by multiple regression using the least-squares method), and x denotes the independent variables.

2.5 Microorganism and culture conditions

Debaryomyces hansenii NRRL Y-7426 was kindly provided by the National Center for Agricultural Utilization Research (Peoria, Illinois, USA). Freeze-dried cells were grown on a basal medium containing 30 g L^{-1} commercial xylose, 3 g L^{-1} yeast extract, 3 g L^{-1} malt extract, and 5 g L^{-1} peptone. The microorganism was maintained in agar slant tubes containing a medium formulated with the same components and concentrations as the previous one plus 20 g L^{-1} agar. Inocula were prepared by solubilization of cells with sterile water and underwent growth during 24 hours in the previous medium without agar. Biomass in inocula was measured by optical density at 600 nm and adjusted by dilution with water, and added to fermentation broth to reach a final concentration in the vicinity of 0.04 g L^{-1}.

Shake flask fermentation experiments were carried out under microaerophilic conditions in 250 mL Erlenmeyer flasks containing 100 mL of culture media (sterilized in autoclave at 100 °C during 60 min) prepared with 85 mL hydrolyzates and 10 mL of the nutrients indicated in Table 3.

Fermentation	Hydrolyzate	Nutrients
1	Raw	Liquid vinasses + stream B (conc.)
2		Corn steep liquor (30 g/L)
3		YE (3 g/L), ME (3 g/L), Peptone (5 g/L)
4		Without nutrients
5	Detoxified	YE (3 g/L), ME (3 g/L), Peptone (5 g/L)

YE: Yeast Extract; ME: Malt Extract.

Table 3. Nutrients employed in fermentations

Fermentation 1 contains the optimal amount of economic nutrients described by Salgado et al. (2010b) consisting in 50 mL liquid vinasses and 25 mL stream B but concentrated 7.5 times in rotavapor at 50 °C to reach a final volume of 10 mL. Fermentation 2 contains another economic nutrient, corn steep liquor. Fermentation 4 was carry out using non supplemented hydrolyzates, meanwhile Fermentations 3 and 5 contains commercial nutrients using raw or dextoxified hyrolyzates.

After inoculation (5 mL), fermentations were carried out in orbital shakers (New Brunswick, Edison, NJ, USA) at 100 rpm and 31 °C for 96 hours. Samples (2 mL) were taken at given fermentation times and centrifuged at 6,000 rpm for 3 min. The supernatants were stored for analyses.

2.6 Analytical methods

Glucose, xylose, arabinose, xylitol, glycerol, acetic acid, ethanol, furfural, and hydroxymethylfurfural (HMF) were measured by a high-performance liquid chromatographic system (Agilent, model 1200, Palo Alto, CA) equipped with a refractive index detector and an Aminex HPX-87H ion exclusion column (Bio Rad) eluted with 0.003 M sulfuric acid at a flow rate of 0.6 mL min^{-1} at 50 °C. Biomass concentration in experiments was measured by centrifugation, washing of the cells, and oven-drying to constant weight at 100 °C.

3. Results and discussion

3.1 Characterization of different agro-industrial wastes

Lignocellulosic materials consists of three types of polymers - cellulose, hemicelluloses, and lignin - that are strongly intermeshed and chemically bonded by non-covalent forces and by covalent cross-linkages. Cellulose and hemicelluloses are macromolecules from different sugars; whereas lignin is an aromatic polymer synthesized from phenylpropanoid precursors (Pérez et al., 2002). They also contain smaller amounts of proteins, pectin, acids, salts, minerals, ashes and extractives, including soluble non-structural sugars, nitrogenous material, chlorophyll, and waxes.

Cellulose is a linear polymer of D-glucose subunits linked to each other by β-(1,4)-glucosidic bonds with fibrous structure with a degree of polymerization of up to 10,000 or higher (Jørgensen et al., 2007). The long-chain cellulose polymers are linked together by hydrogen and van der Waals bonds, which cause the cellulose to be packed into microfibrils. Hemicelluloses and lignin cover the microfibrils (Kumar et al., 2009). Cellulose can appear in crystalline form (crystalline cellulose) and, in smaller percentage, in non-organized cellulose chains (amorphous cellulose) (Pérez et al., 2002).

Hemicelluloses is a complex carbohydrate polymer consisting of short highly branched chains of different five-carbon sugars (xylose, rhamnose and arabinose) and six-carbon sugars (glucose, galactose and mannose) and smaller amounts of non-sugars, mainly acetyl groups, but also uronic acids such as 4-o-methylglucuronic, D-glucuronic, and D-galactouronic acids. The backbone of hemicelluloses is either a homopolymer or a heteropolymer with short branches linked by β-(1,4)-glycosidic bonds and occasionally β-(1,3)-glycosidic bonds, and also can have some degree of acetylation. These polymers do not aggregate, even when they cocrystallize with cellulose chains (Kumar et al., 2009). Glucuronoxylan is the principal component of hardwood hemicelluloses whereas glucomannan is predominant in softwood (Pérez et al., 2002). The degree of polymerization is below 200 (Jørgensen et al., 2007).

Finally, lignin is a complex network formed by polymerization of phenyl propane units and constitutes the most abundant non-polysaccharide fraction in lignocellulose (Jørgensen et al., 2007). Lignin is an amorphous heteropolymer complex. The large molecular structure contains cross-linked polymers of phenolic monomers linked by alkyl-aryl, alkyl-alkyl, and aryl-aryl ether bonds (Kumar et al., 2009). Lignin encrusts the cell walls and cements the cells together (Hamelinck et al., 2005), conferring structural support, impermeability, and resistance against microbial attack and oxidative stress (Pérez et al., 2002).

Table 4 summarizes the quantitative acid hydrolysis of different agroindustrial wastes. Cellulose changes widely from only 10.1 % in vine leaf up to 80-95 % in cotton seed hairs, and 85-99 % in papers, which is an almost exclusively cellulosic material. Furthermore, the composition varies no only among species but also within a single plant depending on

	Cellulose	Hemicelluloses				Lignin	Extractives	Reference
		Total	Xylan	Arabinan	Acetyl groups			
Almond shell	50.7	28.9				20.4	2.5	Demirbas, 2003
	26.8	32.5	26.1	2.4	04.0	27.4	5.0	Nabarlatz et al., 2005
Apple fiber	20.8	10.0-28.4				12.1		Claye et al., 1996
Barley bran husks	23.0	34.3	26.6	6.1	1.6	21.4		Cruz et al., 2000
Chestnut shells	21.1	13.9	10.5	1.4	2.0	46.5	18.6	*
Coastal Bermuda grass	25.0	35.7				6.4		Kumar et al., 2009
Corncob	31.7	38.1	30.9	3.8	3.4	20.3		Cruz et al., 2000
	52.0	32.0				15.1		Demirbas, 2003
	41.2	36.0				6.1		Domínguez et al., 1997
	45	35				15		Kumar et al., 2009
Corn fiber	14.3	16.8				8.4		Mosier et al., 2005
Corn leaves	37.6	37.7	30.3	4.2	3.2	12.6		Cruz et al., 2000
Corn stover	51.2	30.7				14.4		Demirbas, 2003
	37.5	22.4				17.6		Mosier et al., 200517
Cotton seed hairs	80-95	5-20				0		Kumar et al., 2009
Distilled grape marc	10.8	11.2	7.5	2.2	1.6	50.9	14.7	Portilla-Rivera et al., 2008
Flax	34.9	23.6				22.3		Fan et al. 1987
Grasses	25-40	35-50				10-30		Kumar et al., 2009
Hardwood stems	40-55	24-40				18-25		Kumar et al., 2009
Hazelnut kernel shells	29.6	15.7				53.0		Demirbas, 2003
Hazelnut shell	25.9	29.9				42.5	3.3	Demirbas, 2003
	23.7	24.4	19.6	0.7	4.1	40.8	11.1	Portilla-Rivera et al., 2008
Leaf fruit	11.1	14.7	9.2	1.2	4.3	20.5	53.8	*
Leaves	15-20	80-85				0		Kumar et al., 2009
Newspaper	61.3	9.8				12.0		Kim & Moon, 2003
	40-55	25-40				18-30		Kumar et al., 2009
Nut shells	25-30	25-30				30-40		Kumar et al., 2009
Oats	26.6	30.2				21.4		Claye et al., 1996
Oat fiber	26.6	17.0-21.3				21.4		Claye et al., 1996
Oat straw	39.4	27.1				17.5		Fan et al., 1987

Table 4. (continues on next page) Quantitative acid hydrolysis (in %) of agro-industrial wastes

Office paper	68.6	12.4				11.3		Mosier et al., 2005
Olive husk	24.0	23.6				48.4		Demirbas, 2003
Paper	85-99	0				0-15		Kumar et al., 2009
Pistachio shells	15.2	38.5	33.1	0.0	5.4	29.4	17.0	*
Red grape stem	13.3	8.47	6.7	0.77	1.0	35.9	42.4	*
Rice bran	24.4	7.6-24.0				18.4		Claye et al., 1996
Rice hulls	35.6	11.96				15.4		Saha et al., 2005
Softwood stems	45-50	25-35				25-35		Kumar et al., 2009
Solid cattle manure	1.6-4.7	1.4-3.3				2.7-5.7		Kumar et al., 2009
Sorted refuse	60.0	20				20		Kumar et al., 2009
Soybean stems	34.5	24.8				19.8		Fan et al., 1987
Sugarcane bagasse	35.0	35.8				16.1		Sasaki et al., 2003
Sunflower shell	48.4	34.6				17.0	2.7	Demirbas, 2003
Swine waste	6.0	28				-		Kumar et al., 2009
Switchgrass	32.0	25.2	21.1	2.8		18.1	17.5	Hamelinck et al., 2005
	45.0	31.4				12		Kumar et al., 2009
	31.0	20.4				17.6		Mosier et al., 2005
	22.2	13.9	10.4	3.5	0.0	18.0	46.0	*
Thistle	31.1	12.2				22.1		Jiménez & López, 1993
Tomato fiber	19.7	13.2-23.3				13.8		Claye et al., 1996
Vine leaf	10.1	8.37	5.9	1.8	0.67	44.4	37.1	*
Vineshoots	34.1	19.0	12.8	0.90	5.3	27.1	7.1	Bustos et al. 2004
Walnut shell	25.6	22.1				52.3	3.3	Demirbas, 2003
	23.0	21.0	16.0	1.2	3.8	37.4	18.6	Portilla-Rivera et al., 2008
Waste papers from chemical pulps	60-70	10-20				5-10		Kumar et al., 2009
Wheat bran	32.2	16-28				5.2		Claye et al., 1996
Wheat bran leaves	26.2	24.3	15.8	8.5	0.0	7.9	41.5	*
Wheat straw	28.8	39.1				18.6		Demirbas, 2003
	30.0	50				15		Kumar et al., 2009
	38.2	21.2				23.4		Mosier et al., 2005
	48.6	27.7						Saha et al., 2005
White grape stem	20.7	11.37	10.4	0.00	0.97	31.3	36.7	

*Current work.

Table 4. (continued) Quantitative acid hydrolysis (in %) of agro-industrial wastes

several factors such as age, location or stage of growth. For instance, we obtained a percentage of 22.2 % in switchgrass, but Kumar et al. (2009) reported a percentage of 45 %. Meanwhile, hemicelluloses are in general in smaller amounts, although in some particular cases such as leaves can account for values as high as 80-85 % (Kumar et al., 2009), being xylan the main constituent in all the cases indicated. Regarding lignin, Demirbas (2003) reported the highest percentage of lignin (53.0 %) in hazelnut kernel shells, although this percentage varies widely depending on the material. Papers and herbaceous plants show in general the lowest contents; meanwhile softwoods and nuts have the highest contents (see Table 4).

3.2 Prehydrolysis of lignocellulosic materials

Due to the robust structure of lignocellulosic biomass, pretreatment is a prerequisite for hydrolysis into fermentable sugars to be completed within an industrially acceptable time frame (Jørgensen et al., 2007). Although fermentable D-glucose could be produced from cellulose through the action of either acid or enzymes breaking the β-(1,4)- glycosidic linkages (Kumar et al., 2009), the structure of cellulose along with the intermolecular hydrogen bonds gives cellulose high tensile strength, makes it insoluble in most solvents and is partly responsible for the resistance of cellulose against microbial degradation (Ward & Moo-Young 1989). On the contrary, hemicelluloses, because of their branched, amorphous nature, are relatively easy to hydrolyze (Hamelinck et al., 2005), particularly, in contrast to cellulose, the polymers present in hemicelluloses are easily hydrolyzable (Kumar et al., 2009).

Over the years a number of different technologies have been successfully developed for pretreatment of lignocellulose (Jørgensen et al., 2007), including concentrated or dilute-acid hydrolysis. Concentrated acids such as H_2SO_4 and HCl have been used to treat lignocellulosic materials. Nevertheless, although they are powerful agents for cellulose hydrolysis, concentrated acids are toxic, corrosive, hazardous, and thus require reactors that are resistant to corrosion, which makes the pretreatment process very expensive. In addition, the concentrated acid must be recovered after hydrolysis to make the process economically feasible (Kumar et al., 2009).

Among the chemichal hydrolysis methods, dilute-acid hydrolysis is probably the most commonly applied (Taherzadeh & Karimi, 2007). Therefore, dilute-acid hydrolysis (prehydrolysis) where a mineral acid is added to the reaction media to break down polymers, can be a possible first processing step for an integral use of biomass, since hemicelluloses can be almost completely solubilized, whereas little alteration is caused in both lignin and cellulose, which are recovered in the solid phase (Moldes et al., 2007). Among them, sulfuric acid has been of the most interest in such studies as it is inexpensive and effective (Kumar et al., 2009).

The lignocellulosic materials were subjected to prehydrolysis treatments under the same conditions (3 % H_2SO_4, 130 °C, 30 min., liquid to solid ratio = 8 g/g) previously reported by other authors (Portilla-Rivera et al., 2007). Table 5 reports the sugars content and other compounds solubilized during the treatment. Xylose was the sugar released in highest concentration, reaching a maximal xylose content of 41.8 g L^{-1} in pistachio shells. Additionally, glucose and arabinose appear in the lowest amounts of 0.88 g L^{-1} and 0.67 g L^{-1}, respectively, showing a xylose to glucose ratio of 47.6. Nevertheless, some possible toxins of further fermentable broths were generated, including acetate from the deacetylation of xylan and furan dehydration products (furfural from xylose and hydroxymethylfurfural from glucose). Using pistachio shells, 8.0 g L^{-1} of acetic acid and 1.7 g L^{-1} of furfural were released, meanwhile the highest amount of HMF, 1.2 g L^{-1}, was observed after the acid hydrolysis of white grape stems. Other compounds were also present in hydrolyzates, including citric, tartaric and lactic acids, and glycerol. Tartaric acid concentrations were relatively elevated in viticulture wastes (white and red grape stems and vine leafs) with concentrations up to 9.8 g L^{-1}. Nevertheless, these acids are not toxic for further fermentations stages.

Agroindustrial wastes	Citric acid	Tartaric acid	Glucose	Xylose	Xylose/ Glucose ratio	Arabinose	Lactic acid	Glycerol	Acetic acid	HMF	Furfural
Chestnut shells	0.0	6.9	2.6	13.3	5.0	3.2	0.0	0.0	2.8	0.0	0.0
Grass	1.7	1.0	8.4	12.4	1.5	3.4	0.0	0.0	1.5	0.51	0.51
Leaf fruit	0.0	1.4	5.8	9.1	1.6	5.5	0.0	0.0	2.2	0.47	0.0
Pistachio shells	0.0	1.5	0.88	41.8	47.6	0.67	0.0	0.0	8.0	0.0	1.7
Red grape stem	0.79	9.8	5.3	12.6	2.4	2.1	0.0	1.1	2.7	0.0	0.0
Vine leaf	0.0	9.4	4.4	6.8	1.6	3.7	0.0	0.0	1.6	0.0	0.0
Wheat bran leaves	0.0	0.0	23.8	22.5	0.95	9.9	0.0	0.0	0.71	0.69	0.88
White grape stem	1.2	8.2	12.1	13.5	1.1	2.0	0.98	3.5	2.6	1.2	0.0

Table 5. Composition (g/L) of hydrolyzates after prehydrolysis at 130 °C with 3 % H_2SO_4 during 30 min. using a liquid/solid ratio = 8 g g^{-1}

3.3 Optimization of the prehydrolysis of pistachio shells

Considering that pistachio shells provides the highest amount of xylose, the acid hydrolysis of this material was optimized using a 3**(2-0) full factorial design, to produce the highest level of hemicelluloses hydrolysis with the lowest level of degradation of the cellulosic fraction. The use of experimental designs to optimize the hydrolysis conditions is a technology widely used in bibliography (Bustos et al., 2005). In the present work, the independent variables considered and their variation ranges were: H_2SO_4 concentration (1-3 %) and duration of treatments (15-45 min). The standardized (coded) adimensional variables employed, having variations limits (-1,1), were defined as x_1 (coded H_2SO_4 concentration) and x_2 (coded time). The correspondence between coded and uncoded variables was established by linear equations deduced from their respective variation limits:

$$x_1 = ([H_2SO_4] - 2) \tag{2}$$

$$x_2 = (t - 30)/15 \tag{3}$$

The composition of liquors was measured by dependent variables: y_1 (glucose, g L^{-1}), y_2 (xylose, g L^{-1}), y_3 (arabinose, g L^{-1}), y_4 (acetic acid, g L^{-1}), y_5 (formic acid, g L^{-1}) and y_6 (furfural, g L^{-1}). Table 6 shows the set of experimental conditions assayed expressed in terms of coded variables, as well as the experimental data obtained for variables y_1 to y_6. The sequence for the experimental work was randomly established to limit the influence of systematic errors on the interpretation of results.

	Operational conditions		Experimental results					
Experiment	x_1	x_2	y_1	y_2	y_3	y_4	y_5	y_6
1	-1	-1	0.44	34.4	0.62	7.2	0.79	0.21
2	-1	0	0.66	37.9	0.83	7.5	0.75	1.3
3	-1	1	0.62	37.8	0.92	7.9	0.82	1.6
4	0	-1	0.56	39.0	0.60	8.1	0.90	0.65
5	0	0	0.74	41.1	0.64	8.1	0.84	1.3
6	0	1	0.88	38.8	0.65	8.3	1.1	1.9
7	1	-1	0.67	40.6	0.65	8.1	1.2	1.1
8	1	0	0.98	41.8	0.64	8.1	1.1	2.2
9	1	1	1.1	38.0	0.67	8.5	1.3	3.0

Table 6. Operational conditions considered in this study (expressed in terms of the coded independent variables dimensionless H_2SO_4 concentration x_1 and dimensionless time x_2 and experimental results achieved for the dependent variables y_1 (glucose concentration, g/L); y_2 (xylose concentration, g/L); y_3 (arabinose concentration, g/L); y_4 (acetic acid concentration, g/L); y_5 (formic acid concentration, g/L) and y_6 (furfural concentration, g/L). Results are the average of two independent experiments. Standard deviations were in the range of 1-3 % of the mean

Table 7 lists the regression coefficients and their statistical significance (based on a t-test) as well as the statistical parameters (R^2, R^2 adjusted, F and the significance level based on the F test) measuring the correlation and the statistical significance of the models, respectively. It can be noted that all the models showed good statistical parameters for correlation and significance allowing a close reproduction of experimental data. In relation with the influence of independent variables, H_2SO_4 concentration (x_1) caused the strongest effect on the variation of dependent variables considered, as it can be seen from the absolute value of the corresponding coefficients.

Fig. 1 shows a three dimensional representation of the response surface for the predicted dependence of the xylose concentration of samples (y_2) on the sulfuric acid concentration and reaction time as well as a two-dimensional contour plot generated by the model. The response surface shows a slight continuous increase in y_2 with catalyze and an optimum value at intermediate reaction times.

From the coefficients of Table 7, Eq. 4 was deduced for y_2 (xylose concentration):

$$y_2 = 41.06889 + 1.72 * x_1 - 1.21333 * (x_1)^2 - 2.16333 * (x_2)^2 - 1.4975 * x_1 * x_2 \quad (4)$$

Using the "solver" application of Microsoft Excell the maximum xylose concentration predicted for the model (y_1 = 41.8 g L^{-1}) was achieved when x_1 = 0.90 (2.9 %) and x_2 = -0.31 (25.3 min.). This value corresponds to 90.32 % of the theoretical xylose present in the raw material.

a) Regression coefficients and significance level for the dependent variables.

Coeficientes	y_1	y_2	y_3	y_4	y_5	y_6
b_0	0.7822*	41.06889*	0.6422*	8.1289*	0.86*	1.4*
b_1	0.17*	1.72*	-0.06833**	0.3533*	0.1883*	0.5317*
b_{11}	0.01667	-1.21333*	0.091667***	-0.2933***	0.045	0.285
b_2	0.1533*	0.08	0.061667**	0.2483**	0.045***	0.75*
b_{22}	-0.08333***	-2.16333*	-0.018333	0.08167	0.105**	-0.2
b_{12}	0.06***	-1.4975*	-0.07***	-0.0775	0.015	0.1425

b) Statistical parameters (R^2 and F) measuring the correlation and significance of the models.

Variable	R^2	corrected R^2	Fexp.	significance level (based on the F test)
y_1	0.98722	0.96592	540.715	96.69
y_2	0.99356	0.98283	1080.1478	97.66
y_3	0.93434	0.82491	99.6115	92.30
y_4	0.95926	0.89137	164.8324	94.01
y_5	0.97965	0.94574	337.0255	95.81
y_6	0.98035	0.94761	349.3056	95.88

* Significant coefficients at the 90 % confidence level.
** Significant coefficients at the 95 % confidence level.
*** Significant coefficients at the 99 % confidence level.

Table 7. Regression coefficients, significance level for dependent variables and statistical parameters measuring the correlation and significance of the models

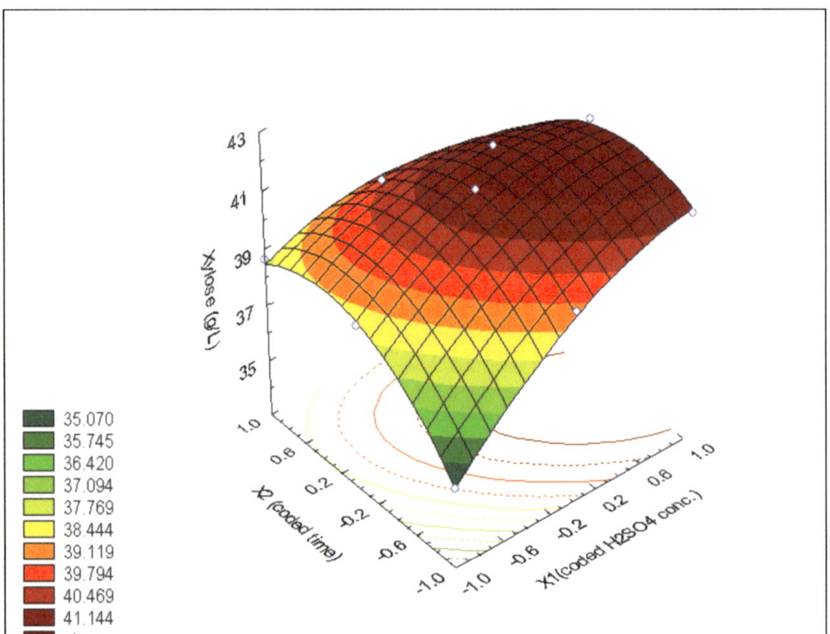

Fig. 1. Response surface and contour plot for the dependence of xylose concentration (y_2) on H_2SO_4 conc. (x_1) and reaction time (x_2)

3.4 Fermentation under the optimal hydrolysis conditions of pistachio shells

Diluted acids were used under the optimal conditions predicted by the solver application to obtain hydrolyzates with high content of xylose. Fig. 2 and Table 8 show the effect of nutrient supplementation on xylitol production by *D. hansenii*. Under all the conditions assayed, a 24h lag phase was observed. Xylose started to be consumed after glucose was firstly depleted within 24h. Conversely to other works with the same yeast (Carvalheiro et al., 2007), it is interesting to note the pronounced diauxic growth pattern obtained for the

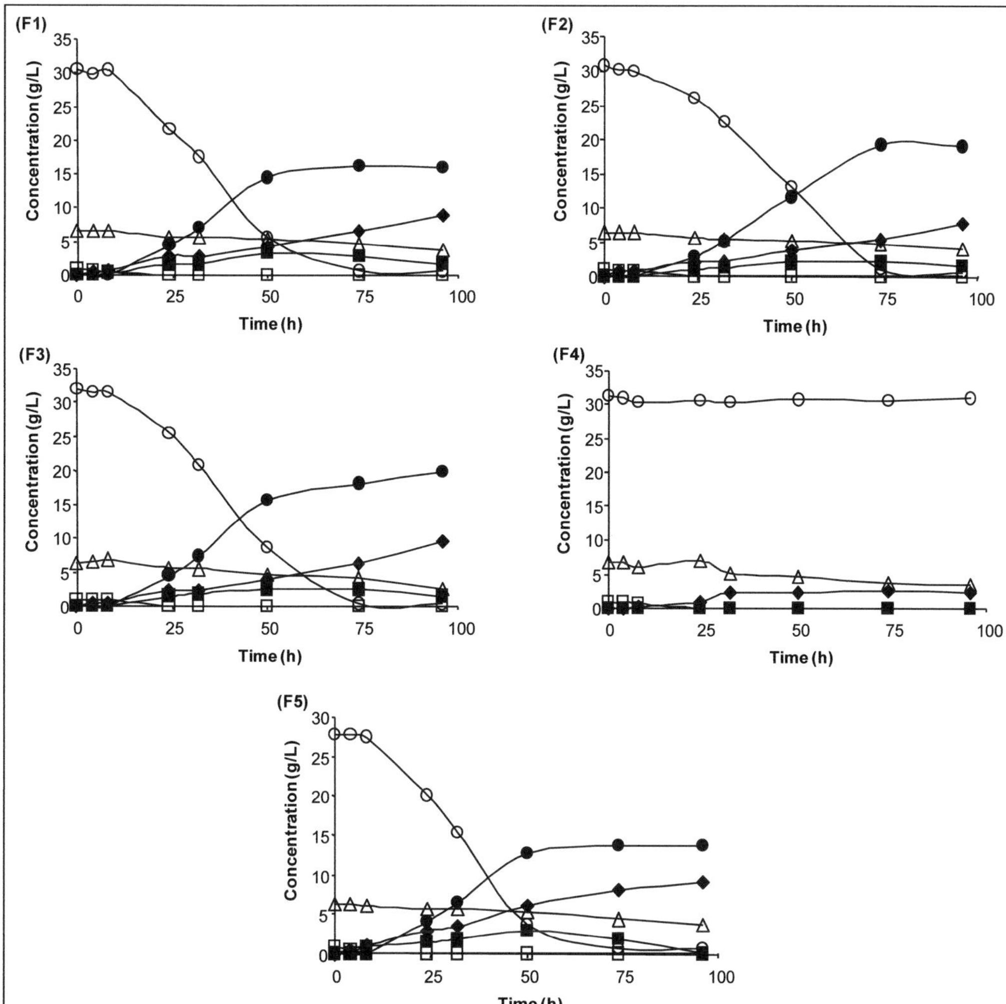

Fig. 2. Course with time for the xylose to xylitol bioconversion by *Debaryomyces hansenii* during fermentations carried out using the nutrients described in Table 3. (O) xylose, (□) glucose, (△) acetic acid, (●) xylitol, (■) ethanol, (◆) biomass. (F1) raw hydrolyzate, liquid vinasses and stream B (conc.); (F2) raw hydrolyzate, corn steep liquor (30 g/L); (F3) raw hydrolyzate, yeast extract (3 g/L), malt extract (3 g/L), peptone (5 g/L); (F4) raw hydrolyzate, without nutrients; (F5) detoxified hydrolyzate, yeast extract (3 g/L), malt extract (3 g/L), peptone (5 g/L)

	Glucose (g/L)		Xylose (g/L)		Arabinose (g/L)		SC (g/L)	Q_S (g/L·h)	Acetic acid (g/L)		Xylitol (g/L)		$Q_{Xylitol}$ (g/L·h)	$Y_{Xylitol/SC}$ (g/g)	Glycerol (g/L)		Ethanol (g/L)		Biomass (g/L)		Q_X (g/L·h)	$Y_{X/SC}$ (g/g)
	T_0	T_f	T_0	T_f	T_0	T_f			T_0	T_f	T_0	T_f			T_0	T_f	T_0	T_f	T_0	T_f		
F1	0.89	0	30.6	0.51	0.57	0.50	31.1	0.420	6.5	4.8	0	16.2	0.219	0.52	0	0.36	0	2.8	0.04	6.7	0.089	0.21
F2	0.93	0	30.8	1.2	1.0	0.93	30.6	0.413	6.5	4.7	0	19.3	0.260	0.63	3.8	3.8	0	2.0	0.04	5.4	0.073	0.18
F3	0.91	0	32.0	0.45	0.64	0.61	32.5	0.439	6.3	4.2	0	18.0	0.244	0.56	2.8	3.1	0	2.5	0.04	6.2	0.084	0.19
F4	0.78	0	31.3	30.4	0.57	0.52	1.7	0.023	6.6	1.9	0	0	0	0	0	0	0	0	0.04	1.5	0.020	0.84
F5	0.76	0	27.8	0.58	0.52	0.41	28.1	0.379	6.4	4.1	0	13.7	0.185	0.49	0	1.5	0	2.0	0.04	8.2	0.111	0.29

SC = total sugars consumed; Q_S = volumetric rate of substrates (xylose, glucose and arabinose) consumption; $Q_{Xylitol}$ = global volumetric productivity of xylitol; $Y_{Xylitol/SC}$ = xylitol yield (xylitol produced/total sugars consumed); Q_X = biomass volumetric productivity; $Y_{X/SC}$ = biomass yield

Table 8. Stoichiometric parameters, productivities and yields after 74h for hemicellulosic pistachio shells hydrolyzates bioconversion in assays carried out by *D. hansenii*. Results are the average of two independent experiments. Standard deviations were in the range of 2-4 % of the mean

mixture of glucose and xylose. A similar pattern was observed for *Candida guilliermondii* (Canilha et al., 2005), and *Candida tropicalis* (Walther et al., 2001). Arabinose and acetic acid were scarcely consumed (see Table 8). Thus, meanwhile no arabinose was consumed in fermentation 4 (raw hydrolyzates without supplementation), the consumption in fermentation 5 (detoxified hydrolyzates supplemented with commercial nutrients) only reached 37 %; whereas acetic acid consumption ranged 39.1-58.8 % regardless of supplementation or detoxification treatment. Supplementation with economic nutrients (vinasses or corn steep liquor) led to an increase in the overall sugars consumption rate in comparison with the use of commercial nutrients.

Xylitol was the main product for all media except for fermentation 4 where no production was observed, meaning that these hydrolyzates require to be supplemented with additional nutrients to stimulate the xylitol production. The necessity to supplement hydrolyzates with nutrients seems to depend on the raw material (Carvalheiro et al., 2007), and while the supplementation was shown to be beneficial for *Candida guilliermondii* during the xylose-to-xylitol bioconversion in *Eucalyptus* hydrolyzates (Canettieri et al., 2001), no addition of nutrients was necessary in rice straw hydrolyzates for the same yeast (Silva & Roberto, 1999). The highest concentration, 19.3 g L^{-1} after 74 hours, was obtained in hydrolyzates supplemented with economic wastes (vinasses and stream B), corresponding to fermentation 2 (Q_P = 0.260 g L^{-1} h^{-1}; $Y_{P/S}$ = 0.63 g g^{-1}), followed by the use of another economic nutrient in fermentation 3, corn steep liquor, where xylitol reached up to 18.0 g L^{-1} (Q_P = 0.244 g L^{-1} h^{-1}; $Y_{P/S}$ = 0.56 g g^{-1}). In fermentations supplemented with commercial nutrients (fermentations 1 and 5), conversely to the expected, the higher xylitol value, 16.2 g L^{-1}, was observed in fermentation without detoxification. Ethanol and glycerol were also produced although in small amounts.

Finally, it is important to mention the amount of biomass generated in all fermentations with the exception of fermentation 4 where only 2.3 g L^{-1} was achieved. Usually, the growth begins with a lag phase in which the microorganism adapts to the enzymatic systems in order to metabolize the new substrate, but without growing (Sánchez et al., 2008), however, the previous adaptation of *D. hansenii* to hydrolyzates minimize this phase, showing a linear growth during the total period of fermentation, producing up to 9.5 g L^{-1} after 96 hours in fermentation 3 (Q_X = 0.099 g L^{-1} h^{-1}; $Y_{X/S}$ = 0.29 g g^{-1}).

4. Conclusions

In this study, the following conclusions can be drawn related to agro industrial wastes evaluation:

1. Based on the characterization and further diluted acid hydrolysis carried out to the agro industrial wastes evaluated, it can be concluded their great potential as feedstocks for the production of industrially relevant food additives.
2. In particular, pistachio shells show the higher xylan content. Optimizing the conditions of hydrolysis by an experimental design reaches 90% of the theorical xylose present in raw material.
3. Hydrolyzate medium obtained with nutrients supplementation can be effectively employed for the xylitol production by *Debaryomyces hansenii*.

5. Acknowledgements

The authors gratefully thank the financial support from the XUNTA DE GALICIA (project 09TAL013383PR).

6. References

Bustos, G., Moldes, A. B., Cruz, J. M., and Domínguez, J. M. (2004). Production of fermentable media from vine-trimming wastes and bioconversion into lactic acid by *Lactobacillus pentosus*. *J. Sci. Food Agr.* 84, 2105–2112.

Bustos, G., Moldes, A. B., Cruz, J. M., and Dominguez, J. M. (2005). Production of lactic acid from vine-trimming wastes and viticulture lees using a simultaneous saccharification fermentation method. *J. Sci Food Agr.* 85, 466–472.

Canettieri, E. V., Almeida e Silva, J. B., and Felipe, M. G. A. (2001). Application of factorial design to the study of xylitol production from eucalyptus hemicellulosic hydrolysate. *Appl. Biochem. Biotechnol.* 24, 159–168.

Canilha, L., Carvalho, W., and Almeida e Silva, J. B. (2005). Influence of medium composition on xylitol bioproduction from wheat straw hemicellulosic hydrolysate. *World J. Microbiol. Biotechnol.* 21, 1087–1093.

Carvalheiro, F., Duarte, L. C., Medeiros, R., and Gírio, F. M. (2007). Xylitol production by *Debaryomyces hansenii* in brewery spent grain dilute-acid hydrolysate: effect of supplementation. *Biotechnol. Lett.* 29, 1887–1891.

Claye, S. S., Idouraine, A., and Weber, C. W. (1996). Extraction and fractionation of insoluble fiber from five fiber sources. *Food Chem.* 57, 305-310.

Cruz, J. M., Domínguez, J. M., Domínguez, H. and Parajó, J. C. (2000). Preparation of fermentation media from agricultural wastes and their bioconversion into xylitol. *Food Biotechnol.* 14, 79-97.

Demirbas, A. (2003). Relationships between lignin contents and fixed carbon contents of biomass samples. *Energy Convers. Manage.* 44, 1481-1486.

Domínguez, J. M., Cao, N., Gong, C. S., and Tsao, G. T. (1997). Dilute acid hemicellulose hydrolyzates from corn cobs for xylitol production by yeast. *Bioresource Technol.* 61, 85-90.

Fan, L. T., Gharpuray, M. M., and Lee, Y. H. (1987). Cellulose Hydrolysis. Springer-Verlag, New York, USA

Hamelinck, C. N., van Hooijdonk, G. and Faaij, A. P. C. (2005). Ethanol from lignocellulosic biomass: techno-economic performance in short-, middle- and long-term. *Biomass Bioenerg.* 28, 384–410.

Jiménez, L., and López, F. (1993). Characterization of paper sheets from agricultural residues. *Wood Sci. Technol.* 27, 468-474.

Jørgensen, H., Kristensen, J. B., and Felby, C. (2007). Enzymatic conversion of lignocellulose into fermentable sugars: challenges and opportunities. *Biofuel, Bioprod. Biorefin.* 1, 119-134.

Kim, S. B., and Moon, N. K. (2003). Enzymatic digestibility of used newspaper treated with aqueous ammonia-hydrogen peroxide solution. *Appl. Biochem. Biotechnol.* 105-108, 365-373.

Kumar, P., Barrett, D. M., Delwiche, M. J., and Stroeve, P. (2009). Methods for pretreament of lignocellulosic biomass for efficient hydrolysis and biofuel production. *Ind. Eng. Chem.* 48, 3713-3729.

Moldes, A. B., Bustos, G., Torrado, A., and Dominguez, J. M. (2007). Comparison between different hydrolysis processes of vine-trimming waste to obtain hemicellulosic sugars for further lactic acid conversion. *Appl. Biochem. Biotechno.* 143, 244–256.

Mosier, N., Wyman, C., Dale, B., Elander, R., Lee, Y. Y., Holtzapple, M., and Ladisch, M. (2005). Features of promising technologies for pre-treatment of lignocellulosic biomass. *Bioresource Technol.* 96, 673–686.

Nabarlatz, D., Farriol, X., and Montane, D. (2005). Autohydrolysis of almond shells for the production of silo-oligosaccharides: product characteristics and reaction kinetics. *Ind. Eng. Chem.* 44, 7746-7755.

Pérez, J., Muñoz-Dorado de la Rubia T., and Martínez, J. (2002). Biodegradation and biological treatments of cellulose, hemicellulose and lignin: an overview. *Int. Microbiol.* 5, 53-63.

Portilla-Rivera, O. M., Moldes, A. B., Torrado, A. M., and Dominguez, J. M. (2007). Lactic acid and biosurfactants production from hydrolyzed distilled grape marc. *Process Biochem.* 42, 1010-1020.

Portilla-Rivera, O., Torrado, A., Domínguez, J. M., and Moldes, A. B. (2008). Stability and emulsifying capacity of biosurfactants obtained from lignocellulosic sources using *Lactobacillus pentosus*. *J. Agr. Food Chem.* 56, 8074–8080.

Prakasham, R. S., Rao, R. S., Hobbs, P. J. (2009). Current trends in biotechnological producion of xylitol and future prospects. Curr. Trends Biotechnol. Pharm. 3 (1), 8-36.

Rivas, B., Dominguez, J. M., Dominguez, H., and Parajó, J. C. (2002). Bioconversion of posthydrolysed autohydrolysis liquors: an alternative for xylitol production from corn cobs. *Enzyme Microb. Technol.* 31, 431–438.

Rodríguez-Couto, S. (2008). Exploitation of biological wastes for the production of value-added products under solid-state fermentation conditions. *Biotechnol. J.* 3, 859–870.

Saha, B. C., Iten, L. B., Cotta, M. A., and Wu, Y. V. (2005). Dilute acid pretreatment, enzymatic saccharification, and fermentation of rice hulls to etanol. *Biotechnol. Progr.* 21, 816-822.

Salgado, J. M., Martínez-Carballo, E., Max, B., and Domínguez, J. M. (2010). Characterization of vinasses from five certified brands of origin (CBO) and use as economic nutrient for the xylitol production by *Debaryomyces hansenii*. *Bioresource Technol.* 101, 2379–2388.

Salgado, J. M., Rodríguez, N., Cortés, S., and Domínguez, J. M. (2010). Improving downstream processes to recover tartaric acid, tartrate and nutrients from vinasses and formulation of inexpensive fermentative broths for xylitol production. *J. Sci. Food Agr.* 90, 2168-2177.

Sánchez, C. (2009). Lignocellulosic residues: Biodegradation and bioconversion by fungi. *Biotecnol. Adv.*, 27, 185-194.

Sánchez, S., Bravo, V., García, J. F., Cruz, N., and Cuevas, M. (2008). Fermentation of D-glucose and D-xylose mixtures by *Candida tropicalis* NBRC 0618 for xylitol production. *World J. Microbiol. Biotechnol.* 24, 709–716.

Sasaki, M. T., Adschiri, T., and Arai, K. (2003). Fractionation of sugarcane bagasse by hydrothermal treatment. *Bioresource Technol.* 86, 301-304.

Silva, C. J. S. M., and Roberto, I. C. (1999). Statistical screening method for selection of important variables on xylitol biosynthesis from rice straw hydrolysate by *Candida guilliermondii* FTI 20037. *Biotechnol. Tech.* 13, 743–747.

Staniskis J. K., and Stasiskiene, Z. (2005). Industrial waste minimization-experience from Lithuania. *Waste Manage. Res.* 23, 282-290.

Taherzadeh M. J., and Karimi K. (2007). Acid-based hydrolysis processes for ethanol from lignocellulosic materials: a review. *Bioresources* 2(3), 472-499.

Walther, T., Hensirisak, P., and Agblevor, F. A. (2001). The influence of aeration and hemicellulosic sugars on xylitol production by *Candida tropicalis*. *Bioresource Technol.* 76, 213-220.

Ward, O. P., and Moo-Young, M. (1989). Enzymatic degradation of cell wall and related plant polysaccharides," *Crit. Rev. Biotechnol.* 8, 237–274.

9

Rosemary Compounds as Nutraceutical Health Products

Silvia Moreno[1], Adriana María Ojeda Sana[1], Mauro Gaya[1], María Verónica Barni[1], Olga A. Castro[1] and Catalina van Baren[2]
[1]Fundación Instituto Leloir, Instituto de Investigaciones Bioquímicas Buenos Aires - CONICET, Patricias Argentinas 435 (1405) CABA
[2]Cátedra de Farmacognosia, (IQUIMEFA - CONICET), FFyB-UBA, Junín 956 2° piso (1113) CABA
Argentina

1. Introduction

The oxidation of lipids in food, cosmetic and pharmaceutical products, together with the growth of undesirable microorganisms results in the development of spoilage, off-flavor, rancidity and deterioration of such products turning them unacceptable for human consumption. In consequence, the addition of exogenous antioxidants such as butylated hydroxytoluene and butylated hydroxyanisole is frequently required to improve the stability of these products. As synthetic butylated derivatives may have toxic effects there is an increasing interest in the use of natural antioxidants, such as phenols isolated from plants to avoid undesired food borne diseases (Shylaja & Peter, 2004).

In recent years, a greater emphasis has been placed on the link between the prevention of chronic diseases and the human diet. The present popularity of natural antioxidants, as the dietary polyphenols, on human health has promoted the surge of *in vitro* studies examining the effects of these physiological active components (Williams et al., 2004).

During the past decade, consumers began to view food in a new way. The antioxidant properties of polyphenols have been widely studied, but it has become clear that the mechanisms of action of these compounds go beyond the modulation of oxidative stress (Scalbert et al., 2005). These compounds have great potential in the emerging nutritional industry, because they are often considered as food and medicines as well, therefore they may be used in the prevention and curative treatments (Sies, 2010). This issue has been of interest from ancient times, Hippocrates, 400 B.C. said, "Let food be your medicine and medicine your food".

Different foods have been identified as containing health-promoting properties beyond their basic nutritional value, and stimulating innovation in the field of nutrition and health is the search for nutraceuticals. A nutraceutical is defined as a food or part of a food that provides medicinal benefits or health, including prevention and/or treatment of diseases.

In order to use nutraceuticals in the prevention and treatment of human pathologies, many questions still remain unanswered:

Which natural source is to be used? Which is a good candidate to be used as nutraceutical?.

Rosemary (*Rosmarinus officinalis* L., Lamiaceae) is considered one of the most important sources for the extraction of phenolic compounds with strong antioxidant activity. This specie grows worldwide and has been cultivated since long ago, in ancient Egypt, Mesopotamia, China and India (Bradley, 2006). Rosemary extracts, enriched in phenolic compounds are effective antioxidants due to their phenolic hydroxyl groups but they also possess plenty of other beneficial effects like antimicrobial, antiviral, anti-inflammatory, anticarcinogenic activities and is also known to be an effective chemopreventive agent (al-Sereiti et al., 1999; Aherne et al., 2007).

Therefore, this specie contains bioactive ingredients which provide a complementary value other than the nutritional one, to be applied in the food industry. However, particular bioactives of rosemary responsible for some biological activities, as antimicrobial, have not been deeply characterized. A lesser amount of information exists about their mechanism of action. The present chapter focuses on the most significant rosemary biological properties, reviewing the free radical scavenging and antibacterial actions of non-volatile constituents and essential oils, the antibacterial activity of main rosemary bioactives in combination with antibiotics, as well as possible antibacterial mechanism of action is proposed, among other topics. In addition, a toxicity assay using the nematode *Caenorhabditis elegans* is covered.

2. Rosemary the best natural antioxidant

2.1 Antioxidant action of different fraction of rosemary

Rosemary plants have many phytochemicals which constitute potential sources of natural compounds as phenolic diterpenes, flavonoids phenolic acids and essential oils. About 90% of the antioxidant activity is attributed mainly to a high content of non-volatile components as carnosic acid and carnosol (phenolic diterpenes) and rosmarinic acid (Bradley, 2006). It is clear that *R. officinalis* constituents have antioxidant activity according to traditional use and scientific evidence, although little information is available on the relationship between chemical composition and antioxidant activity of the essential oils and non-volatile extracts.

We investigated the antioxidant activity of volatile and non-volatile fractions isolated from two leaf phenotypes of rosemary plants growing in the same farm in Argentina. Plants showing a wide (W) and a narrow (N) leaves phenotypes were collected from Jardín Botánico Arturo E. Ragonese from National Institute of Agricultural Technology-INTA Castelar, Argentina, January 2008. The essential oils were obtained by hydrodistillation of dried leaves using a Clevenger-type apparatus and samples were analyzed by high resolution gas chromatography coupled with mass spectrometry. Ethanol extracts were prepared according to the method previously reported (Moreno et al., 2006) and stored at -20 °C. To determine the dry weight of each extract, 1 ml of the sample was dried in an oven to constant weight. The extracts were centrifuged using a 5804 Eppendorf centrifuge at 5000 rpm for 15 min at room temperature before HPLC analysis. Quantification of phenolic compounds and identification was performed on an HPLC (LKB Bromma) equipped with a diode array detector, using a 250 mm × 4 mm C18 Luna analytical column (Phenomenex, USA), as previously described (Moreno et al., 2006).

The antioxidant activity was tested using a stable radical 2,2-diphenyl-1-picrylhydrazyl (DPPH) as described by Brand-Williams et al., 1995. The percentage of DPPH radical was calculated measuring the change in absorbance at 517 nm and EC_{50} (extract concentration necessary to decolorate DPPH radical in a 50%) was determined for each fraction.

Results showed that different amounts of key bioactive compounds were present in the essential oils of both phenotypes. The main constituents of W phenotype essential oil were α-pinene and 1,8-cineole, while the N phenotype contained a comparable amount of 1,8 cineole and high contents of myrcene (Table 1). Therefore, the first is referred as an α-pinene chemotype and the other one as a myrcene chemotype. The essential oil of the myrcene chemotype exhibited approximately a double fold antioxidant activity than the α-pinene phenotype. Recently, other authors reported that pure myrcene showed antioxidant activity and eliminated oxidative stress in rats in a time-dependent manner (Ciftci et al., 2011).

Essential oil	EC_{50} (μl/ml)	Main constituents	(%)
W Phenotype	25.0 ± 1.0	α-Pinene	31.2 ± 2.5
		1,8 -Cineole	21.6 ± 1.6
N Phenotype	10.0 ± 0.4	Myrcene	31.1 ± 2.3
		1,8-Cineole	18.7 ± 1.4

Table 1. Antioxidant activity and main constituents of essential oils isolated from two phenotypes of rosemary plants. Values ± SD

Chemical analysis of ethanol rosemary extracts indicated that W phenotype contained about 36.8% of diterpenes (carnosic acid plus carnosol) and 8.4% of rosmarinic acid, while the other phenotype contained a higher amount of diterpenes and lesser amounts of rosmarinic acid (Table 2). The ethanol extract isolated from the N phenotype, containing a high amount of diterpenes, exhibited approximately double fold antioxidant activity than plants of W phenotype.

Our results showed that the volatile and non-volatile fractions isolated from the same phenotype had similar antioxidant activity.

Ethanol extract	EC_{50} (μg/ml)	Main constituents	(%)
W Phenotype	20.0 ± 0.8	Carnosic acid + carnosol	36.8 ± 2.7
		Rosmarinic acid	8.4 ± 0.5
N Phenotype	10.8 ± 0.4	Carnosic acid + carnosol	50.0 ± 4.5
		Rosmarinic acid	2.9 ± 2.5

Table 2. Antioxidant activity and main constituents of ethanol extracts isolated from two phenotypes of rosemary plants. Values ± SD

2.2 Rosemary as protective agent against oxidative protein damage

A large number of reports have shown rosemary constituents to be an efficient antioxidant against lipid peroxidation and DNA damage induced by radical oxygen species in rat liver mitochondria and microsomes at concentrations of 3 - 30 μM, demonstrating their ability to protect tissues and cells against oxidative stresses (Bradley, 2006). On the other hand, it is well-known that several antioxidants exhibited pro-oxidant effect producing protein damage under certain conditions as in the presence of transition metals such as Fe and Cu. In order to study the protection of rosemary compounds against protein damage in comparison with ascorbate and 6-Hydroxy-2,5,7,8-tetramethylchroman-2-carboxylic acid (Trolox), hydroxyl radical-mediated oxidation experiments, were carried out using a metal-catalyzed reaction. Bovine serum albumin (4 μg) was incubated with or without Cu^{+} (100 μM) and H_2O_2 (1 mM) in the absence or presence of ascorbate, Trolox or the methanol rosemary extract (obtained as described in Moreno et al., 2006). Reactions were performed in opened tubes at 37°C, mixed with loading buffer and loaded in 12.5% dodecyl sulfate-polyacrylamide gel electrophoresis as reported by Mayo et al., 2003. Figure 1 shows that 20 μg of the plant extract used containing a concentration of 18 μM diterpenes, reduced significantly protein damage compared with 20 μM of ascorbate (compare the intensity of the protein's monomer in lane 8 vs. lane 6). Figure 1 also shows that Trolox and ascorbate only protected protein modifications when they were present at lower concentrations, and no protection of the protein was observed at higher concentrations. Ascorbate reveals a higher pro-oxidant action than Trolox.

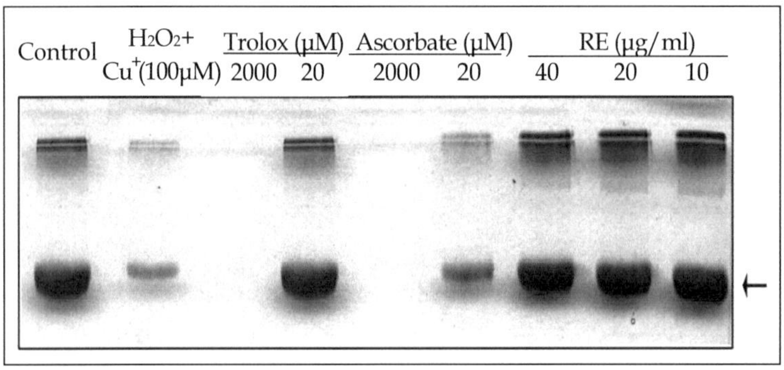

Fig. 1. Gel electrophoresis of bovine serum albumin protein after treatment with H_2O_2 + $CuSO_4$ in the presence of Trolox, ascorbate and methanol rosemary extract (RE) at the concentrations indicated. Control was performed with the addition of ethanol 0.2%. Arrow indicates the bovine serum albumin monomer

3. Antibiotics and antioxidants of rosemary working together

Due to the recent trend in green consumers, there is an increasing interest in the antimicrobial properties of rosemary compounds. Essential oils and organic or aqueous extracts of isolated from this specie not only have antioxidant activity but also present antibiotic effects, therefore they have gained acceptance in industry to replace existing synthetic preservatives in foods (Davidson, Sofos, & Branen, 2005).

In the area of health, the near-term interest of plant products as antimicrobial agents is related to the development of new strategies/therapies for infections caused by bacterial species (Cowan, 1999; Lewis and Ausubel, 2006). Recently, it was reported that natural plant products can potentiate the activity of antibiotics in combination (Coutinho et al., 2009). Moreover, the use of bacterial resistance modifiers derived from natural sources, mainly from plants, such as efflux pump inhibitors, was suggested to be useful to suppress the emergence of multidrug resistant strains (Stavri et al., 2007).

To determine the validity of rosemary compounds as nutraceuticals, a rigorous analysis of the biological activities of their bioactives is required as well as the further study of their antibacterial mechanism of action.

3.1 Antibiotic action of different fractions of rosemary

We previously reported the effective antimicrobial action of non-volatile rosemary extracts containing 33 - 46% of diterpenes (carnosic acid plus carnosol) against common food pathogenic Gram positive bacteria as *Staphylococcus aureus* and *Enterococcus faecalis* as well as the Gram negative bacteria *Escherichia coli* (Moreno et al., 2006). These microorganisms cause severe problems in human health (NNIS, 2004). According to a 2011 research study, *S. aureus* was found in ~50% of beef, pork, and poultry products throughout the United States, 96% of these isolates were resistant to at least 1 antibacterial agent relevant in human medicine and 52% were resistant to three or more types (Waters et al., 2011).

Here, we show the antibacterial performance of a methanol rosemary extract (obtained as described in Moreno et al., 2006) in comparison with common food preservatives. The results are expressed in percent of inhibition of bacterial growth (see Equation 1)

$$\% \text{ Inhibition of bacterial growth} = \frac{(A_{595}\text{Control} - A_{595}\text{Sample})}{A_{595}\text{Sample}} \times 100 \quad (1)$$

Where, A_{595} Control is the absorbance of the bacterial culture without compounds.

Rosemary extracts presented a higher antimicrobial efficacy than benzoic acid, butylated hydroxytoluene (BHT) and butylated hydroxyanisole (BHA) to inhibit *S. aureus* growth (Fig. 2A), while a similar action than BHT and benzoic acid and a minor antimicrobial activity in relation to BHA was seen against *E. coli* (Fig. 2B).

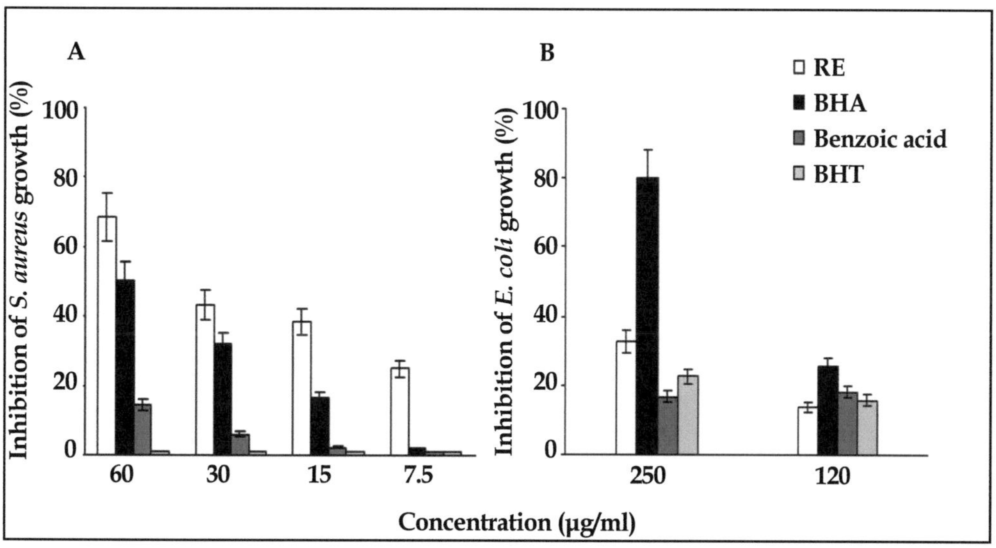

Fig. 2. Effect of the methanol rosemary extract (RE) on the *S. aureus* growth (A) or on the growth of *E. coli* (B), in comparison with BHA, BHT and benzoic acid. Values are shown as the mean of three independent experiments ± SD

The performance of rosemary oils and ethanol extracts isolated from the W phenotype and N phenotype plants against the human pathogen *E. faecalis,* was also compared (Table 3). Results showed that the essential oil from the α-pinene chemotype (W phenotype) exhibited a higher antibacterial activity than the essential oil isolated from N phenotype. By contrast, the ethanol extract isolated from the N phenotype containing higher amounts of carnosic acid and carnosol exhibited the highest antimicrobial action (Table 4).

Phenotype	Essential oil	Inhibition of *E. faecalis* growth (%)
W	26 μl/ml	100
N	26 μl/ml	45.5 ± 3

Table 3. Antibiotic activity of essentials oils isolated from two phenotypes of rosemary plants. Values are shown as the mean of three independent experiments ± SD

Phenotype	Ethanol extract	Inhibition of *E. faecalis* growth (%)
W	0.25 mg/ml	50.0 ± 6
N	0.25 mg/ml	100

Table 4. Antibiotic activity of ethanol extracts isolated from two phenotypes of rosemary plants. Values are shown as the mean of three independent experiments ± SD

All findings suggested that carnosic acid and α-pinene are key ingredients responsible that confer the antibacterial properties in the non-volatile and volatile fractions isolated from rosemary plants, respectively.

The essential oils isolated from N phenotype exhibited minor antibacterial activity than the other one, although it presented a higher performance as antioxidant.

We are investigating the antibacterial action against several Gram positive and negative microorganisms. Others authors demonstrated antibacterial activity of volatile compounds against *Listeria monocytogenes, Salmonella typhimurium, Escherichia coli, Shigella dysenteria, Bacillus cereus* (For review see Burt, 2004).

3.2 Antibiotic potentiation by rosemary bioactives

The use of combined antioxidants has gained acceptance in industry and has been applied to different aspects of food preservation (Davidson, Sofos & Branen, 2005). Although, up to date, a rational basis for the use of phytochemicals enhancing and/or broadening the biological antioxidant and antimicrobial activities against food-borne pathogens, is still poorly explored (Wei & Shibamoto, 2007). In the health's area, natural compounds are usefull strategies for the development of therapies against infections caused by bacterial species, and they are used in combination with common antibiotics potentiating their activity (Coutinho et al., 2009).

We previously reported a synergistic antioxidant effect between the methanol rosemary extract and BHT and a synergistic interaction with BHA to inhibit *E. coli* and *S. aureus* growth (Romano et al., 2009). Here, we reported the *in vitro* antibiotic interaction of rosemary extracts with common antibiotics using the broth microdilution method against *S. aureus* and *E. faecalis.*

Figure 3A shows the dose-response curve, in which the addition of 6.25 μg/ml of the plant extract clearly displaced the curve to the left, meaning that an increment in the antimicrobial action against *S. aureus* took place in the binary mixture at all gentamicin concentrations tested. It can be extrapolated from the curve that 0.035 μg/ml of pure gentamicin is needed to achieve 50% of inhibition, while the same effect can be achieved with 0.015 μg/ml of the aminoglycoside in the presence of rosemary extract.

Then, to study the type of interactions binary mixtures with different concentrations of rosemary extract and pure carnosic acid with gentamicin were analyzed by isobolograms (Fig. 3B).

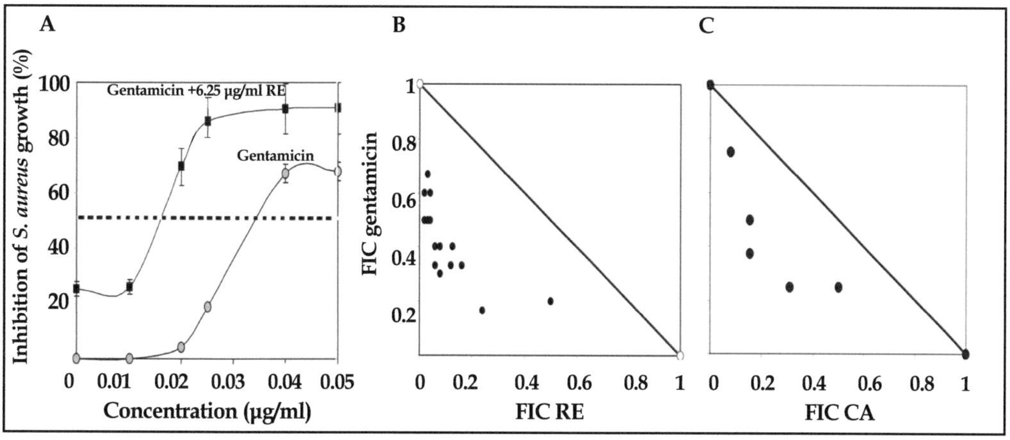

Fig. 3. Rosemary extract (RE) and carnosic acid (CA) acts synergistically with gentamicin against *S. aureus*. Dose-response curve (A), isobologram of gentamicin in combination with RE (B) or pure carnosic acid (C). Inhibition of *S. aureus* 25923 growth was determined after 24 h of incubation. FIC, fractional inhibitory concentration, normalized to the unit of RE, gentamicin or pure carnosic acid, corresponds to the MIC_{50} = 12 μg/ml, 0.035 μg/ml and 8 μg/ml, respectively. Values are shown as the mean of three independent experiments ± SD

Compounds concentrations lower than their MIC_{50} values were prepared for every combination tested as described previously (Romano et al., 2009) and isobolograms were performed. When compounds in combination are more effective than what might be expected from their dose-response curves (synergy), smaller amounts will be needed to produce the effect under consideration, and a concave-up isobole results (Tallarida, 2001).

The Fractional inhibitory concentration (FIC index) was determined (see Equation 2).

$$\text{FIC index} = FIC_A + FIC_B = [A]/MIC_A + [B]/MIC_B \quad (2)$$

FIC_A, FIC_B: Fractional inhibitory concentration of drug A & B respectively. MIC_A, MIC_B: Minimum inhibitory concentration of drug A & B respectively. [A], [B]: Concentration of drug A & B respectively.

FIC index obtained by checkerboard method is interpreted as follows: ≤ 0.5- synergy; > 0.5 and ≤ 4- additivity; and > 4- antagonism.

Figure 3B shows that different combinations of rosemary extracts corresponding to FIC 0.01 to 0.25 with gentamicin (FIC of 0.2 to 0.4) resulted in fixed-ratio points below the additive line and exhibited values of FIC index ≤ 0.5, verifying synergistic antimicrobial effects of the plant extract with gentamicin.

As carnosic acid was shown to be the main antimicrobial compound of non-volatile rosemary extracts against *S. aureus* (Moreno et al., 2006), this compound was assayed in combination with gentamicin at concentrations equivalent to those found in the extract (1 to 10 µg/ml). The minimal inhibition concentration (MIC) of carnosic acid against the bacteria was 16 µg/ml and was referred as a FIC of 1. Values of FIC index ≤ 0.5 (FIC 0.3 of carnosic acid in combination with FIC 0.2 to 0.3 of gentamicin) were observed (Fig. 3C).

On the other hand, rosmarinic acid exhibited a minor antibacterial effectiveness than carnosic acid and exhibited additive antibacterial actions with gentamicin (Data not shown). Therefore, data showed that the addition of both the extract and carnosic acid allows a reduction of the gentamicin amount in approximately 3 - 4 folds to obtain the same antibiotic action. The bactericidal action of rosemary extracts and pure carnosic acid were confirmed by time-kill curves.

Table 5 shows that carnosic acid plus tetracycline, tobramycin, kanamycin, ciprofloxacin and gentamicin exhibits significant synergistic antibiotic activity against *S. aureus.*

Antibiotics	MIC values of antibiotic in combination with MIC of carnosic acid[a] (µg/ml)				FIC Index
	0	1/8	1/4	1/2	
Tetracycline	0.50	<u>0.06</u>	0.06	0.06	0.25
Tobramycin	0.50	<u>0.12</u>	0.12	0.06	0.37
Kanamycin	2.00	1.00	<u>0.50</u>	0.25	0.50
Vancomycin	2.00	<u>1.00</u>	1.00	0.50	0.62
Ciprofloxacin	0.50	<u>0.06</u>	0.12	0.12	0.25
Penicillin	0.06	-	<u>0.03</u>	0.03	0.75
Gentamicin	0.50	<u>0.06</u>	0.06	0.06	0.25

Table 5. Antibiotic activity of carnosic acid in combination with several antibiotics against *S. aureus* ATCC 25923. [a]Underlined values: combinations of minimum FIC used to calculate the FIC index. Values are the mean of three independent experiments

In order to confirm the role of carnosic acid as the main bioactive compound of rosemary extracts involved in the synergistic effect with gentamicin, additional experiments were carried out, examining carnosic acid in combination with gentamicin against *E. faecalis* (Fig. 4). Results showed that 30 µg/ml of the diterpene in combination with 0.25 µg/ml of gentamicin resulted in an strong antibiotic synergistic effect as the expected value for the sum corresponding to individual effects was 42% of inhibition, whereas experimental data showed a100 % of inhibition on bacteria growth for the combination mixture.

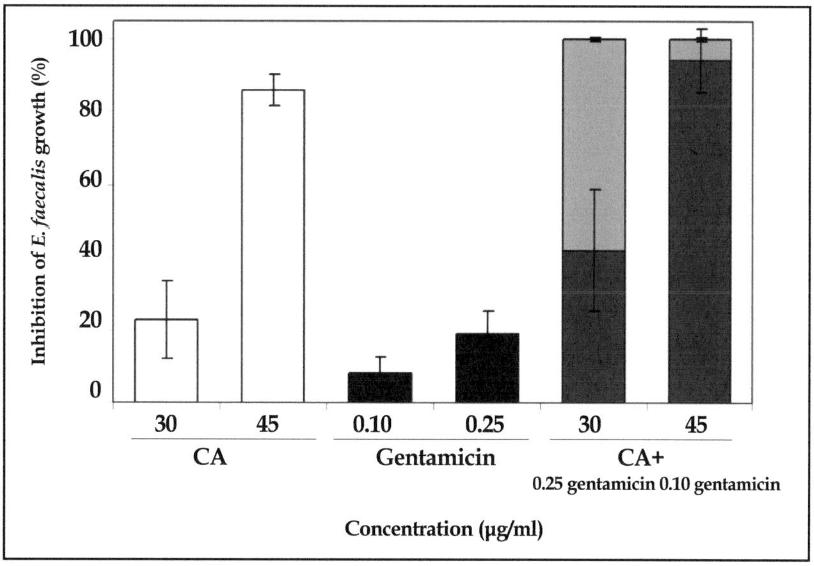

Fig. 4. Antibiotic activity of carnosic acid in combination with gentamicin against *E. faecalis* ATCC 29212. Dark grey bars represent the expected results if an additive interaction took place and light grey bars show the experimental data

3.3 Mechanism of antibacterial action of rosemary bioactives

R. officinalis plants exhibited antibiotic activity although less is known regarding the mechanisms of action of the particular bioactives. Polyphenols still constitute a promising source of new drugs and there is a high interest in understanding their mechanisms of action (Scalbert, 2005). We carried out studies to determine the antibacterial effect of the carnosic acid. This liphophilic diterpene has been found associated with chloroplasts membranes (Perez-Fons et al., 2006). Herein the cell membrane permeabilization effect by carnosic acid using the membrane potential sensitive cyanine dye DiS-C3-(5). This fluorescent probe is a caged cation, which distributes between cells and medium depending on the cytoplasmic membrane potential. Once it is inside the cells, it becomes concentrated and self-quenches its own fluorescence. The fluorescence monitored with an excitation wavelength of 610 nm and an emission wavelength of 670 nm. A blank with only cells and the dye was used to subtract the background and compounds as the proton motive force inhibitor carbonyl cyanide-m-chlorophenylhydrazone (CCCP) and Polimycyn B were used as controls for their ability to decreased or increase the membrane potential, respectively. *S aureus* suspension was incubated with 1.6 µM DiS-C3-(5) until a stable reduction in fluorescence and 100 mM KCl was added to equilibrate the cytoplasmic and external K^+ concentration. Figure 5 shows time decreases of the fluorescence intensities upon addition of 32 µg/ml of carnosic acid to *S. aureus* cells in a similar way than the inhibitor CCCP, while the cationic antimicrobial peptides Polimyxyn B produced an increment on the fluorescence. CCCP is a small amphipathic molecule which dissolves in phospholipid bilayers, providing a polar environment for the ion and an hydrophobic face to the outside world, it is an uncoupling agent that specifically increases the proton permeability, and disconnects the electron transport chain from the formation of ATP, discharging the pH gradient, and destroying the membrane potential.

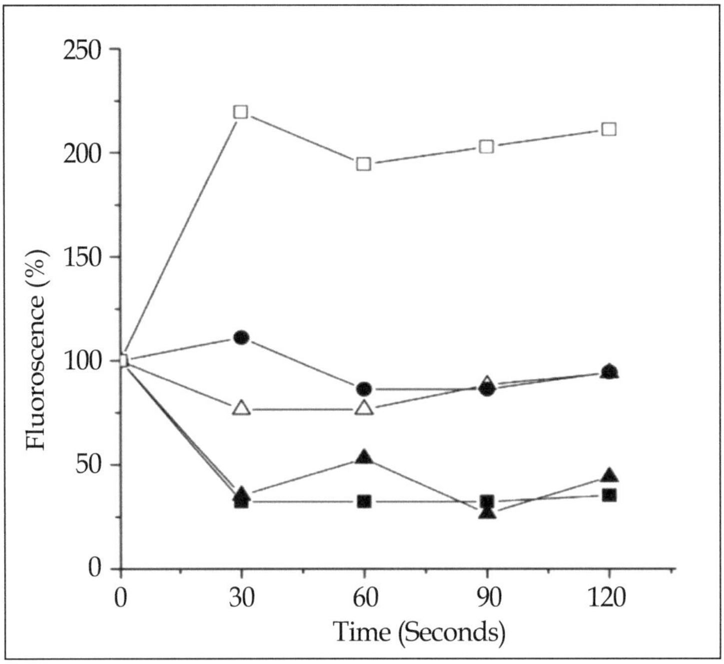

Fig. 5. Time changes of the fluorescence intensities of DiS-C_3-(5) upon addition of carnosic acid to *S. aureus* cells: 32 µg/ml (▲), 16 µg/ml (△), cells treated with CCCP (■), cells treated with Polimyxyn B (□) and untreated cells (●). Representative records from two similar independent experiments are shown

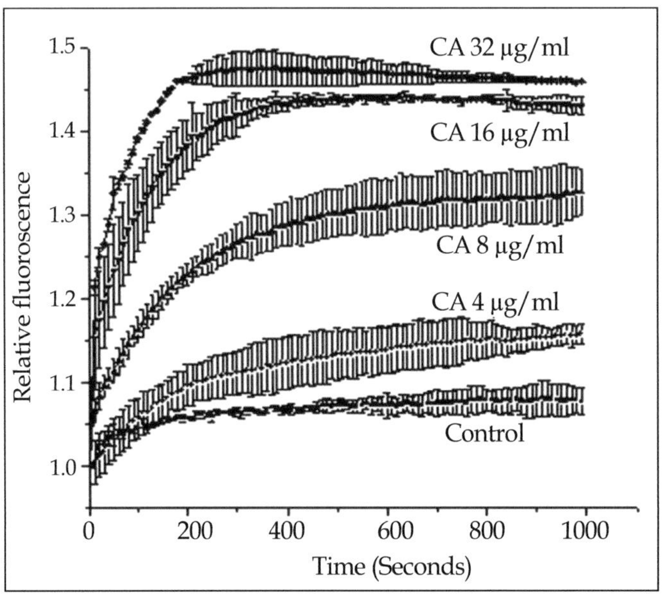

Fig. 6. Effect of carnosic acid on EtBr efflux in *S. aureus* cells. Control corresponds to the initial intracellular fluorescence of EtBr at time 0. Representative efflux records from two similar independent experiments are shown

As it can be seen above, carnosic acid at 1/4 - 1/8 MIC value (2 to 4 µg/ml) showed a selective synergistic interaction with tetracycline and aminoglycosides (Table 5) that may be explained by selective modulation of the antibiotics accumulation possibly by inhibition of their efflux in *S. aureus*. To assess whether this compound may have an efflux pump inhibitor activity in *S. aureus,* the efflux substrate ethidium bromide (EtBr) was used. EtBr shows a characteristic fluorescence when it enters the cell and binds to DNA and its uptake was determined by monitoring the fluorescence using a multiwell plate reader (DTX 880 Multimode detector, Beckman Coulter) at 29ºC every 10 sec intervals for 3 min with excitation and emission wavelengths of 535 nm and 485 nm, respectively. Figure 6 shows that carnosic acid rapidly increased EtBr uptake and this effect was dose dependent indicating a positive effect on the intracellular accumulation of the fluorescent molecule.

Whether carnosic acid is a general disruptor of the transmembrane electrochemical potential, it should display a synergistic effect with chloramphenicol which is effluxed out of the cell by secondary transport relying in the membrane potential. Additionally, an antagonist effect with ampicillin and penicillin G, which enter inside cells by a proton symport mechanism, should be displayed however, none of these effects were observed (Table 5).

Other studies from our laboratory on the antibiotic action of carnosic acid against *E. faecalis* showed similar results. In time-kill studied, carnosic acid displayed a bacteriostatic effect at its MIC value (64 µg/ml), whereas a bactericidal effect was achieved at 2 x MIC (128 µg/ml) increasing the permeability of cell membrane effects. At sub MIC values, the diterpene inhibited the drug efflux pumps of secondary transporters (Repetto, 2009). Other authors reported the antibacterial and resistance modifying activity of *R. officinalis* and suggested that the carnosic acid has the capacity to modify the resistance pattern of strains of *S. aureus* expressing multidrug efflux pumps (Oluwatuyi et al., 2004). Regarding this feature, we decided to investigate several bioactive compounds of *R. officinalis* with antimicrobial activity per se, or as modulators of bacterial resistance, against multidrug-resistant clinical strains of *S. aureus* isolated from pediatric patients (Ojeda Sana et al., 2011).

Our findings, together with results from other authors, suggest that this diterpene is a potential antibacterial agent to be used in combinational therapy (at least with aminoglycosides, tetracyclin and fluoroquinolones) against sensitive as well as multidrug resistant, vancomicyn and/or methicillin resistant Gram-positive cocci. Moreover, as an antibacterial compound, carnosic acid can not only target membrane permeability and enhance drug uptake, but also this compound, at sub MIC values inhibited the drug efflux transport probable by altering the cell membrane potential. We proposed a model for the antibacterial action of carnosic acid (Fig. 7).

3.4 Rosemary bioactives kill inthaphagocytic *S. aureus* cells

We examined the intracellular antibacterial activity of carnosic acid against *S. aureus*. A model of *S. aureus* infected RAW 654.7 mouse macrophages has been monitored for long-term (24 h) experiments. This bacterium adheres to phagocytes and easily invades them and tends to restrict the phagolysosomal compartment, where it largely escapes destruction and survives in a semiquiescent state for prolonged periods (Maurin et al., 2001). These intraphagocytic forms are considered responsible for the well-known recurrent character of staphylococcal infections as well as for the many failures of apparently appropriate antibiotic treatments.

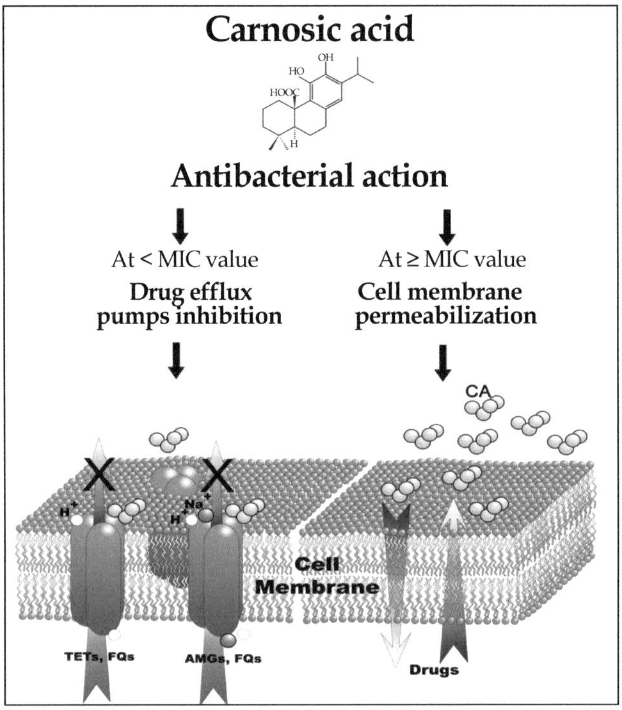

Fig. 7. Proposed model for the antibacterial action of carnosic acid. NMGs, aminoglycosides; TETs, Tetracyclins and FGs, fluoroquinolones

Confocal microscopy was used to ascertain the localization of the bacteria inside RAW 264.7 macrophages observed 24 h after phagocytosis of opsonized *S. aureus.* As shown in Fig. 8 A and B *S. aureus* clearly appeared intracellular. Pericellular membrane, located mostly on the inner face was labeled with FITC- phalloidin (green signal), and bacteria were labeled with DAPI (Blue signal). Later, macrophages were incubated for 24 h with carnosic acid from the MIC (16 µg/ml, determined in broth at pH 7.3). Carnosic acid caused a marked reduction in CFU, diminishing 3-4 log the bacterial growth at a concentration of 1 ½ MIC (24 µg/ml) (Fig. 8C). This effect was comparable to the control cells incubated with 0.5 µg/ml ciprofloxacin, a fluoroquinolone antibiotic that has excellent antibacterial activity against gram positive bacteria and intracellular penetration.

Earlier we demonstrated the *in vivo* antibacterial efficacy of a rosemary extract containing high amounts of carnosic acid against *S. aureus* in two skin infection models in mice (Barni et al, 2009).

Clinically effective antimicrobial agents exhibit selective toxicity towards the bacterium rather than the host. It is this characteristic that distinguishes antibiotics from disinfectants. The basis for selectivity will vary depending on the particular antibiotic. Carnosic acid at the concentrations that kill *S. aureus* has a high selectivity to bacteria and is not toxic to macrophages.

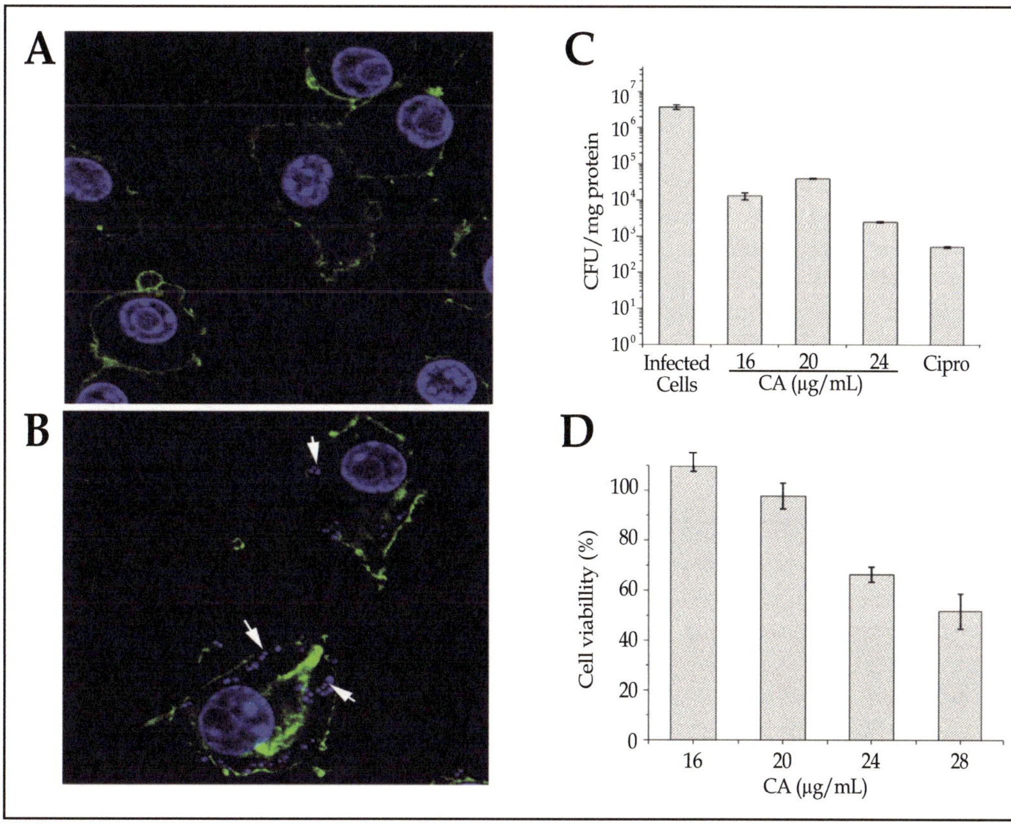

Fig. 8. Intraphagocitic killing of *S. aureus* by carnosic acid. Confocal microscopy of uninfected RAW 264.7 macrophages (A) or after phagocytosis of *S. aureus* (B). Arrowheads indicate intraphagocytic bacteria. (C) variations in the number of CFU per mg of protein (± SD; n= 2) after 24 h of incubation with carnosic acid or with 0.5 µg/ml of ciprofloxacin (Cipro). (D) Effect of carnosic acid following a 24-h exposure using a standard MTS cell viability assay

4. Safety of rosemary: Toxicity evaluation

Due to the potential use in human nutrition and health in foods of carnosic acid, it is important to determine its toxicological effects in a living organism.

In this study, we used *Caenorhabditis elegans* as a model system to examine the toxic effects of rosemary bioactives. This nematode, present in soil and found in temperate regions of the world, has emerged as an important animal model in various fields including toxicological research and rapid toxicity assessment for new chemicals (Moy T et al., 2006). *C. elegans* is easily cultured on agar medium plates with *Escherichia coli* OP50 as a food source. The ease of laboratory cultivation, its small size, large brood size, short development time (the entire complete life cycle from egg to egg producing hermaphrodite occurs over 3 days at 20°C under normal laboratory conditions) and the well-studied biology, makes the *C.elegans* an ideal model organism for biological studies. The transparency of this nematode allows for high quality microscopic images to be taken. Toxicity assays will be used in this study to test how carnosic acid affects *C. elegans* survival by exposing them to various concentrations of this compound. The *C. elegans* life cycle is comprised of an embryonic stage, four larval stages designated L1-L4 and adulthood. As it had been previously described that some nematode killing observed in this assays was a consequence of eggs being retained in the uterus and hatching internally, we substituted wild type worms by the *glp-4* temperature-sensitive sterile mutants to prevent internal hatching of progeny. *glp-4* does not make a germline at the restrictive temperature and survive without a bacterial food source, whereas WT *C. elegans* die by internal hatching of progeny when transferred to bacteria-free media. Worms were synchronized by isolating eggs from gravid adults, hatching the eggs overnight in M9 buffer, and plating L1-stage worms onto lawns of OP50 *E. coli* on nematode growth medium (NGM) agar media. Worms were grown to sterile, young adults by incubation at 25°C for 48 - 52 h, washed off the plates with M9 buffer. Approximately 40 worms were transferred to 24 well plates containing 500 microliters of M9 medium and the compounds to be tested. Each compound was tested in individual wells, and the screen was performed by using triplicate 24-well plates. To score for survival, the plates were shaken by hand, worms were considered to be dead if they did not move or exhibit muscle tone when viewed using a stereo microscope at 20X magnification. Worms were observed every two days and scored for survival analysis at day 8 after transfer. As carnosic acid was dissolved in ethanol, we first tested if the 1.6% final ethanol concentration found in the vehicle of this assay had any effect over the survival of worms. We didn't find significative differences between the survival of ethanol treated and untreated worms. Our results demonstrated that up to 250 µg/ml of carnosic acid tested the compound does not adversely affect the normal physiology of the nematodes, while the highest concentration tested (500 µg/ml) of carnosic acid had moderate worm mortality (30%) (Fig. 9).

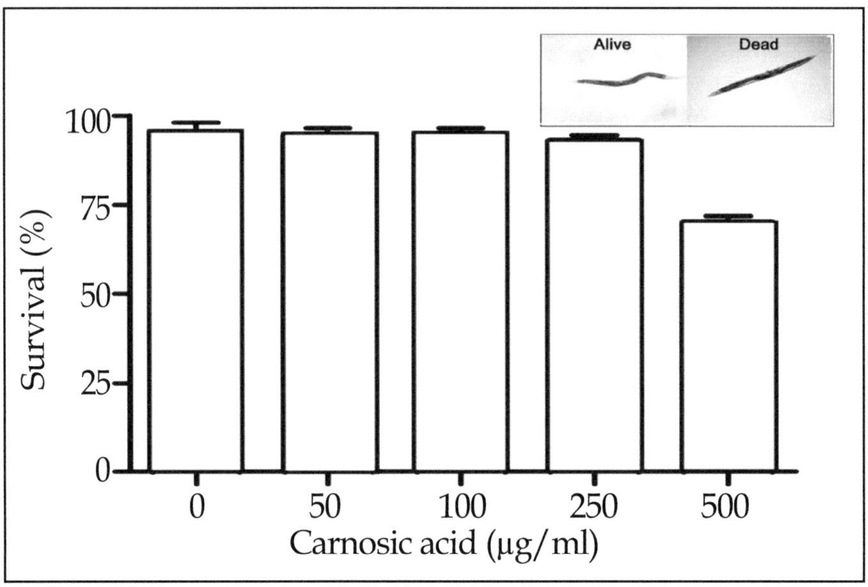

Fig. 9. Effect of carnosic acid over *C.elegans* survival. (Inset) typical morphology of living nematodes maintaining a sinusoidal shape (left) and dead ones appearing as straight, rigid rods (right) (at ×20 magnification). Values are shown as the mean of three independent experiments ± SD

5. Conclusions

The present chapter summarizes the most significant rosemary biological properties, reviewing its free radical scavenging, antibacterial actions of essential oils and non-volatile pure constituents and in combination with antibiotics.

Both volatile and non-volatile secondary metabolites isolated from rosemary plants growing in the same place exhibited comparable antioxidant activities. Non-volatile constituents exhibited at least a minor pro-oxidant property than ascorbic acid and Trolox, at the concentration assayed. The non-volatile fractions containing high content of carnosic acid as well as the volatile oils rich in α-pinene exhibited the highest antibiotic activity.

Rosmarinus officinalis extract as well pure carnosic acid, showed synergistic effect in combination with aminoglycosides and ciprofloxacin against *S. aureus*. We proposed a model for the antibacterial action of carnosic acid which may be useful for future applications.

Unexpectedly, carnosic acid showed a clear antistaphylococcal action towards extracellular and intraphagocytic forms of *Staphylococcus aureus*, pointing out a potential use of this compound in the treatment of *S. aureus* infections, without significant effect on the macrophage viability. Moreover, carnosic acid at bactericide concentrations did not show adverse effects on the viability of a living organism as the nematode *C. elegans*.

6. Rosemary in human nutrition and health: Future and prospect

All together, the available evidence indicates that rosemary compounds might be of therapeutic benefit in bacterial infections and be an ideal candidate for nutraceutical health products. Non-volatile extracts of rosemary containing approximately an amount of 20 μg/ml (18 μM) of carnosic acid as the key compound, killed several bacteria and represent a therapeutic alternative against extracellular-intracellular *S. aureus* infections. This compound did not show pro-oxidant effects and its use is safe at least until a concentration of 250 μg/ml (750 μM). The *in vivo* antibacterial efficacy of an ethanol extract of *Rosmarinus officinalis* L. containing high amounts of carnosic acid against the pathogenic bacteria *S. aureus* has been demonstrated previously in mouse (Barni et al, 2009). Even though, prospective controlled clinical studies are still lacking.

Rosemary is the only spice commercially available for use as an antioxidant in Europe and the United States. This specie has the advantage to contain different antioxidant molecules (lipophilic monoterpenes and diterpenes, as well as hydrophilic derivatives of caffeic acid as rosmarinic acid) that could be effective in both, aqueous fluids as well as in lipophilic parts of the body as a very effective antioxidant to scavenge free radicals. In addition, non-volatile extracts of rosemary can also be used to decrease 4.4 - to 17-fold the amounts of the synthetic butyl derivatives used as food or cosmetic preservatives (Romano et al., 2009).

Although not discussed in this chapter, due to the associated bioactivities of carnosic acid (anti-inflammatory-and anticarcinogenic effects), rosemary polyphenols can be considered as a potential source of promising new nutraceuticals formulations (Mengoni et al., 2011).

Improvements in the processes of regulation of rosemary bioactives are needed, and the general tendency is to perpetuate the German Commission E experience, which combines scientific studies and traditional knowledge (monographs).

There is still work to be done regarding how to use rosemary derivatives inside the human diet. Particular attention needs to be given to stability studies and interactions of rosemary constituents with other food constituents. Further investigations will be directed towards the application of phenolic compounds in various food matrices. The use of functional foods enriched with rosemary compounds need technologies for incorporating health-promoting ingredients into food without reducing their bioavailability or functionality. In this sense, we are also working on the addition of rosemary bioactives into edible films (Proyect CYTED, 309AC0382 action: Getting additive materials from plant by-products of the region and its application in the development of biodegradable packaging food processing and nutraceutical use). Finally, the production and biotechnological studies and genetic improvement of rosemary plants will offer great advantages, since it will be possible to obtain uniform and high quality raw materials which will ensure the efficacy and safety of rosemary products.

7. Acknowledgment

We would like to thank the ANCyT, Argentina for the financial support (Grants PICT 2005-35401 and PICT 2008-1969); to the University of Buenos Aires B014 and 20020090200401 and CYTED 309AC0382. Moreno S. is a career researcher of the National Research Council of Argentina (CONICET) and Ojeda-Sana A.M. and Barni M.V. doctoral fellowship. Thanks also to Ing. Elechosa and Tec. Juarez for the provision of plant materials and to Lic Contartese for his invaluable assistance in the preparation of rosemary extracts.

8. References

Aherne, S.A., Kerry, J.P. & O'Brien, N.M. (2007). Effects of plant extracts on antioxidant status and oxidant-induced stress in Caco-2 cells. *British J Nutr* 97: 321-328. URL: *www.nutritionsociety.org.uk*

al-Sereiti, M.R., Abu-Amer K.M. & Sen P. (1999) Pharmacology of Rosemary (*Rosmarinus officinalis* Linn.) and its therapeutic potentials. *Indian J Exp Biol* 37(2): 124-30. URL: *www.pubget.com*

Barni, M.V., Fontanals, A. & Moreno, S. (2009). Study of the antibiotic efficacy of an ethanolic extract from *Rosmarinus officinalis* against *Staphylococcus aureus* in two skin infection models in mice. *Boletín Latinoamericano y del Caribe de Plantas Medicinales y Aromáticas* 8 (3): 219 - 223.

Bradley P. (2006) *British herbal compendium, A handbook of scientific information on widely used plant drugs,* British herbal Medicine Association, Bournemouth.

Brand-Williams, W., Cuvelier, M.E. & Berset, C. (1995). Use of free radical method to evaluate antioxidant activity. *Lebensmittel-Wissenschaft und Technologic* 28: 25–30. URL: *www.academicpress.com*

Burt, S. (2004). Essentials oils: their antibacterial properties and potential applications in foods - a review. *Intl J Food Microbiol* 94: 223-253. URL: *www.scimagojr.com*

Ciftci, O., Tanyildizi, S. & Godekmerdan A. (2011). Curcumin, myrecen and cineol modulate the percentage of lymphocyte subsets altered by 2, 3, 7, 8-Tetracholorodibenzo-p-dioxins (TCDD) in rats. *Hum Exp Toxicol* [Epub ahead of print]. URL: *www.unboundmedicine.com*

Clinical and Laboratory Standards Institute (1999). Methods for determining bactericidal activity of antimicrobial agents; approved standard. (6th Ed.). *CLSI document M26-A.* Clinical and Laboratory Standards Institute, Wayne, Pa.

Coutinho, H.D.M., Costa, J.G.M. & Lima, E.O. (2009). Herbal therapy associated with antibiotic therapy: Potentiation of the antibiotic activity against methicillin - Resistant *Staphylococcus aureus* by *Turnera ulmifolia* L. *BMC Complement Altern Med* 9: 35. URL: *www.liebertpub.com*

Cowan, M.M. (1999). Plant products as antimicrobial agents. *Clinical Microbiol Reviews* 12: 564–582. URL: *www.cmr.asm.org*

Davidson, P.M., Sofos, J.N. & Branen, A.L. (2005). *Antimicrobials in Food.* (3rd Ed.). New York: CRC Press, pp. 1-8.

Erkan, N. Ayranci, G. & Ayranci, E. (2008). Antioxidant activities of Rosemary *(Rosmarinus Officinalis* L.) extract, blackseed (Nigella sativa L.) essential oil, carnosic acid, rosmarinic acid and sesamol. *Food Chem* 110: 76–82. URL: www.elsevier.com

ESCOP Monographs. (2009). The Scientific Foundation for Herbal Medicinal Products, Rosmarini Folium, Thieme, 2nd ed., pp. 429–436

Lewis, K. & Ausubel, F.M. (2006). Prospects for plant-derived antibacterials. *Nat Biotechnol* 24: 1504 - 1507. URL: *www.nature.com*

Maurin, M. & Raoult. D. (2001). Use of aminoglycosides in treatment of infections due to intracellular bacteria. *Antimicrob Agents Chemother* 45: 2977–2986. URL: *aac.asm.org*

Mayo, D.X., Tan, R.M., Sainz, M., Natarajan, Lopez-Burillo S. & Reiter, R.J. (2003). Protection against oxidative protein damage induced by metal-catalyzed reaction or alkylperoxyl radicals: comparative effects of melatonin and other antioxidants. *Biochim Biophys Acta* 1620: 139–150. URL: *www.pubget.com*

Mengoni, E.S., Vichera G., Rigano L.A., Rodriguez-Puebla M.L., Galliano S.R., Cafferata E.E., Pivetta, O.H., Moreno, S. & Vojnov, A.A. (2011). Suppression of COX-2, IL-1beta and TNF-alpha expression and leukocyte infiltration in inflamed skin by bioactive compounds from *Rosmarinus officinalis* L. *Fitoterapia* 82 (3): 414-421. URL: *www.elsevier.com*

Moreno, S., Scheyer, T., Romano, C., & Vojnov, A.A. (2006). Antioxidant and antimicrobial activities of Rosemary extracts linked to their polyphenol composition. *Free Rad Res* 40: 223-231. URL: *tandf.informaworld.com*

Moy, T., Ball, A.R. & Anklesaria, Z. (2006). Identification of novel antimicrobials using a live-animal infection model. *Proc Natl Acad Sci* U.S.A. 103: 10414-10419. URL: *www.pnas.org*

National Nosocomial Infections Surveillance (2004). National Nosocomial Infections Surveillance (NNIS) System Report, data summary from January 1992 through June 2004, issued October 2004. *Am J Infect Control* 32: 470-485. URL: *www.ajicjournal.org*

Ojeda Sana, A.M., Blanco, A., Lopardo, H., Cáceres Guido, P.A., Macchi, A., van Baren, C., Moreno, S. (2011). Antibiotic effectiveness of *Rosmarinus officinalis* bioactive compounds against multidrug-resistant bacteria of difficult clinical treatment. *Proceedings VII.* Argentine Congress of General Microbiology. Tucumán, Argentina. May 18 - 20, 2011.

Oluwatuyi, M., Kaatz, G.W. & Gibbons, S. (2004). Antibacterial and resistance modifying activity of *Rosmarinus officinalis. Phytochem* 65: 3249-3254. URL: *www.elsevier.com*

Perez-Fons, L., Garzon, M.T. & Micol, V. (2010). Relationship between the antioxidant capacity and effect of Rosemary (*Rosmarinus officinalis* L.) polyphenols on membrane phospholipid order. *J Agric Food Chem* 58(1): 161-71. URL: pubs.acs.org

Piddock, L.J.V. (2006). Clinically relevant chromosomally encoded multidrug resistance efflux pumps in bacteria. *Microbiol Mol Biol Rev* 19: 382-402. URL: *mmbr.asm.org*

Repetto, M.V. (2009). Bachelor Thesis, Facultad Ciencias Exactas y naturales, Universidad de Buenos Aires.

Romano, C.S., Abadi, K., Repetto, V., Vojnov, A.A. & Moreno, S. (2009). Synergistic antioxidant and antibacterial activity of Rosemary plus butylated derivatives. *Food Chem* 115: 456-461. URL: *www.elsevier.com*

Scalbert, A., Johnson, I.T. & Saltmarsh, M. (2005). Polyphenols: antioxidants and beyond. *Am J Clin Nutr* 81(1 Suppl): 215S-217S. URL: *www.ajcn.org*

Shylaja, M.R. & Peter, K.V. (2004). *The functional role of herbal spices.* In: Peter, K.V. ed. Handbook of herbs and spices, Vol. 2. Boca Raton: CRC Press.

Sies, H. (2010) Polyphenols and health: update and perspectives. *Arch Biochem Biophys* 501(1): 2-5. URL: *www.sciencedirect.com*

Stavri, M., Piddock, L.J.V. & Gibbons, S. (2007). Bacterial efflux pump inhibitors from natural sources. *J Antim Chemother* 59: 1247-1260. URL: *www.jchemother.it*

Tallarida, R.J. (2001). Drug synergism: Its detection and applications. *J Pharmacol Exp Ther* 298: 865–872. URL: *jpet.aspetjournals.org*

Waters, A.E., Contente-Cuomo, T., Buchhagen, J., Liu, C.M., Watson, L., Pearce, K., Foster, J.T., Bowers, J., Driebe, E.M., Engelthaler, D.M., Keim, P.S. & Price, L.B. (2011). Multidrug-Resistant *Staphylococcus aureus* in US Meat and Poultry. *Clin Infect Dis* 52:1227-30. URL: *www.idsociety.org*

Wei, A., & Shibamoto, T. (2007). Antioxidant activities and volatile constituents of various essential oils. *J Agric Food Chem* 55: 1737–1742. URL: *pubs.acs.org*

Williams, R.J., Spencer, J.P.E. & Rice-Evans, C. (2004). Flavonoids: Antioxidants or signalling molecules?. *Free Rad Biol Med* 36(7): 838-849. URL: *www.elsevier.com*

10

Potential of Probiotic *Lactobacillus* Strains as Food Additives

Sheetal Pithva[1], Padma Ambalam[1],
Jayantilal M. Dave[2] and Bharat Rajiv Vyas[1]
[1]*Department of Biosciences, Saurashtra University, Rajkot*
[2]*201 Shivam, Vrindavan Society, Kalawad Road, Rajkot*
India

1. Introduction

The increasing consumer awareness that diet and health are linked is stimulating innovative development of novel products by the food industry. Lactic acid bacteria (LAB) have received much attention over recent decades due to the health-promoting properties of certain strains, called probiotics. The concept probiotics has been redefined over time. Fuller defined it as "A live microbial feed which beneficially affects the host animal by improving intestinal microbial balance" (Fuller, 1989). The probiotic products traditionally incorporate intestinal species of *Lactobacillus* because of their long history of safe use in the dairy industry and their natural presence in the human intestinal tract, which is known to contain a myriad of microbes, collectively called the microbiota. Intestinal LAB in humans are intimately associated with the host's health because they are an important biodefense factor in preventing colonization and subsequent proliferation of pathogenic bacteria in the intestine. In fact, probiotics have been used for as long as people have eaten fermented foods. However, it was Metchnikoff at the turn of the 20th century who first suggested that ingested bacteria could have a positive influence on the normal microbial flora of the intestinal tract (Metchnikoff, 1907). He hypothesized that lactobacilli were important for human health, longevity, and promoted yogurt and other fermented foods as healthy. Food derived from plants, animals, or their products often contain many types of microbes. These microbes from natural and external sources colonize food by contact, which can occur anytime between production and consumption. Microbial contamination of food (i.e. the colonization by unwanted microorganisms) can have many undesirable consequences ranging from spoilage to food borne illness. However, some microbes possess properties that are beneficial for food production or conversion or storage. These food grade microorganisms are used to produce a variety of fermented foods (with improved storage capability) from raw animal and plant material. Having natural preservatives in mind, LAB and their metabolites are good alternatives. The increasing consumer awareness of the risks derived not only from food-borne pathogens, but also from the artificial chemical preservatives used to control them (Abee *et al.*, 1995), has led to renewed interest in so-called "green technologies" including novel approaches for a minimal processing and exploitation of bacteriocins for biopreservation (Papagianni, 2003). Biopreservation can be explained as the link between fermentation and preservation, and refers to extension of the shelf-life and improvement of the safety of food using microorganisms and/or their metabolites (Kao & Frazier, 1966; Klaenhammer, 1988; Holzapfel *et al.*, 1995). Furthermore, the use of LAB and or their metabolites for food preservation is generally accepted by consumers as something "natural" and "health-promoting" (Montville & Winkowski, 1997). Among LAB, addition of *Lactobacillus* culture to food is an approach to food preservation, it also contributes to taste, texture and also inhibits food spoilage bacteria by producing growth inhibiting substances like bacteriocins, lactic acid etc. Strategies utilized to study incorporation of biopreservatives into food include: direct use of LAB-strains with proven antimicrobial activity as starter cultures or food additives, use of biopreservatives preparation in the form of previously fermented product, or use of partially-purified, purified or chemically synthesized bacteriocins (De Vuyst & Vandamme, 1994).

2. Food-associated Lactic acid bacteria

The first essential step in food fermentation is the catabolism of carbohydrates by the LAB. LAB as a group exhibit an enormous capacity to degrade different carbohydrates and related compounds. LAB are Gram-positive, non-spore forming cocci, coccobacilli or rods and most genera have a DNA base composition of less than 50% G+C, lack catalase, grow under microaerophilic or anaerobic conditions, and typically ferment glucose mainly to lactic acid (homo-fermentative), but can also have lactic acid, CO_2, and ethanol/acetic acid as end products (hetero-fermentative). In nature, species of the LAB are found in gastro-intestinal tract (GIT) of mammals and also in fermented food products (dairy, meat, vegetables, fruits and beverages). LAB associated with foods are generally restricted to the genera *Lactobacillus, Lactococcus, Leuconostoc, Pediococcus* and *Streptococcus*. Orla-Jensen (1919) proposed a classification of lactic acid bacteria, which was based on morphology, temperature range of growth, nutritional characteristics, carbon sources utilization and agglutination effects. Orla-Jensen (1919) differentiated three major groups. The first group contained *Thermobacterium, Streptobacterium* and *Streptococcus* which were all catalase negative and produce mainly lactic acid besides traces of other by-products. The second group contained *Betabacterium* and *Betacoccus,* which also lack catalase but as a rule formed detectable amounts of gas and other by-products, besides lactic acid. The third group consisting of *Microbacterium* and *Tetracoccus* show a positive catalase reaction. In 1960, Van den Hammer showed that representative of *Betabacterium* did not possess fructose-1,6-bisphosphate aldolase, in contrast to *Thermobacterium* and *Streptobacterium*. These findings supported the discrimination of the three physiological groups: (i) the obligately homo-fermentative lactobacilli, lacking both glucose-6-phosphate dehydrogenase and 6-phosphogluconate dehydrogenase (*Thermobacterium*), (ii) the facultatively homo-fermentative lactobacilli having both dehydrogenases but degrading glucose preferably via the Embden-Meyerhof-Parnas pathway (*Streptobacterium*) and (iii) the obligately hetero-fermentative lactobacilli lacking fructose 1,6-bisphosphatealdolase (*Betabacterium*). *Thermobacterium, Streptobacterium* and *Betabacterium* were considered to be the three subgenera within the genus *Lactobacillus*.

The genus *Lactobacillus* belongs to the large group of lactic acid bacteria. The genus *Lactobacillus* belongs phylogenetically to the phylum *Firmicutes* (Garrity *et al.*, 2004). The family *Lactobacillaceae* comprises the main family in the order *Lactobacillales* which itself belongs to the class *Bacilli*. Lactobacilli can be found in a variety of ecological niches, such as plants (fruits, vegetables, cereal grains) or plant-derived materials, silage, fermented foods (yogurt, cheese, olives, pickles, salami, etc.), as well as in the oral cavities, GIT, and vaginas of human and animals. The bacteria that occupy a niche in the GIT are true residents or autochthonous (i.e., found where they are formed). Other bacteria are just "get a lift" through the gut and are allochthonous (i.e., formed in another place). Autochthonous strains have a long-term association with a particular host, and they form stable populations of a characteristic size in a particular region of the gut. It is often difficult to determine whether or not a particular microorganism is truly autochthonous to a particular host (Tannock, 2004).

3. The role of lactic acid bacteria in the functional food concept

3.1 The functional food concept

Functional food is food that promotes human health above the provision of basic nutrition. The term "functional food" was first proposed in Japan two decades ago and legally approved there as Food for Specified Health Use (FOSHU). A relatively recently proposed working definition describes functional food as "food that can be satisfactorily demonstrated to affect beneficially one or more target functions in the body, beyond adequate nutritional effects, in a way relevant to an improved state of health and well-being and/or reduced risk of diseases" (Contor, 2001). Functional foods are also known as designer foods, medicinal foods, nutraceuticals, therapeutic foods, super foods, foodiceuticals, and medifoods (Shah, 2001). Functional food has a significant and growing global market, of which the largest segment in Europe, Japan and Australia comprises food containing probiotics, prebiotics and synbiotics (Stanton *et al.*, 2005).

3.2 Probiotics

The idea that LAB prevents intestinal disorders and diseases is nearly as old as the science of microbiology (Molin, 2001). Therefore, in the development of probiotic food intended for human consumption, strains of LAB have most commonly been used. The term "probiotic" (Greek: for life) was first used by Lilly and Stillwell (1965). "Probiotic" was later more widely used and defined by Parker (1974), and further improved by Fuller (1989) with the following definition: "A live microbial food supplement which beneficially affects the host animal by improving its intestinal microbial balance". This definition has later been slightly revised (Schaafsma, 1996; Schrezenmeir & de Vrese, 2001) to "Foods containing live and defined bacteria, which when given in sufficient numbers, exert beneficial effects by altering the microflora in the host" or as expressed by Salminen *et al.*, (1998) "Viable preparation in food or dietary supplements to improve the health of humans and animals". According to these definitions, an impressive number of microbial species and genera can be considered as probiotics. However, only strains classified as LAB are (due to their traditional use in food) currently considered of importance in regard to food and nutrition.

3.3 Prebiotics

Since the viability of the live bacteria in food products and during transit through the GIT may be variable, the "prebiotic" concept has been developed. A prebiotic is defined as a "non-digestible food ingredient that beneficially affects the host by selectively stimulating the growth and/or activity of one or a limited number of bacteria in the colon that can improve the host health" (Gibson & Ruberfroid, 1995). Thus, selective growth of certain indigenous gut bacteria is improved by the administration of the prebiotic and thereby any viability problems of orally administered bacteria in the upper GIT can be overcome. Some oligosaccharides, due to their chemical structure, are resistant to digestive enzymes and therefore pass into the large intestine where they become available for fermentation by saccharolytic bacteria. Compounds that are either partially degraded or not degraded by the host and are preferentially utilized by probiotic bacteria as a carbon and/or energy source. The criteria which allow classification of a food ingredient as a prebiotic, are defined by Fooks and Gibson (2002), and include the following statements, ex. fructo-oligosaccharides, xylo-oligosaccharides, lactose derivatives such as lactulose, lactitol, galacto-oligosaccharides and soyabean oligosaccharides.

i. It must be neither hydrolysed, nor absorbed in the upper part of the GIT.
ii. It should be selectively fermented by one or a limited number of potentially beneficial bacteria in the colon.
iii. Its presence should alter the colonic microbiota towards a healthier composition.
iv. It should induce effects which are beneficial to the host's health.

3.4 Synbiotics

A further possibility in microflora management procedures is the use of synbiotics, i.e., the use in combination of probiotics and prebiotics (Gibson & Roberfroid, 1995). The live microbial additions may be used in conjunction with a specific substrate for growth and the end result should be improved survival of the probiotic, which has a readily available substrate for its fermentation, as well as the individual advantages that each may offer (Fooks & Gibson, 2002). Many studies suggest that the consumption of synbiotic products has higher beneficial effects on the human health than probiotic or prebiotic products (Gmeiner *et al.*, 2000), leading to improved survival of probiotic bacteria during the storage of the product and during the passage of the intestinal tract. Moreover, the synbiotic product may allow an efficient implantation of probiotic bacteria in colonic microbiota, because the prebiotic has a stimulating effect on the growth and/or activities of both the exogenous and the endogenous bacteria (Champagene & Gardner, 2005).

3.5 Important aspects for selection of probiotic strains

When selecting a probiotic strain, a number of aspects should be considered, and the theoretical basis for selection should involve safety, functional as well as technological aspects (Salminen *et al.*, 1998; Adams, 1999; Saarela *et al.*, 2000). When selecting a preferable probiotic strain, several aspects of functionality have to be considered, as specified below:

i. Strains for human use are preferably of human origin and isolated from a healthy human GIT and non-pathogenic
ii. must survive through upper GIT and arrive alive at its site of action and able to function in the gut environment
iii. adhere to the intestinal epithelium cell lining and colonize the lumen of the intestinal tract
iv. strains should not carry transmissible antibiotic resistance genes

v. must be able to survive GI transit (acid and bile salt tolerant)
vi. must have good technological properties so that it can be manufactured and incorporated into food products without loosing viability and functionality or creating unpleasant flavors or textures
vii. functional aspects include viability and genetic stability

3.5.1 Acid and bile tolerance

Probiotic lactobacilli encounter various environmental conditions upon ingestion by the host and during transit in the GIT. Firstly, they need to survive the harsh conditions of the stomach. Humans secrete approximately 2.5 litres of gastric juice each day, generating a fasting pH of 1.5, increasing to pH 3 to 5 during food intake and that the food transit time through the human stomach is about 90 minutes. The aggregation of cells could possibly be explained by an increased hydrophobicity of the cell surface at low pH. The cell envelope of gram-positive bacteria consists of an inner plasma membrane and a thick outer layer of peptidoglycan. In contrast to gram-negative bacteria, cell walls of gram-positive bacteria contain large amounts of negatively charged teichoic acids (polymers of glycerol or ribitol joined by phosphate groups). Hence, one can assume that the teichoic acids become protonated at low pH, leading to a more hydrophobic surface. Ingested microorganisms must endure numerous environmental extremes to survive in the human GIT. Bile tolerance is one of the most essential criteria for the selection of a probiotic strain. Bile acids are synthesized in the liver from cholesterol and are secreted from the gall-bladder into the duodenum, where they play an important role in the digestion of fat. Bile acids are conjugated to either glycine or taurine. Bile is a digestive secretion that plays a major role in the emulsification of lipids. Bile acids are surface active, amphipathic molecules with potent antimicrobial activity and act as detergents, disrupting biological membranes. It has the ability to affect the phospholipids, proteins of cell membranes and disrupt cellular homeostasis. Therefore, the ability of pathogens and commensals to tolerate bile is likely to be important for their survival and subsequent colonization in the GIT (Begley *et al.*, 2005).

In our study, we have isolated *Lactobacillus rhamnosus* Fb from healthy human infant feces, is a gram-positive, catalase negative, non-motile and non-spore forming rod-shaped organism (De Man *et al.*, 1960). Its ability to ferment ribose, rhamnose and growth at 15°C and 45°C indicate that it belongs to the group *Streptobacterium* (Orla-Jensen, 1943). The identity of the *L. rhamnosus* Fb was confirmed by 16S rDNA sequence analysis. The primary requirement for potential probiotic organisms is to survive during the passage through the acidic (pH 1-3) environment of stomach. *L. rhamnosus* survives at pH 2 for 2 h, 87% of total cells remain viable, a sufficiently long time for the cells to pass through the stomach and reach their site of action in the intestine. *L. rhamnosus* show high survival which is satisfactory especially as probiotic strains can be buffered by food or other carrier molecules and in fact are not directly exposed to such a low pH in the stomach. After passage through acidic condition cells were exposed to bile salt (0.1% pancreatin, 0.5% bile salt, pH 8) viability increase after 3 h of incubation to 95%. Mimicking gastro-intestinal transit, we observed that the bacterial stress originated by low pH may be overcome after the subsequent treatment in presence of bile (Charteris *et al.*, 1998). A bile concentration of 0.3% is usually used for screening of bile tolerant strains, as this is considered as an average intestinal bile concentration of the human GIT (Gilliland *et al.*, 1984). *L. rhamnosus* possesses the ability to grow in the presence of 0.4% phenol and remain viable in 0.6% phenol, a toxic metabolite produced by intestinal bacteria during putrefaction in the GIT (Khedekar, 1988). *L. rhamnosus* also possesses the ability to grow in the presence of 6% NaCl (Jacobsen *et al.*, 1999). The ability of *L. rhamnosus* cells to survive in the presence of bile, NaCl and phenol can help them to survive, grow, colonize and elicit the beneficial effects to the host.

4. Health benefits of functional probiotic culture

A number of health benefits are claimed in favour of products containing probiotic organisms including antimicrobial activity, gastrointestinal infections, improvement in lactose metabolism, antimutagenic properties, anticarcinogenic properties, reduction in serum cholesterol, anti-diarrhoeal properties, immune system stimulation, improvement in inflammatory bowel disease and suppression of *Helicobacter pylori* infection (Ambalam *et al.*, 2009; 2011; Kurmann & Rasic, 1991; Shah, 2007). Some of the health benefits are well established, while other benefits have shown promising results in animal models. However, additional studies are required in humans to substantiate these claims. Health benefits imparted by probiotic bacteria are strain specific, and not species- or genus-specific. It is important to note that no strain will provide all proposed benefits, not even strains of the same species, and not all strains of the same species will be effective against defined health conditions. The strains of *Lactobacillus* and *Bifidobacterium* are able to restore the normal balance of microbial populations in the intestine and most commonly used as probiotics (Shah, 2006).

4.1 Antimicrobial activity of probiotic *Lactobacillus rhamnosus*

Many mechanisms have been postulated by which Lactobacilli could produce antimicrobial activity. In addition to their competitive inhibition of the epithelial and mucosal adherence of pathogens and inhibition of epithelial invasion by pathogens, lactobacilli and bifidobacteria show antimicrobial activity by producing antimicrobial substances and/or stimulating mucosal immunity (Servin, 2004). Probiotic bacteria produce organic acids, hydrogen peroxide and bacteriocins as antimicrobial substances that suppress the multiplication of pathogenic and putrefying bacteria. Lactic and acetic acids account for over 90% of the organic acids produced. Lowering of pH due to lactic acid or acetic acid produced by these bacteria in the gut has a bacteriocidal or bacteriostatic effect. *Lactobacillus rhamnosus* has shown antimicrobial activity against *Escherichia coli, Enterobacter aerogenes, Salmonella typhi, Shigella sp., Proteus vulgaris, Pseudomonas aeruginosa, Serratia marcescens, Staphylococcus aureus, Bacillus subtilis, Bacillus megaterium, Bacillus cereus, Helicobacter pylori, Campylobacter jejuni, and Listeria monocytogenes* (Ambalam *et al.*, 2009). *L. rhamnosus* produces other antimicrobial metabolites, as evidenced from the antimicrobial activity of Cell Free Culture filtrate (CFC) even when Extracellular Protein Concentrate (EPC) (independently) was less active. This evidence suggests the multifactorial nature of the antimicrobial activity and possibly a synergistic effect. Roles of other metabolites remain to be identified.

4.2 Classification of bacteriocins from Lactic acid bacteria

Bacteriocins of LAB are a heterogeneous group of bacterial antagonists and several classification criteria have been used to group them. The bacteriocins have been divided into three main categories: Class I Lantibiotics ribosomally synthesized peptides that undergo extensive post-translational modifications for ex. Nisin. Class II Non-lantibiotic peptides ribosomally synthesized peptides that undergo minimal post-translational modification. They have diverse chemical and genetic characteristics. They have been sub-divided into three main groups: (a) single peptides, often with a characteristic YGNGVXC amino acid motif near the N-terminus (b) two-peptide bacteriocins (c) In Nes' classification bacteriocins produced by the cell's general secretory (sec) pathway and in Klaenhammer's classification, thiol-activated peptides. Class III Non-lantibiotic, heat labile proteins. These are relatively uncommon among the antibacterial compounds of LAB, Class IV A fourth category of complex bacteriocins containing protein and lipid or carbobydrate moieties was included in Klaenhammer's classification (Klaenhammer, 1993) but Nes and coworkers (Nes *et al.*, 1996) excluded this category because these compounds have not been purified and evidence for them is based on loss of activity following treatment with carbohydrate-or lipid-hydrolysing enzymes and Class V bacteriocins with circular, unmodified structure ex. enterocin (Eijsink, *et al.*, 1998; Guyonnet *et al.*, 2000). *L. acidophilus* produces various bacteriocins and antibacterial substances such as Lactocidin, Acidolin, Acidophilin, Lactacium-B and inhibitory protein (Shah, 1999).

4.3 Characteristics of antimicrobial protein(s) of *Lactobacillus rhamnosus*

Antimicrobial activity of Cell Free Culture filtrates (CFC) against the test organisms increase with the culture age of *L. rhamnosus* Fb and became stable when culture reached the stationary phase. A similar antimicrobial spectrum of CFC filtrate was observed against all the test organisms (Fig. 1, b & d). The antimicrobial spectrum of Extracellular Protein Concentrate (EPC) against the test organisms changed with culture age. Antimicrobial activity against *E. coli* was observed in the initial growth phase, activity increases with culture age upto 18 h, it decreases later with increasing culture age (Fig. 1, a & c). Antimicrobial activity against *Ent. aerogenes, B. subtilis, B. megaterium* and *Staph. aureus* appeared after 6 h of growth and increased with culture age became stable in stationary phase. Whereas activity against *Shigella sp., Ps. aeruginosa* and *B. cereus* was observed after 12 h of growth and increased with culture age upto 30 h before decreasing marginally. Antimicrobial activity against *S. typhi* and *P. vulgaris* was observed in stationary phase and it did not change much with the increase in culture age. The antimicrobial extracellular proteins are produced during exponential and stationary phases. Changes in the antimicrobial activity spectrum of the EPC during different growth phases provide evidence that the EPC is a mixture of antimicrobial peptides and its composition changes with the culture age (Ambalam *et al.*, 2009). Antimicrobial activity spectrum changes with culture age, indicates that the antimicrobial activity is attributed to the mixture of antimicrobial peptides. Antimicrobial activity of EPC shows activity over broad pH range (2-9) but the activity varies with test organisms. At pH 2 to 5 and 8 mode of inhibition was bactericidal against *E. coli, Ent. aerogenes, S. typhi, Shigella sp., P. vulgaris, Ser. marcescens, Ps. aeruginosa, Staph. aureus, B. megaterium, B. cereus and B. subtilis*. While at pH 6, 7 and 9 the activity was bacteriostatic. Antimicrobial activity of EPC was thermostable (60 min at 100°C), thermostability was evidenced from the bactericidal activity of heat treated EPC, heat treatment caused complete loss of activity against *Ps. aeruginosa* and *Bacillus spp.* Heat stability of antimicrobial proteins has been suggested to be the major feature of low molecular weight bacteriocins and arises from complex pattern of disulphide intramolecular bonds that stabilize secondary structures by reducing the number of possible unfolded structures (Cintas *et al.*, 1995). Currently we do not know the reasons for the stability of

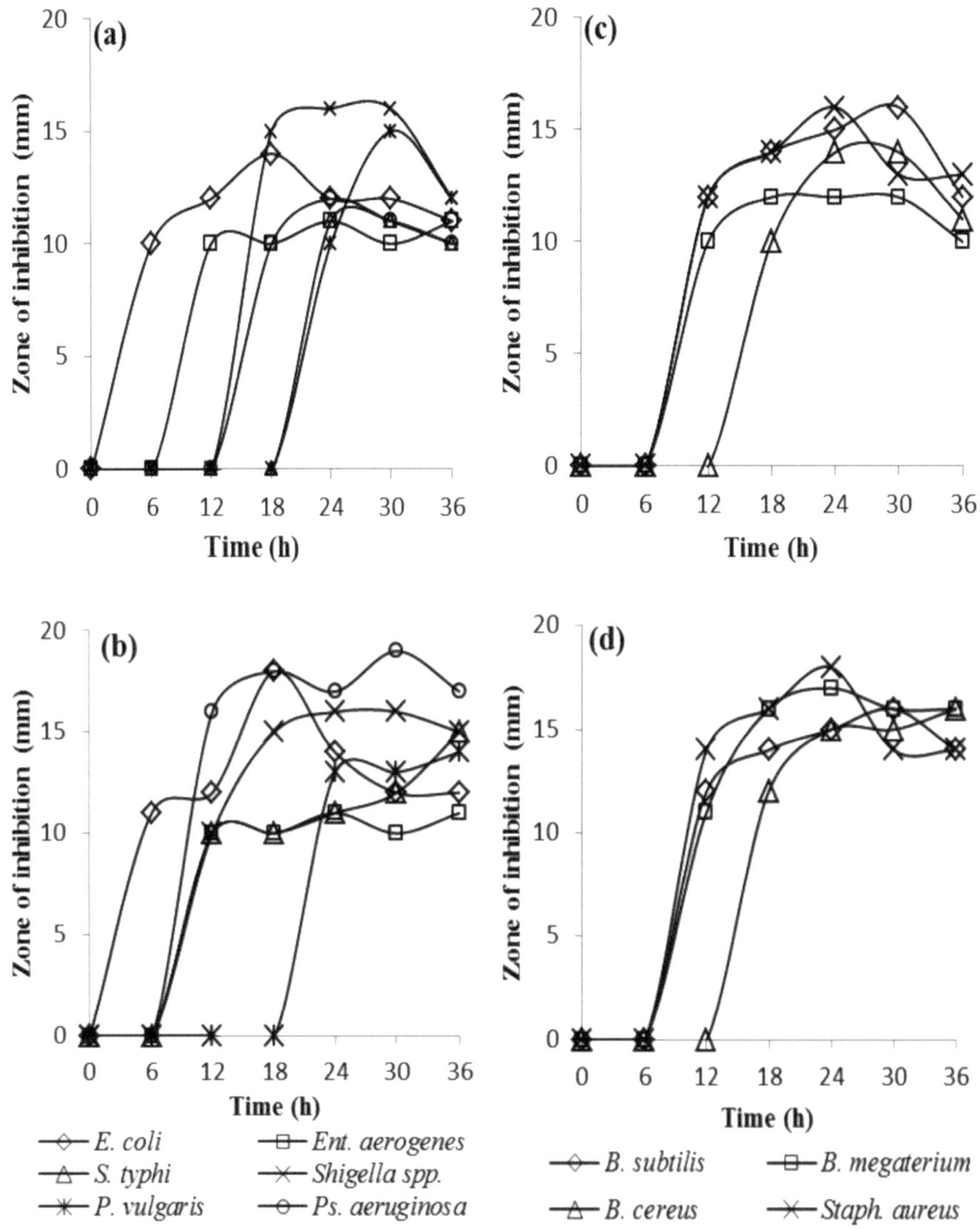

Fig. 1. Association of antimicrobial spectrum of EPC produced *by L. rhamnosus* Fb culture during logarithmic and idiophasic growth. (a) & (c) Antimicrobial activity of Extracellular Protein Concentrate (EPC) and (b & d) Cell Free Culture Filtrate (CFC) of *Lactobacillus rhamnosus* produced at the different growth phases (6, 12, 18, 24, 30 and 36 h) determined by well-diffusion assay against *E. coli, Ent. aerogenes, S. typhi, Shigella sp., P. vulgaris, Ps. aeruginosa, B. subtilis, B. megaterium, B. cereus* and *Staph. aureus*

antimicrobial peptides but the work is in progress to further characterize the structure and functions of EPC. Sensitivity of EPC to proteolytic enzymes like Proteinase K, Trypsin and Pepsin shows strain specificity, it varies with test organism which also further indicates the proteinaceous nature of the active agent. EPC treated with Proteinase K cause reduction in antimicrobial activity against *B. megaterium* (80%), *Staph. aureus* (79%), *B. cereus* (71%), *P. vulgaris* (60%), *Ser. marcescens* (53%), *S. typhi* (46%), *E. coli* (41%), *Ent. aerogenes, Shigella sp.* and *Ps. aeruginosa* marginally (<10%). Proteinase K exhibits broad substrate specificity. Proteinase K degrades many proteins in the native state even in the presence of detergent. The predominant site of cleavage is the peptide bond adjacent to the carboxyl group of aliphatic and aromatic amino acids which block alpha amino group. It is commonly used for its broad specificity (Ebeling *et al.,* 1974). Trypsin cleaves peptide chains mainly at the carboxyl site of the amino acids lysine or arginine, except when either is followed by proline. Antimicrobial activity of EPC treated with trypsin was completely lost against *Ps. aeruginosa, Staph. aureus* and *Bacillus spp.* indicates the presence of the active site for antimicrobial action may be present at the carboxyl site of protein. Pepsin is most efficient in cleaving peptide bonds between hydrophobic and preferably aromatic amino acids such as phenylalanine, tryptophan, and tyrosine (Dunn, 2001). EPC treated with pepsin results in complete loss of antimicrobial activity against *S. typhi* and *Staph. aureus* while partially reduced against *Bacillus spp.* Variable sensitivity of antimicrobial activity of EPC against test organisms imply the presence of a more than one antimicrobial peptides active against different test organisms. Gel permeation chromatography (Sephadex G-25) of EPC provides additional evidence of inhibitory protein is low molecular protein. Gel electrophoresis (Tricine SDS PAGE) of EPC shows that proteins present in EPC resolved into three bands, one diffuse band representing low molecular weight peptides (4 kDa) and the other marking the presence of higher molecular weight proteins (Schagger & Von Jagow, 1987). Gel over-layer confirmed that inhibition of the test organisms was due to the diffuse band of low molecular weight protein(s). EPC is a dynamic mixture of antimicrobial peptides, since the antimicrobial spectrum of the EPC is intimately associated with the growth phase. The following experimental evidences related to heat stability, sensitivity to proteolytic enzymes, and gel permeation chromatography further implicate the presence and involvement of more than one antimicrobial peptides in the EPC. Purification and characterization of antimicrobial peptides of *Lactobacillus rhamnosus* Fb strain provides novel approach as anti-infective drug, as it shows wide spectrum of antimicrobial action against human pathogens and food spoilage organisms. It has also potential for food additives, treatment of antibiotic resistant organisms. Probiotic formulation derived from this culture can be used to treat gastrointestinal problems including various forms of dysbacteriosis. However further studies are necessary to investigate the possibility of using this novel antimicrobial peptides as an anti-infective agent and *in vivo* study.

4.4 Present approaches and future prospects for bacteriocins in food application

The increasing demand for high-quality 'safe' foods that are not extensively processed has created a niche for natural food preservatives. The ideal natural food preservative should fulfil the following criteria (Hill *et al.,* 2002), acceptably low toxicity, stability to processing and storage, efficacy at low concentration, economic viability, no medical use, and no deleterious effect on the food. While most bacteriocins fulfil all these criteria, to date nisin is the only bacteriocin to be commercially exploited on a large scale, having gained Food and

Drug Administration (FDA) approval in the USA in 1988, although it had been in use in Europe for some time (the WHO approved the use of nisin in 1969). Its success has stimulated further research targeted towards identifying new bacteriocins from LAB which potentially could be used in a similar manner. Many bacteriocins have now been characterized that exhibit antibacterial activity against a range of pathogenic and food spoilage bacteria. It is to be expected that bacteriocins and bacteriocin-producing LAB (used as starters or protective cultures) will find many roles in both fermented and nonfermented foods as a means of improving food quality, naturalness and safety. Three approaches are commonly used in the application of bacteriocins for biopreservation of foods (Schillinger *et al.*, 1996). Inoculation of food with LAB that produce bacteriocin in the products. The ability of the LAB to grow and produce bacteriocin in the products is crucial for its successful use, Addition of purified or partially-purified bacteriocins as food preservatives, use of a product previously fermented with a bacteriocin producing strain as an ingredient in food processing.

4.5 Antimutagenic properties

Humans are continually exposed to a variety of natural and artificial mutagens generated by industrial and environmental activities (Vorobjeva *et al.*, 2002). One of the possible ways of the lowering of mutation pressure on animals and human is the increasing antimutagens levels and of the antimutagenic activity of bacteria, predominantly those inhabiting the intestine of mammals being the ingredients of probiotic, used in food processing and ensilage. Probiotic organisms are reported to bind mutagens to the cell surface (Orrhage *et al.*, 1994). Probiotic *L. rhamnosus* 231 cells has ability to bind, biotransform and detoxify different mutagens like acridine orange (AO), *N*-methyl-*N'*-nitro-*N*-nitrosoguanidine (MNNG), 2-amino-3,8-dimethylimidazo-[4,5-*f*]-quinoxaline (MeIQx) (Ambalam *et al.*, 2011). Binding of AO by Lr 231 is due to adsorption, thereby leading to removal of mutagen in solution and is instantaneous, pH- and concentration-dependent. Whereas, binding of MNNG and MeIQx by Lr 231 results into biotransformation leading to detoxification with subsequent loss of mutagenicity as determined by spectral analysis, thin layer chromatography and Ames test. Lr 231 exhibits ability to bind and detoxify potent mutagens, and this property can be useful in formulating fermented foods for removal of potent mutagens. Similar results were also observed by Sreekumar & Hosono (1998) instantaneous binding of the mutagen Trp-P-1 by *Lactobacillus gasseri*. Lankaputhra and Shah (1998) studied the antimutagenic activity of organic acids produced by probiotic bacteria against several mutagens and promutagens. In their study, butyric acid showed a broad-spectrum antimutagenic activity against all mutagens or promutagens studied and live bacterial cells showed higher antimutagenicity than killed cells. Probiotic bacteria are reported to reduce faecal enzymatic activities including β-glucuronidase, azoreductase, and nitroreductase, which are involved in activation of mutagens (Goldin & Gorbach, 1984).

4.6 Anticarcinogenic properties

Bacteria and metabolic products such as genotoxic compounds (nitrosamine, heterocyclic amines, phenolic compounds, and ammonia) are responsible for colorectal cancer. The consumption of cooked red meat, especially barbequed meat, and low consumption of fibre are reported to play a major role in causing colorectal cancer. The colonic flora is also reported to cause carcinogenesis mediated by microbial enzymes such as β-glucuronidase, azoreductase, and nitroreductase, which convert procarcinogens into carcinogens. Certain strains of *L. acidophilus* and *Bifidobacterium* spp. are reported to decrease the levels of enzymes such as β-glucuronidase, azoreductase, and nitroreductase responsible for activation of procarcinogens and consequently decrease the risk of tumour development (Yoon *et al.*, 2000). Short chain fatty acids produced by *L. acidophilus* and *Bifidobacterium, L. plantarum* and *L. rhamnosus* are reported to inhibit the generation of carcinogenic products by reducing enzyme activities (Cenci *et al.*, 2002). The anticarcinogenic effect of probiotic bacteria reported is due to the result of removal of sources of procarcinogens (or the enzymes that lead to their formation) improvement in the balance of intestinal microflora, normalized intestinal permeability (leading to prevention or delaying of toxin absorption), strengthening of intestinal barrier mechanisms, and activation of non-specific cellular factors (such as macrophages and natural killer cells) via regulation of γ-interferon production. Orally administered *Bifidobacterium* is also reported to play a role in increasing production of IgA antibodies and functions of Peyer's patch cells (Singh *et al.*, 1997).

4.7 Reduction in serum cholesterol

The level of serum cholesterol is a major factor for coronary heart disease, and elevated levels of serum cholesterol, particularly LDL-cholesterol, have been linked to an increased risk (Liong & Shah, 2006). There is a high correlation between dietary saturated fat or cholesterol intake and serum cholesterol level. Feeding of fermented milk containing very large numbers of probiotic bacteria to hypercholesterolaemic human subjects has resulted in lowering cholesterol from 3.0 to 1.5 g/L. Probiotic bacteria are reported to de-conjugate bile salts: deconjugated bile acid does not absorb lipid as readily as its conjugated counterpart, leading to a reduction in cholesterol level. *L. acidophilus* is also reported to take up cholesterol during growth and this makes it unavailable for absorption into the blood stream (Klaver & Meer, 1993).

5. Conclusions

Lactobacillus strains play an important role in food fermentation processes. Modern concepts or perspectives of the application of *Lactobacillus* strains include the following selections; the best adapted and safe for human application as it is an important biodefense factor in human intestinal tract, non-pathogenic, with probiotic effects and/or health-promoting effects and food protective activities. From our study we conclude that *L. rhamnosus* is a potential candidate for probiotic product preparation or as food additives as evidenced by its ability to acid-bile tolerance, salt, antimicrobial activity against human pathogens and food spoilage organisms. The antimicrobial extracellular low molecular weight proteins are produced during exponential and stationary phases. Changes in the antimicrobial activity spectrum of the EPC during different phases of growth provide evidence that the EPC is a mixture of antimicrobial peptides and its composition changes with the culture age. Strain-specific thermostability and sensitivity to proteolytic enzymes of antimicrobial peptides provide further evidence that the antimicrobial activity of EPC is due to mixture of peptides that are heat sensitive and/or resistant. Thermostability of antimicrobial peptides confers additional advantage which can survive the thermal processing cycle of foods and can also work over a broad pH range and could therefore be used in acidic food condition. These broad antimicrobial spectra produced by *L. rhamnosus* are potentially valuable in topical treatment, bio-control, food additives, and other applications that aim at eradicating gram-positive and gram-negative pathogens or non-pathogenic contaminants in the targeted environment. Further work on the purification, characterization of these antimicrobial peptides and mode of action is in progress.

6. Acknowledgements

Meritorious fellowship of University Grants Commission to Sheetal Pithva is gratefully acknowledged

7. References

Abee, T.; Krockel, L. & Hill, C. (1995). Bacteriocins: mode of action and potential in food preservation and control of food poisoning. *International Journal of Food Microbiology*, vol no. 28, pp.169-185

Adams, M.R. (1999). Safety of industrial lactic acid bacteria. *Journal of Biotechnology, vol no.* 68, pp. 171-178

Ambalam, P.; Prajapati, J.B.; Dave, J.M.; Nair, B.M.; Ljungh, A. & Vyas, B.R.M. (2009). Isolation and Characterization of Antimicrobial Proteins Produced by a Potential Probiotics strain of Human *Lactobacillus rhamnosus* 231 and Its Effect on Selected Human Pathogens and Food Spoilage Organisms. *Microbial Ecology in Health and Disease,* vol no. 21, pp. 211-220

Ambalam, P; Dave J. M.; Nair, B.M. & Vyas B.R.M. (2011*). In vitro* Mutagen Binding and Antimutagenic Activity of Human *Lactobacillus rhamnosus* 231. *Anaerobe,* vol no. 17, pp. 217-222

Barefoot, S.F & Klaenhammer T.R. (1984). Purification and characterization of the *L. acidophilus* bacteriocin lactacin B. *Antimicrobial Agents & Chemotherapy,* vol no. 26, pp. 328-334

Begley, M.; Gahan, C. G. & Hill, C. (2005). The interaction between bacteria and bile. *FEMS Microbiology Review,* vol no. 4, pp. 625-51

Cenci, G.; Rossi, J.; Throtta, F. & Caldini, G. (2002). Lactic acid bacteria isolated from dairy products inhibit genotoxic effect of 4-nitroquinoline-1-oxide in SOS-chromotest. *Systematic and Applied Microbiology,* vol no. 25, pp. 483–490

Champagne, C.P. & Gardner, N.J. (2005). Challenges in the addition of probiotic cultures to foods. *Critical Reviews in Food Science and Nutrition,* vol no. 45, pp. 61-84

Charteris, P. W.; Kelly, M. P.; Morelli, L. & Collins, K. J. (1998). Development and application of an *in vitro* methodology to determine the transit tolerance of potentially probiotic *Lactobacillus* and *bifidobacterium* species in the upper human gastrointestinal tract. *Journal of Applied Microbiology,* vol no. 84, pp. 759-768

Cintas, L.M.; Rodriguez, J.M.; Fernandez, M.F.; Sletten, K.; Nes, I. F.; Hernandez, P.E. & Holo, H., (1995). Isolation and characterization of pediocin L50, a new bacteriocin from *Pediococcus acidiliactici* with a broad inhibitory spectrum. *Applied & Environmental Microbiology,* vol no. 61, pp. 2643-2648

Contor, L. (2001). Functional food science in Europe. *Nutritional Metabolites and Cardiovascular Diseases,* vol no. 11, pp. 20-23

De Man, J.C.; Rogosa, M. & Sharpe, M.E. (1960). A medium for the cultivation of Lactobacilli. *Journal of Applied Bacteriology,* vol no. 23,pp. 130–5

De Vuyst, L. & Vandamme, E. J. (1994). Antimicrobial potential of lactic acid bacteria. In bacteriocins of Lactic acid bacteria, Microbiology, Genetics and Applications (ed.) De Vuyst, L., and Vandamme, E. J. pp. 99-141. Glasgow, UK: Blackie Academic and Professional

Dunn B. M. (2001). "Overview of pepsin-like aspartic peptidases". *Current Protocols in Protein Science,* Chapter 21, Unit 21.3.

Ebeling, W., *et al.* (1974). Proteinase K from *Tritirachium album* Linder. *European Journal of Biochemistry,* vol no. 47, pp. 91

Eijsink, V.; Skeie, M.; Middelhoven, P.; Brurberg, M. & Nes, I. (1998). Comparative studies of class IIa bacteriocins of lactic acid bacteria. *Applied & Environmental Microbiology,* vol no. 64, pp. 3275–3281

Fooks, L.J. & Gibson, G.R. (2002). Probiotics as modulators of the gut flora. *British Journal of Nutrition,* vol no. 88, S39-S49

Fuller, R. J. (1989). Probiotics in man and animals. *Applied Biotechnology,* vol no. 66, pp. 365-378

Garrity, G.M; Bell, J.A. & Lilburn, T.G. (2004). Taxonomic outline of Prokaryotes. Bergey's Manual of Systematic Bacteriology, 2nd edition Release 5, Springer-verlag New York: 1-399

Gibson, G.R. & Roberfroid, M.B. (1995). Dietary modulation of the human colonic microbiota: introducing the concept of prebiotics. *Journal of Nutrition,* vol no. 125, pp. 1402-1412

Gilliland, S.E.; Stanlely, E. T. & Bush, J. L. (1984). Importance of bile tolerance of *Lactobacillus acidophilus* used as a dietary adjunct. *Journal of Dairy Science,* vol no. 67, pp. 3045-3051

Gmeiner, M.; Kneifel, W.; Kulbe, K.D.; Wouters, R.; De Boever, P.; Nollet, L. & Verstraete, W. (2000). Influence of a synbiotic mixture consisting of *Lactobacillus acidophilus* 74-2 and a fructooligosaccharide preparation on the microbial ecology sustained in a simulation of the human intestinal microbial ecosystem. *Applied Microbiology and Biotechnology,* vol no. 53, pp. 219-223

Goldin, B. R. & Gorbach, S. L. (1984). The effect of milk and *Lactobacillus* feeding on human intestinal bacterial enzyme activity. *American Journal of Clinical Nutrition,* vol no. 39, 756–761

Guyonnet, D.; Fremaux, C.; Cenatiempo, Y. & Berjeaut, J. (2000). Method for rapid purification of class IIa bacteriocins and comparison of their activities. *Applied & Environmental Microbiology,* vol no. 66, pp. 1744–1748

Hill, C.; Keeffe, T. & Ross, R.P. (2002). Bacteriocins-antimicrobial agents with potential in food preservation and food safety. In: *Encyclopaedia of Dairy Science.* Academic Press, London, UK

Holzapfel, W. H.; Geisen, R. & Schillinger, U. (1995). Biological preservation of foods with reference to protective cultures, bacteriocins and food grade enzymes. *International Journal of Food Microbiology,* vol no. 24, pp. 343-362

Jacobsen, C.N.; Rosenfeldt Nielsen, V.; Hayford, A.E.; Moller, P.L.; Michaelsen, K.F.; Paerrgaard, A. et al. (1999) Screening of probiotic activities of 47 strains of *Lactobacillus* spp. by *in vitro* techniques and evaluation of the colonization ability of

5 selected strains in humans. *Applied & Environmental Microbiology*, vol no. 65, pp. 4949–56

Kao, C. T. & Frazier, W. C. (1966). Effect of lactic acid bacteria on growth of *Staphylococcus aureus*. *Applied & Environmental Microbiology*, vol no. 14, pp. 251-255

Khedekar, C.D. (1988). Characterization of human strains of *Lactobacillus acidophilus* isolates for their suitability in preparation of milk beverage and their antibacterial cum therapeutic ability. MSc (Dairy Microbiology) thesis, Gujarat Agriculture University, Anand

Klaenhammer, T. R. (1988). Bacteriocins of lactic acid bacteria. *Biochemie*, vol no. 70, pp. 337-349

Klaenhammer, T. R. (1993) Genetics of bacteriocins produced by lactic acid bacteria. *FEMS Microbiology Reviews*, vol no. 12, pp. 39-85

Klaver, F. M. & Meer, R. V. D. (1993). The assumed assimilation of cholesterol by lactobacilli and Bifidobacterium bifidum is due to their bile salt deconjugating activity. *Applied & Environmental Microbiology*, vol no. 59, pp. 1120–1124

Kurmann, J. A. & Rasic, J. L. (1991). The health potential of products containing bifidobacteria. In R. K. Robinson (Ed.), Therapeutic properties of fermented milks (pp. 117–157). London, UK: Elsevier Applied Science Publishers

Lankaputhra, W. E. V. & Shah, N. P. (1998). Antimutagenic properties of probiotic bacteria and of organic acids. *Mutation Research*, vol no. 39,pp. 169–182

Lilly, D.M. & Stillwell, R.H. (1965). Probiotics, growth promoting factors produced by microorganisms. *Science*, vol no. 147, pp. 747-748

Liong, M. T. & Shah, N. P. (2006). Effects of *Lactobacillus casei* synbiotic on serum lipoprotein, intestinal microflora and organic acids in rats. *Journal of Dairy Science*, vol no. 89, pp. 1390–1399

Metchnikoff, I.I. (1907). The prolongation of life: Optimistic studies. Springer Published Comp. New York

Molin, G. (2001). Probiotics in foods not containing milk or milk constituents, with special reference to *Lactobacillus plantarum* 299v. *American Journal of Clinical Nutrition*, vol no. 73, pp. 380S-385S

Montville, T. J. & Winkowski, K. (1997). Biologically based preservation systems and probiotic bacteria. In: Doyle, M. P., Beuchat, L. R., Montville, T. J. (Eds.), Food Microbiology. Fundamentals and Frontiers. ASM press, USA, pp. 557-576.

Nes, I.F.; Bao Diep, D.; Havarstein, L.S.; Brurberg, M.B.; Eijsink, V. & Holo, H. (1996). Biosynthesis of bacteriocins in lactic acid bacteria. *Antonie Van Leeuwenhoek*, vol no. 70, pp. 113-128

Orla-Jensen, S. (1919). The Lactic Acid Bacteria. Copenhagen: Andr. Fred Host and Son. Mem. Acad. Roy. Sci., *Danemark, Sect. Sci.* 8, 5, 81-197

Orla-Jensen, S.(1943). The lactic acid bacteria. Copenhagen: Munksgaard. Ergaenzungsband. Mem. Acad. Roy. Sci. *Danemark, Sect. Sci. Biol.* Vol no. 2, 3, pp. 1-145

Orrhage, K.; Sillerstro¨ m, E.; Gustafsson, J. A.; Nord, C. E. & Rafter, J. (1994). Binding of mutagenic heterocyclic amines by intestinal and lactic acid bacteria. *Mutation Research*, vol no. 311, pp. 239–248

Parker, R.B. (1974). Probiotics, the other half of the antibiotic story. *Animal Nutrition and Health*, vol no. 29, pp. 4-8

Papagianni, M. (2003). Ribosomally synthesized peptides with antimicrobial properties: biosynthesis structure, function and application. *Biotechnology Advances*, vol no. 21, pp. 465-499

Saarela, M.; Mogensen, G.; Fonden, R.; Matto, J. & Mattila-Sandholm, T. (2000). Probiotic bacteria: safety, functional and technological properties. *Journal of Biotechnology*, vol no. 84, pp. 197-215

Salminen, S.; Deighton, M.A.; Benno, Y. & Gorbach S.L. (1998). Lactic acid bacteria in health and disease. In: Salminen S, Von Wright A, eds. Lactic acid bacteria: microbiology and functional aspects, 2nd ed. New York: Marcel Dekker Inc, 211-253

Schaafsma, G. (1996). State of the art concerning probiotic strains in milk products. *IDF Nutritional News Letters*, vol no. 5, 23-24

Schagger, H.; Von Jagow, G. (1987). Tricine-SDS-PAGE for separation of 1-100 kDa proteins. *Analytical Biochemistry*, vol no. 166, pp. 368–79

Schillinger, U.; Schillinger G. R. & Holzapfel, W.H. (1996). Potential of antagonistic microorganisms and bacteriocins for the biological preservation of foods. *Trends in Food Science and Technology*, vol no. 7, pp.158-164

Schrezenmeir, J. & de Vrese, M. (2001). Probiotic, prebiotic and symbiotic-approaching a definition. *American Journal of Clinical Nutrition*, vol no. 73, pp.S361-S364

Servin, A. L. (2004). Antagonistic activities of lactobacilli and Bifidobacteria against microbial pathogens. *FEMS microbiology Reviews*, vol no. 28, pp. 405-440

Shah, N. P. (1999). Probiotic bacteria: Antimicrobial and antimutagenic properties. *Probiotica*, vol no. 6, pp. 1–3

Shah, N. P. (2001). Functional foods, probiotics and prebiotics. *Food Technology*, vol no. 55, pp. 46–53

Shah, N. P. (2006). Health benefits of yogurt and fermented milks. In R. C. Chandan (Ed.), Manufacturing yogurt and fermented milks (pp. 327–340). Iowa, USA: Blackwell Publishing Professional

Shah, N. P. (2007). Functional cultures and health benefits. *International Dairy Journal*, vol no. 17,pp. 1262–1277

Singh, J.; Rivenson, A.;Tomita, M.; Shimamura, S.; Ishibashi, N. & Reddy, B. S. (1997). *Bifidobacterium longum* and *Lactobacillus acidophilus* producing intestinal bacteria inhibits colon cancer and modulates the intermediate biomarkers of colon carcinogenesis. *Carcinogenesis*, vol no. 18, pp. 833–841

Sreekumar, O. & Hosono, A. (1998). Antimutagenicity and the influence of physical factors in binding *Lactobacillus gasseri* and *Bifidobacterium longum* cells to amino acid pyrolysates. *Journal of Dairy Science*, vol no.81, pp. 1508–1516

Stanton, C.; Paul Ros, R.; Fitzgerald, G.F. & Sindern, D.V. (2005). Fermented functional foods based on probiotics and their biogenic metabolites. *Current Opinion in Biotechnology*, vol no. 16, pp. 198-203

Tannock, G. W. (2004). A Special Fondness for Lactobacilli. *Applied & Environmetal Microbiology*, Vol. 70, No. 6 pp. 3189–3194

Van der Hammer, E. (1960). The carbohydrate metabolism of the lactic acid bacteria. Thesis, University of Utrecht, The Netherlands

Vandamme, P.; Pot, B.; Gillis, M.; de Vos, P.; Kersters, K. & Swings, J. (1996). Polyphasis taxonomy, a consensus approach to bacterial systematics. *Microbiological Reviews*, vol no. 60, pp. 407-438

Vorobjeva, L. I. & Abilev, S. K. (2002). Antimutagenic Properties of Bacteria: Review. *Applied Biochemistry and Microbiology,* Vol. 38, No. 2, pp. 97–107

WHO (World Health Organization) (1969). Specification for identity and purity of some antibiotics. *World Health Organization/Food Add,* 69, 34, 53-67

U.S. Food & Drug Administration (1988). Nisin preparation: Affirmation of GRAS status as direct human food ingredient. *Federal Register,53,* April 6

Yoon, H.; Benamouzig, R.; Little, J.; Francois-Collange, M. & Tome, D. (2000). Systematic review of epidemiological studies on meat, dairy products, and egg consumption and risk of colorectal adenomas. *European Journal of Cancer Prevention,* vol. 9, No. 3, pp. 151–164

11

Earth Food *Spirulina* (*Arthrospira*): Production and Quality Standarts

Edis Koru
Ege University Fisheries Faculty,
Dept. of Aquaculture Algae Culture Lab.
Bornova, Izmir
Turkey

1. Introduction

According to researchers, in the first photosynthetic life forms known on earth 3.6 billion years ago, was created by the God. Blue-green algae, cyanobacteria, is the evolutionary bridge between bacteria and green plants. It contained within it everything life needed to evolve. This immortal plant has renewed itself for billions of years, and has presented itself to us in the last 40 years. *Spirulina* has 3.6 billion years of evolutionary wisdom coded in its DNA. *Spirulina,* or what was most likely *Arthrospira,* is a photosynthetic, filamentous, spiral-shaped, multicellular and blue-green microalga that has a long history of use as food. For this microorganism contain chlorophyll *a,* like higher plants, botanists classify it as a microalga belonging to *Cyanophyceae* class; but according to bacteriologists it is a bacterium due to its prokaryotic structure (Fig. 1). *Spirulina,* essentially an exceptional simple extract of blue-green algae, has been extensively studied and is now in widespread usage throughout the world as a food product and as a dietary supplement (Fox, 1996; Paleaz, 2006).

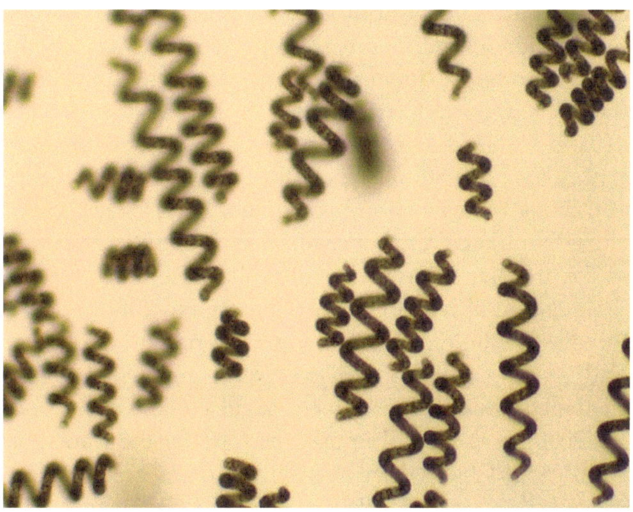

Fig. 1. Microscopic view of microalgae *Spirulina* (Cyanobacteria), Photo by E. Koru

The goals of this paper are: to review the present literature on the historical evolution and potential nutritional use of *Spirulina* and its products; to provide insight into the potential implications of the studies reviewed in the context of possible nutritional and therapeutic applications in health management, and; to identify areas of interest for future research.

2. Historical background on the use of *Spirulina*

2.1 Evolutionary history

The cyanobacteria are believed to have evolved 3.5 billion years ago. Fossils discovered in the 3.5-Ga-old Apex chert in northwestern Western Australia crude filamentous cyanobacteria with strikingly similar morphologies to present-day filamentous cyanobacteria (*Oscillatoriacea*). The occurrence of aerobic respiration and oxygenic photosynthesis, photosynthetic carbon dioxide fixation like that of extant cyanobacteria, cell division more similar to the extant cyanobacterial and recent rRNA analyses showing that the *Oscillatoriacea* are among the earliest evolved also lend further evidence to the fossil record. *Arthrospira* belongs to the Class *Oscillatoriacea* and therefore has a very old lineage. Despite their old lineage, the fossil cyanobacteria are morphologically very similar to their extant forms, suggesting a slow evolutionary process. *Spirulina* (*Arthrospira*) is a ubiquitous organism. After the first isolation by Turpin in 1827 from a freshwater stream, species of *Spirulina* have been found in a variety of environments: soil, sand, marshes, brackish water, seawater, and freshwater. Species of *Spirulina* have been isolated, for instance, from tropical waters to the North Sea, thermal springs, salt pans, warm waters from power plants, fish ponds, etc. Thus, the organism appears to be capable of adaptation to very different habitats and colonizes certain environments in which life for other microorganisms is, if not impossible, very difficult (Ciferri, 1983; Tomaselli, 2003).

2.2 Use of *Spirulina* as human food and animal feed

It is not known with accuracy when human began to use microalgae. The current use of these resources has three precedents: tradition, scientific and technological development, and the so-called, "green tendency". There are reports that it was used as food in Mexico during the Aztec civilization approximate 400 years ago. Bernal Díaz del Castillo, a member of Hernán Cortez´s troops, reported in 1521, that *S. maxima* was harvested from the Lake Texcoco, dried and sold for human consumption in a Tenochtitlán (today Mexico City) market (Sánchez, et al., 2003). In 1940, a French phycologist Dangeard published a report on the consumption of dihé by the Kanembu people near Lake Chad. Scientist Dangeard, also noted these same algae populated a number of lakes in the Rift Valley of East Africa, and was the main food for the flamingos living around those lakes. Twenty-five years later during 1964-65, a botanist on a Belgian Trans-Saharan expedition, Jean Léonard, reported finding a curious greenish, edible cakes being sold in native markets of Fort-Lamy (now N'Djamena) in Chad. When locals said these cakes came from areas near Lake Chad, Léonard recognized the connection between the algal blooms and dried cakes sold in the market (Habib et al., 2008). It is still being used as food by the Kanembu tribe in the Lake Chad area of the Republic of Chad where it is sold as dried bread called "*dihe*" (Fox, 1996; Belay, 2002). In 1967 *Spirulina* was established as a "wonderful future food source" in the International Association of Applied Microbiology (Sasson, 1997). This traditional food

(Fig. 2), by a European scientific mission, and is now widely cultured throughout the world. In its commercial use, the common name, *Spirulina*, refers to the dried biomass of the cyanobacterium, *Arthrospira platensis*, and is a whole product of biological origin. *A. platensis* and *A. maxima*, that are commonly used as food, dietary supplement, and feed supplement (Wikfors & Ohno, 2001). The re-introduction of *Spirulina* as a health food for human consumption in the late 1970s and the beginning of the 1980s was associated with many controversial claims which attribute to *Spirulina* a role of a 'magic agent' that could do almost everything, from curing specific cancer to antibiotic and antiviral activity. Since most claims were never backed up by detailed scientific and medical research, they will not be discussed in this chapter. Nevertheless, one cannot ignore the fact that more than 70 per cent of the current *Spirulina* market is for human consumption, mainly as health food. Primarily concern in *Spirulina* focused mainly on its rich content of protein, essential amino acids, minerals, vitamins, and essential fatty acids. *Spirulina* is 60-70% protein by weight and contains a rich source of vitamins, for example vitamin B12 and provitamin A (β-carotene), and minerals, especially iron. One of the few sources of dietary γ-linolenic acid (GLA), it also contains a host of other phytochemicals that have potential health benefits. This first data was enough to launch many research projects for industrial purposes in the 1970s, because micro-organisms (*Chlorella*, *Spirulina*, yeast, some bacteria and moulds) seemed at that time to be the most direct route to inexpensive proteins - the iconic "single cell proteins". In 1970's, a request was received from a company named Sosa-Texcoco Ltd by the "Institut français du pétrole" to study a bloom of algae occurring in the evaporation ponds of their sodium bicarbonate production facility in a lake near Mexico City. As a result, the first systematic and detailed study of the growth requirements and physiology of *Spirulina* was performed. This study, which was a part of Ph.D. thesis by Zarrouk (1966), was the basis for establishing the first large-scale production plant of *Spirulina* (Sasson, 1997). *Arthrospira* has been produced commercially for the last 30 years for food and specialty feeds. Commercial algae like *Spirulina*, are normally produced in large outdoor ponds under controlled conditions (Fig. 3). Some companies also produce directly from lakes. Current production of *Spirulina* worldwide is estimated to be about 3,000 metric tons.

The incidence of toxic episodes in freshwaters due to the presence of toxic cyanobacteria has considerably increased over the last few years. There is also increasing evidence that low levels of exposure may have chronic effects in humans and therefore strict and reliable control of these toxins should be carried out in order to prevent serious public health problems. It is known that cyanobacteria have been implicated in a number of episodes of human illnesses worldwide (Martinez, 2007; Šejnohová, 2008). For this reason toxicological tests obligation done so the *Spirulina* culture and biomass. There is no report of cyanobacterial toxins in *Arthrospira* species to date. Although inadvertent harvest of these toxic species is a risk when harvesting algae from natural bodies of water with mixed phytoplankton populations, it is very unlikely to be a problem in properly controlled and properly managed monoculture of *Arthrospira* (Belay, 2008). So, sold widely in health food stores and mass-market outlets throughout the world, *Spirulina*'s safety as food has been established through centuries of human use and through numerous and rigorous toxicological studies (Belay, 1997; Cevallos, et al., 2008).

The nutritive value of *Arthrospira* is amplified in that it has a relatively low percentage of nucleic acids (~4%) as comperad with the high content of nucleic acids in bacteria. Also, it is

Fig. 2. Harvesting, traditionally drying *Spirulina* dihé in a sand filter, and selling local market (www.AlgaeIndustryMagazine.com, Algae in Historical Legends March 10, 2011, by Robert Henrikson)

Fig. 3. *Spirulina* production ponds under controlled conditions (İzmir-Aydın/Turkey), Photo by Edis Koru

extremely high in vitamin B_{12}, the mucoprotein cell walls are easy to digest, unlike the cellulose cell wall found in many other nutritional algae, it is completely non-toxic, and its lipids are made up of unsaturated fatty acids that do not form chlesterol. This makes *Spirulina* a potential food item for persons suffering from coronary illness and obesity (Richmond, 1992). In addition to being exeptionally high in protein, *Arthrospira* appears to have the highest vitamin B_{12} content of any unprocessed plant or animal food, representing a boon to vegetarian diets. Two or three heaped tablespoons of *Spirulina* (~20-25 g), provide all the daily body requirements of vitamin B_{12}, as well as significant quantities of other B-complex vitamins, including 70% of the recommended daily allowance for vitamin B_1 (thiamine), 50% for B_2 (riboflavin), and 12% for B_3 (niacin). Also, 10 g. of *Spirulina* contains about 25000 international units of vitamin A, representing over 500% of the recommended daily allowance. Other nutritional attributes of *Spirulina* include essential unsaturated fatty acids, the most important of which is γ-linolenic acid (6,9,12-octadecatrienoic-acid) of which it has a high content (Becker, 1992, Richmond, 1992). Also, researchers have reported the therapeutic effects of *Spirulina* as a growth promoter, probiotic, and booster of the immune system in animals including fishes (Venkataraman, 1993; James et al., 2006).

Some of the best worldwide known *Spirulina* producing companies are: Earthrise Farms (USA), Cyanotech (USA), Hainan DIC Microalgae Co., Ltd (China), Marugappa Chettir Research Center (India), Genix (Cuba) and Solarium Biotechnology (Chile) (Belay, 1997).

Nowadays, *Spirulina* has been marketed and consumed as a human food and has been approved as a food for human consumption by many governments, health agencies and associations of these countries: Argentina, Australia, Austria, Bahrain, Bahamas, Bangladesh, Belarus, Belgium, Brazil, Bulgaria, Canada, Chad, Chile, China, Colombia, Costa Rica, Croatia, Czech Republic, Denmark, Ecuador, Egypt, Ethiopia, Finland, France, Germany, Greece, Guam, Gulf States Haiti, Hungary, India, Iceland, Indonesia, Ireland, Israel, Italy, Jamaica, Japan, Kenya, Korea, Kuwait, Liechtenstein, Luxembourg, Macedonia, Malaysia, Mexico, Myanmar, Monaco, Netherlands, New Zealand, Nigeria, Norway, Peru, Philippines, Poland, Portugal, Romania, Russia, Saudi Arabia, Serbia, Singapore, Slovenia, South Africa, Spain, Sweden, Switzerland, Taiwan, Thailand, Togo, Turkey, Ukraine, United Kingdom, United States, Venezuela, Vietnam, Zaire, Zimbabwe. (Becker & Venkataraman, 1984; Vonshak, 2002; Koru, 2009, Henrikson, 2010).

3. Biological properties of *Spirulina*

3.1 Morphology and taxonomy

Spirulina (*Arthrospira*) is symbiotic, multicellular and filamentous blue-green microalgae with symbiotic bacteria that fix nitrogen from air. *Spirulina* can be rod- or disk-shaped. Their

Fig. 4. Morphology of *Spirulina (Arthrospira)*, Scanning electron micrograph of a trichome of axenic *S. platensis*. (Ciferri, 1983), Photo by R. Locci

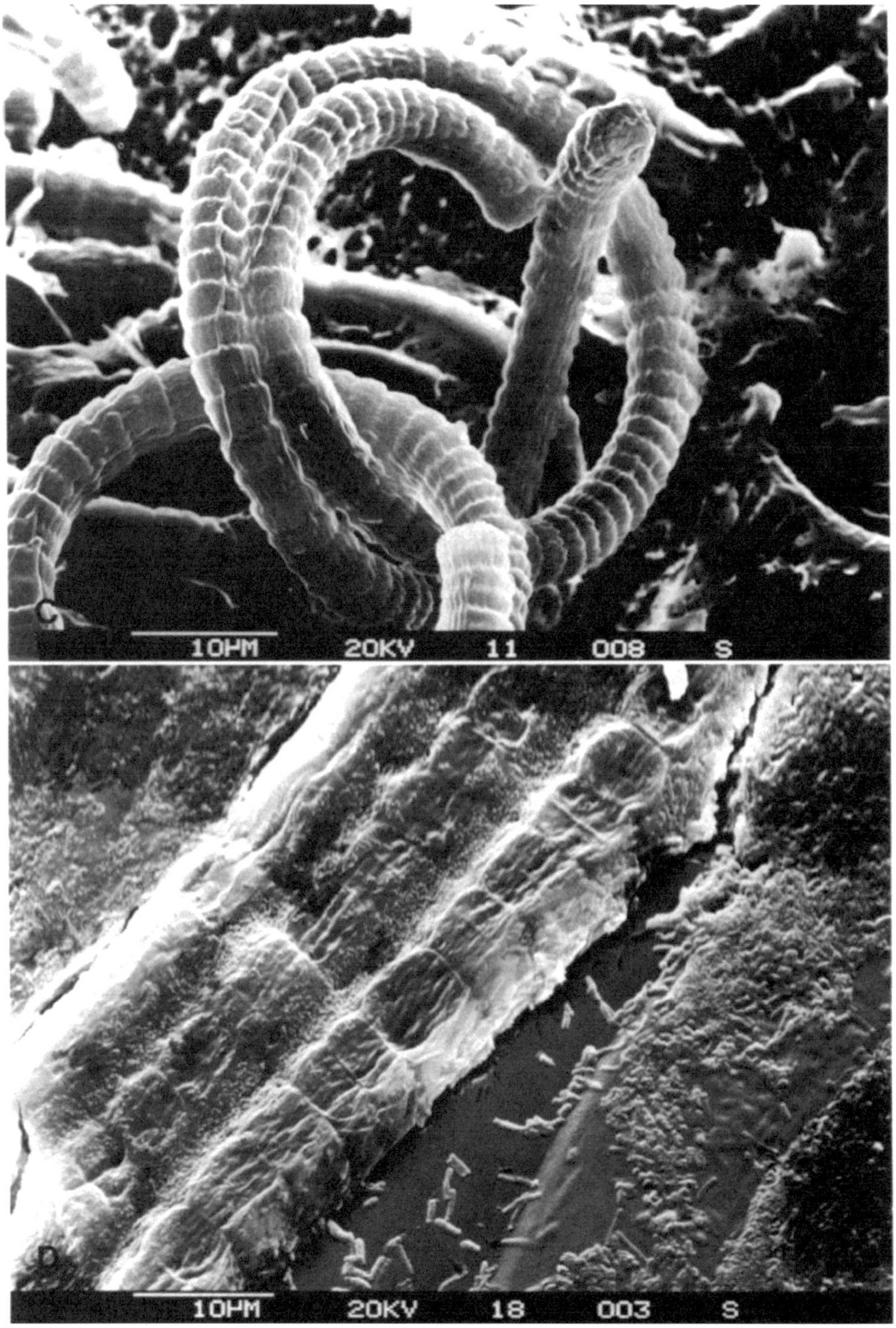

Fig. 5. Morphology of *Spirulina(Arthrospira)*, (C). Scanning electron micrograph of a portion of a trichome of axenic *S. platensis*; (D). Scanning electron micrograph of non axenic trichomes of *S. maxima*. (Ciferri, 1983), Photo by R. Locci

main photosynthetic pigment is phycocyanin, which is blue in colour. These bacteria also contain chlorophyll *a* and carotenoids. Some contain the pigment phycoythrin, giving the bacteria a red or pink colour. *Spirulina* are photosynthetic and therefore autotrophic. *Spirulina* reproduce by binary fission. *Arthrospira* (*Spirulina*) species show great plasticity in morphology. This is attributed to environmental factors like temperature and other physical and chemical factors and possibly also due to genetic change. Transmission Electron Microscope observations show for *Spirulina* prokaryotic organization, capsule, pluri-stratified cell wall, photosynthetic or thylakoid lamella system, ribosomes and fibrils of DNA region and numerous inclusions (Fig. 4,5). The capsule has fibrillar structure and covers each filament protecting it. The irregular presence of capsule around the filaments in *S. platensis* is a differentiating morphological characteristic to compare with *S. maxima* (Ciferri, 1983; Fox, 1996; Belay, 1997; Tomaselli, 2002).

3.2 Spirulina biochemical composition

Microalgae could be utilized for the production of several chemicals which are either unique to the algae or found at relatively high concentrations and command a high market value. In this respect, *Spirulina* is one of the more promising microalgae. It is especially rich, relative to other sources, in the polyunsaturated fatty acid γ- linolenic acid (GLA), and in pigments such as phycocyanin, myxoxanthophyl and zeaxanthin. The biochemical composition of *Spirulina* has been analyzed since 1970, showing high protein concentration, 60–70% of its dry weight, whose nutritive value is related to the quality of amino acid. *Spirulina* contains essential amino acids, including leucine, isoleucine and valine. It also contains a relative high concentration of provitamin A, vitamin B_{12} and β-carotene. *Spirulina* have 4–7% lipids, essential fatty acids as linolenic and γ-linolenic acid, and ω-3 and ω-6 polyunsaturated fatty acids (Wu, L.C. et al., 2005) (Table 1,2,3). *Cyanobacteria* and algae possess a wide range of colored compounds, including carotenoids, chlorophyll, and phycobiliproteins. C-phycocyanin is the principal phycobiliprotein. A selenium-containing phycocyanin has been isolated from *S. platensis* (Shekharam et al., 1987). *A. platensis* contains about 13.5% carbohydrates, the sugar composition is mainly composed of glucose, along with rhamnose, mannose, xylose, galactose, and two unusual sugars: 2-*O*-mehtyl-l-rhamnose and 3-*O*-methyl-l-rhamnose. Nowadays the antiviral activity of *Spirulina* has been attributed to three groups of substances: sulfated polysaccharides, sulfoglycolipids, and a protein-bound pigment, the allophycocianin (Barrón et al., 2008). *Spirulina* contains 2.2%-3.5% of RNA and 0.6 %-1% of DNA, which represents less than 5% of these acids, based on dry weight.

Food Type	Crude Protein (%)
Spirulina powder	65
Chicken Egg	47
Beer Yeast	45
Chicken meat	24
Skimmed Powdered milk	37
Cheese	36
Beef meat	22
Fish	22

Table 1. Quantity of *Spirulina* proteins and other foods (Henrikson, 1994)

Vitamins	mg 100/g
Provitamin A	2.330×10^3IU/kg
(β-carotene)	140
Vitamin E	100 a-tocopherol equiv.
Thiamin B_1	3.5
Riboflavin B_2	4.0
Niacin B_3	14.0
Vitamin B_6	0.8
Vitamin B_{12}	0.32
Biotin	0.005
Folic acid	0.01
Phantothenic acid	0.1
Vitamin K	2.2

Table 2. Vitamins in *Spirulina* powder (Belay, 1997)

Composition	Quantity (per 100 g dry wt)
General composition	
Moisture	3.5 g.
Protein	63.5 g.
Fat (Lipids)	9.5 g.
Fibre	3.00 g.
Ash	6.70 g.
N-free extract	15 g.
Colorants	
Phycocyanin	15.6 g.
Carotenoids	456.00 mg
Chlorophyll-*a*	1.30 g
Vitamins	
Provitamin A	213.00 mg
Thiamin (V.B_1)	1.92 mg
Riboflavin (V.B_2)	3.44 mg
Vitamin B_6	0.49 mg
Vitamin B_{12}	0.12 mg
Vitamin E	10.40 mg
Niacin	11.30 mg
Folic acid	40 μg
Panthothenic acid	0.94 mg
Inositol	76.00 mg
Minerals	
Phosphorus	916.00 mg
Iron	53.60 mg
Calcium	168 mg
Potassium	1.83 g
Sodium	1.09 g
Magnesium	250 mg

Table 3. A typical analysis of *Spirulina* product. Based on sample of dried *Spirulina* powder of Turkish Algae Company analyzed by Republic of Turkey Ministry of Agriculture and Rural Affairs Izmir Province Control Laboratory and Ege University (Koru et al., 2008)

Composition	**Fatty acids (%)**	
(C14) Myristic acid	0.23[a]	1.6-1.7[b]
(C16) Palmitic acid	46.07[a]	18-38[b]
(C16:1)Δ9 Palmitoleic acid	1.26[a]	5-7[b]
(C18:1)Δ9 Oleic acid	5.26[a]	5-6[b]
(C18:2)Δ9,12 Linoleic acid	17.43[a]	6-15[b]
(C18:3)Δ9,12,15 γ-Linolenic acid	8.87[a]	4 22[b]
Others	20.88[a]	10-68[b]

Table 4. Fatty acid composition of *Spirulina platensis* powder ([a] Othes and Pire, 2001; [b] Diraman et al., 2009)

3.3 General quality and safety assurance for *Spirulina*

In 1970's, *Spirulina* underwent extensive safety studies with animals and fish. Independent feeding tests in France, Mexico and Japan showed no undesirable results and no toxic side effects on humans, rats, pigs, chickens, fish and oysters. Many independent rat feeding trials were conducted in Japan and no negative effects at all were found for acute or chronic toxicity or reproduction. The re-introduction of *Spirulina* as a health food for human consumption in the late 1970s and the beginning of the 1980s was associated with many controversial claims which attribute to *Spirulina* a role of a 'magic agent' that could do almost everything, from curing specific cancer to antibiotic and antiviral activity. Since most

Spirulina food standards: quality requirements of France, Sweden, Japan and Earthrise Farms (USA)				
Standard	France[a]	Sweden[b]	Japan[c]	USA[d]
Protein	%55-65	%55-65	≥%50	%55-65
Total Carotenoids	♦	♦	>100 mg %	300 mg/100g
Chlorophyll-a	>500 mg %	♦	>500 mg %	900 mg/100g
Phycocyanin	♦	♦	>2000 mg %	8,000 mg/100g
Moisture	♦	♦	<7 %	<7 %
Standard Plate Count	<100,000/g	1,000,000/g	<200,000/g	<200,000/g
Mold	♦	<1000/g	♦	<100/g
Saccharomyces sp.	♦	♦	♦	< 40/g
Coliform bacteria	<10/g	<100/g	negative	negative
Staphylococcus aureus	<100/g	<100/g	♦	negative
Salmonella enteritidis	negative	negative	♦	negative
total heavy metals (Lead, Mercury, Cadmium, Arsenic)	♦	♦	< 20.0 ppm	2.1 ppm
Insect fragmen	♦	♦	♦	<30 pcs per 10 g
Rodent hair	♦	♦	♦	< 1.5 pcs per 150 g

[a] Superior Public Hygiene Council of France, 1984, 1986.
[b] Ministry of Health, Sweden.
[c] Japan Health Foods Association, auth. by Ministry of Health and Welfare.
[d] Earthrise Farms, 1995.
definitions: < = less than, ≥ more than, ♦ = no set Standard, /g = per gram, parts per million (ppm)

Table 5. General Quality and Safety Standards for *Spirulina* (Shimamatsu, 2004, Henrikson, 2010)

claims were never backed up by detailed scientific and medical research, they will not be discussed in this chapter. Nevertheless, one cannot ignore the fact that more than 70 per cent of the current *Spirulina* market is for human consumption, mainly as health food. (Vonshak, 2002, Shimamatsu, 2004, Henrikson, 2010). So, there are established national and international quality standards for *Spirulina* products. Tables 5 show the typical analysis of the contents of *Spirulina* product, the quality standard in Europe, Japan and United States Food and Drug Administration's (FDA) requirement for *Spirulina* product respectively. Recently,cyanobacterial toxins have become a major issue in public health due to the increased occurrence of toxic cyanobacterial blooms. These toxic blooms contain algae that produce hepatotoxins called microcystins (Carmichael, 1994). *Spirulina* companies like Earthrise Farms have already developed methods for the determination of these toxins and actually certify each lot of their product to be toxin free. *Spirulina* does not normally contain microcystins but contamination of outdoor culture by other cyanobacteria is a possibility.

4. Conclusion and outlook

A bibliographical review on *Spirulina* (*Arthrospira*) identifies this microorganism as microalgae or bacteria, by bacteriologist and botanists respectively. This study has revealed many research studies done on its properties, some of these are related to human and animal food uses. *Spirulina* is claimed as a non-toxic, nutritious food, with some corrective properties against viral attacks, anemia, tumoral growth and low prostaglandins production in mammals; and as a source of the yellow coloration of egg yolk when consumed by hens, and a growth, sexual maturation and fertility factor, in poultry and bovines. This material contains proteins, carbohydrates, essential fatty acids, vitamins, minerals, carotenes, chlorophyll *a* and phycocyanin. *Arthrospira* may be produced in rather simple pilot plants or industrial installations if good conditions and quality controls are assured. *Spirulina* has been used as food for centuries in human history. Moreover, it has been produced commercially over the past 30 years and consumed by thousands of people without problem. Many research show that it has a good nutritional profile in addition to containing some phytonutrients that have potential health benefits. The technology of *Spirulina* production has also advanced in the past 30 years, resulting in higher-quality of product at relatively lower cost. *Spirulina* or *Arthrospira* has attracted the attention of researchers for many years, as shown by the hundreds of publications in its various aspects. There are more and more information is being made available about its biology, biotechnological and nutrition applications, and health application and even in what seem to be remote applications like biofuel production and as a life support system in space studies. This interest will no doubt continue. Research along these lines will be rewarding both socially and professionally.

5. References

Becker, E. W. & Venkataraman, L. V. (1984). *Production and utilization of the blue-green alga Spirulina in India,* Biomass,Volume 4, Issue 2, pp. 105-125.

Becker, E. W. (1992). Micro-algae for human and animal consuptions, in Borowitzka, Michael A. & Borowitzka, Lesley J. (ed.), *Micro-algal Biotechnology*, Cambridge University Press, Cambridge, UK, p.223-256.

Belay A. (1997). Mass culture of *Spirulina* outdoors: the Earthrise Farms experience. in Vonshak, A. (ed.), *Spirulina platensis (Arthrospira) Physiology, Cell Biology and Biotechnology*, London: Taylor & Francis, pp.131-158.

Belay, A. (2002). The Potential Application of *Spirulina* (*Arthrospira*) as a Nutritional and Therapeutic Supplement in Health Management, *The Journal of the American Nutraceutical Association*, Vol. 5, No. 2, pp.1-24, JANA 27.

Belay, A. (2008). *Spirulina* (*Arthrospira*): Production and Quality Assurance, in, M.E. Gershwin &Amha Belay (ed.), *Spirulina in human nutrition and health*, CRC Pres, Taylor & Francis, p:16-40.

Barrón, L.B., Torres-Valencia,M.J., Chamorro-Cevallos, G.&Zúñiga-Estrada, A. (2008). *Spirulina* as an Antiviral Agent, in, M.E. Gershwin &Amha Belay (ed.), *Spirulina in human nutrition and health*, CRC Pres, Taylor & Francis, pp.227.

Ciferri, O. (1983). *Spirulina*, the Edible Microorganism, *Microbiological reviews*, American Society for Microbiology Vol.47, No:4, pp. 551-578.

Carmichael, W. (1994). The toxins of cyanobacteria, *Scientific American* 279: 78–86.

Cevallos, C.G., Lilia Barrón, B. & Vázquez-Sánchez, J. (2008). Toxicologic Studies and Antitoxic Properties of *Spirulina*, in M.E. Gershwin & Amha Belay (ed.), *Spirulina in Human Nutrition and Health*, CRC Press Taylor & Francis Group Boca Raton, FL 33487-2742, pp. 42-65.

Diraman, H., Koru, E. & Dibeklioglu, H. (2009). Fatty Acid Profile of *Spirulina platensis* Used as a Food Supplement, The Israeli Journal of Aquaculture - Bamidgeh 61(2), 134-142.

Fox, D.R. (1996). SPIRULINA: Production&Potential, Edisud, France, pp.232.

Habib, M.A.B., Parvin, M., Huntington, T.C. & Hasan, R.M. (2008). *A Review on Culture, Production and Use of Spirulina as Food for Humans and Feeds for Domestic Animals and Fish*, FAO Fisheries and Aquaculture Circular. No. 1034, Rome-Italy, ISBN 978-92-5-106106-0, pp. 1-41.

Henrikson, R. (1994). Microalga *Spirulina*, superalimento del futuro, Ronore Enterprises. 2ª ed. Ediciones Urano, Barcelona, España. pp. 222.

Henrikson, R. (2010). *Spirulina*: World Food, How this micro algae can transform your health and our planet, Published by Ronore Enterprises, Inc.PO Box 909, Hana, Maui, Hawaii 96718 USA, ISBN 1453766987, pp.195.

Henrikson, R. (2011). Algae in Historical Legends, Algae Industry Magazine March 10, URL:www.AlgaeIndustryMagazine.com.

James R., Sampath K., Thangarathinam R. & Vasudhevan, I. (2006). Effect of dietary *Spirulina* level on growth, fertility, coloration and leucocyte count in red swordtail, *Xiphophorus helleri*. *Isr. J. Aquacult. - Bamidgeh*, 58:97-104.

Koru, E., Cirik, S. & Turan, G. (2008). *The use of Spirulina for fish feed production in Turkey, University-Industry Co-Operation Project (USIGEM)*, Project principal investigator and consultant by Edis Koru, pp.100, Bornova-İzmir/Turkey (in Turkish).

Koru, E. (2009). *Spirulina* Microalgae Production and Breeding in Commercial, *Turkey Journal of Agriculture*, - May-June 2008, Issue:11, Year:3, pp:133-134, Çamdibi-Izmir/Turkey (in Turkish).

Martinez, G.A. (2007). Hepatotoxic Cyanobacteria, in Botana, M.L. (ed.), *Phycotoxins: Chemistry and Biochemistry*, Blackwell Publishing, p.259.

Othes, S. & Pire, R. (2001). Fatty acid composition of *Chlorella* and *Spirulina* Microalgae species, *J. AOAC Int.*, 84: 1708-1714.

Pelaez, F. (2006*). The historical delivery of antibiotics from microbial natural products – Can history repeat?, Biochem. Pharmacol.*, 71 (7), p.981.

Richmond, A. (1992). *Spirulina*, in Borowitzka, Michael A. & Borowitzka, Lesley J. (ed.), *Micro-algal Biotechnology*, Cambridge University Press, Cambridge,UK, p.86-121.

Shekharam, K.M., Venkataraman, L.V.& Salimath, P.V. (1987). Carbohydrate composition and characterization of two unusual sugars from the blue green alga *Spirulina platensis, Phytochemistry*, 26, 2267.

Sasson, A. (1997). *Micro Biotechnologies: Recent Developments and Prospects for Developing Countries*. BIOTEC Publication 1/2542. pp. 11–31. Place de Fontenoy, Paris. France. United Nations Educational, Scientific and Cultural Organization (UNESCO).

Sánchez, M., Castillo, B.J., Rozo, C. & Rodríguez, I. (2003). *Spirulina (Arthrospira): An edible microorganism. A review*, Universitas Scientiarum, Vol. 8, No.1, pp.1-16, URL: www.javeriana.edu.co/universitas.docs../Vol 8_htm.

Shimamatsu, H. (2004). Mass production of *Spirulina*, an edible microalga, *Hydrobiologia* 512: 39–44.

Šejnohová, L. (2008). *Microcystis*. New findings in peptide production, taxonomy and autecology by cyanobacterium *Microcystis, Masaryk University in Czech Republic, Faculty of Science Department of Botany and Zoology&Institute of Botany Czech Academy of Sciences,* Ph. D. Thesis, ISBN:978-80-86188-27-0.

Tomaselli, L. (2002). Morphology, Ultrastructure and Taxonomy of *Arthrospira (Spirulina) maxima* and *Arthrospira (Spirulina) platensis,* in: Vonshak, A. (ed.), *Spirulina platensis (Arthrospira) Physiology, Cell Biology and Biotechnology*, Taylor & Francis, ISBN 0-203-48396-0, London, pp.1-17.

Tomaselli, L. (2003). The microalgal cell, In: *Handbook of Microalgal Culture, in*: Richmond, A. (ed.), Blackwell Science, ISBN: 0632059532, pp.2-10.

Venkataraman L.V. (1993). *Spirulina* in India. In: *Proc. Natl. Seminar Cyanobacter Res. - The Indian Scene*. Natl. Facility for Marine Cyanobacteria, Bharathidasan Univ., Tiruchirapalli, India, pp. 92-116.

Vonshak, A. (2002). Use of *Spirulina* Biomass, In: Vonshak, A. (ed.), *Spirulina platensis (Arthrospira) Physiology Cell Biology and Biotechnology,* Taylor & Francis, ISBN 0-203-48396-0, London, pp. 159-173.

Wikfors, H.G. & Ohno, M. (2001). *Impact Of Algal Research In Aquaculture,* Journal of Phycology, 37, pp.968-974.

Wu, L.C., Ho, J.A., Shieh, M.C., & Lu, Y.M. (2005). Antioxidant and antiproliferative activities of *Spirulina* and *Chlorella* water extracts, *J. Agric. Food Chem.*, 53(10): 4207.

Zarrouk, C. (1966). Contribution à l'étude d'une cyanophycée influencée de divers facteurs physiques et chimiques sur la croissance et la photosynthèse de *Spirulina maxima* (Setch. et Gardner) Geitler, University of Paris, Paris, France. (PhD Thesis).

12

Aluminum in Food – The Nature and Contribution of Food Additives

Robert A. Yokel
Pharmaceutical Sciences, University of Kentucky
USA

1. Introduction

Aluminum (Al) is distributed throughout the environment because of its presence as the third most abundant element on earth. Concern about Al toxicity to humans, including from food sources, has persisted since the demonstration that it has the potential to be a neurotoxicant (Wiley 1928, 1929; Schaeffer et al. 1928; Döllken 1898; Gies 1911; Anon. 1913; Yokel and Golub 1997; WHO 1997; Krewski et al. 2007; ATSDR 2008). Exposure of humans to Al is mainly from food, water, airborne dust, antiperspirants, immunizations, allergy injections and antacids (Table 1). Foods and beverages are the single largest contributor of Al intake for the typical human, providing ~ 3.5 to 10 mg/day. Food additives provide a significant percentage of the daily intake. Among the food additives, sodium aluminum phosphates (SALPs) are the main contributors. Drinking water provides ~ 0.1 mg. Other sources for some humans, their daily exposures and intakes, estimated percentage absorbed, and amount of Al that enters the blood are summarized in Table 1. The history of the use and regulation of Al food additives in the US, approved Al-containing food additives in the US and some other countries, primary food types contributing Al to the diet, typical daily dietary Al intake, dietary intake in relation to established tolerable intake and minimal risk levels, and some discussion of Al absorption, distribution, excretion, and toxicity are presented in this chapter.

Aluminum is present in food naturally, as a food additive, and taken up through contact with Al used in food preparation and storage. The Al content of foods is highly variable, depending upon the food product, its processing or lack thereof, the site of growth of the food stuff, and for grains if they are Al-tolerant varieties. Contamination of food with soil, that typically contains 5 to 10% Al, can significantly increase the Al content of the food, although the Al may not be in a readily absorbable form.

2. The history of aluminum-containing food additives in the US

Smith (1928) reviewed the Report of the Referee Board of Consulting Scientific Experts, Created by Executive Order of President Roosevelt in February, 1908, to address the safety of the use of Al compounds in food. He also reviewed the published literature and some unpublished studies up to 1925 and court cases relevant to the use of Al in baking powders and conducted some studies himself. He concluded, in agreement with the Referee Board created by President Roosevelt and the court ruling in State of Missouri vs. Whitney Layton,

Source	Al concentration	Daily Al exposure	Estimated percentage absorbed	Al absorbed daily ($\mu g/kg$)[a]
Typical Exposures				
Water	Average ~ 70 $\mu g/l$	100 μg	0.3 [b]	0.004
Food - total diet		3500-10,000 μg [c]	0.1 to 0.3 [d]	0.05-0.4
Air-office	0.15 $\mu g/m^3$ [e]	1 μg	1 to 2 from lungs [f] 0.1 to 0.3 from GI tract	0.0002 0.00003
Air-outside	0.2 - 1 $\mu g/m^3$ [e,g]	4 μg [h]	1 to 2 from lungs [f] 0.1 to 0.3 from GI tract	0.001 0.0001
Antiperspirants	5-7.5% [i]	50,000-75,000 μg	up to 0.012 [j]	up to 0.1
Vaccines, pediatric patient	125-330 μg/dose	1.4 μg [k]	100 eventually [l]	0.07
Elevated Exposures				
Antacids/phosphate Binders		up to 5,000,000 μg	0.1	80
Industrial Air	25-2500 $\mu g/m^3$	250-25,000 μg per work day	1 to 2 from lungs [f] 0.1 to 0.3 from GI tract	0.6-8 0.008-1
Allergy immunotherapy	150-850 μg/dose	7-40 μg [m]	100 eventually [l]	0.1-0.6
Dialysis solution	If tap water 50 $\mu g/l$	2400 μg	25 [n]	9
Total Parenteral Nutrition Solutions	Neonatal/pediatric	9-23 $\mu g/kg$ [o]	100	9-23
	Adult	1.5 $\mu g/kg$ [p]	100	1.5

[a]Based on a 70 kg adult except for vaccines (20 kg child) and total parenteral nutrition solutions. [b](Yokel et al. 2001a; Zhou, Harris, and Yokel 2008; Stauber et al. 1999; Priest et al. 1998). [c]Based on reports cited in the text. [d](Stauber et al. 1999; Yokel and Florence 2006; Yokel, Hicks, and Florence 2008). [e](Horemans et al. 2008). [f]Based on Al exposure in an industrial setting: (Sjögren et al. 1997; Gitelman et al. 1995; Pierre et al. 1995; Riihimäki et al. 2000; Priest 2004). [g](Jones and Bennett 1986). [h](Priest 2004). [i]Based on 20% Al zirconium glycine complex or 25% Al chlorohydrate in a topical product, which are typical concentrations (POISINDEX information system, Micromedex, Inc, Englewood, CO). [j]Based on (Flarend et al. 2001), assuming that the percentage of Al absorbed does not change with repeated exposure. [k]Based on 30 injections in the first 6 years of life, an average weight of 20 kg, 0.75 mg from RECOMBIVAX HB®, 1.32 mg from Pentacel®, 0.5 mg from Prevnar®, 0.5 mg from HAVRIX® and none from RotaTeq®, Fluzone®, M-M-R® II, or Varivax®, assuming absorption over 6 years. [l](Flarend et al. 1997). [m]Based on a typical allergen extract treatment schedule and maintenance injections for 3.5 years of one allergen extract. [n](Kovalchik et al. 1978). [o](Speerhas and Seidner 2007; Poole et al. 2010; Bohrer et al. 2010). [p](Speerhas and Seidner 2007).

Table 1. Sources of Al exposure, Al concentration in the source, resultant daily Al exposure from the source, estimated percentage absorbed from the source, and calculated amount of Al absorbed daily, normalized to body weight. Modified from (Yokel and McNamara 2001)

1899, that baking powder containing sodium aluminum sulfate had not been shown to cause functional disorders or disease or impairment of the digestion and general health to humans.

Al salts were used in food prior to 1958. The U.S. Congress amended the Food, Drug and Cosmetic Act (FDC Act) in 1958 (Food Additives Amendment) to require pre-market approval of any substance intentionally added to food, that becomes a component of food or otherwise affects the characteristics of food, unless the use of the substance is generally recognized as safe (GRAS) or otherwise excepted from the definition of food additive, e.g., a color additive. A temporal summary of the development of food regulation in the US is available at: (http://www.fda.gov/Food/FoodIngredientsPackaging/GenerallyRecognizedasSafeGRAS/ucm094040.htm). It legislated exemptions for many substances with a history of safe use as food. For substances used in foods prior to 1958, the FDA permits expert opinion to be based on a reasoned judgment founded in experience with common food use, taking into account reasonably anticipated patterns of consumption, cumulative effects in the diet, and safety factors appropriate for the utilization of animal experimentation data. Appropriate regulation is also required that is published in the Federal Register and codified in the Code of Federal Regulations (CFR). The food additive petition process of the FDA is based on Section 409 of the FDC Act. Corresponding regulations are described in CFR Title 21. The GRAS substance list appears in 21 CFR Parts 171, 182, 184, and 186. These regulations are accessible on the Internet, e.g., 21 CFR 133.169 at: http://frwebgate.access.gpo.gov/cgi-bin/get-cfr.cgi?TITLE=21&PART=133&SECTION=169&TYPE=TEXT.

In 1959 the US FDA published a list of substances it considered to be GRAS for use in foods. GRAS is defined in CFR 121.1. Legal authority for Al as GRAS is codified in 21 CFR 182. The list included Al salts. Food additive approval does not require human clinical testing and there is no risk-benefit analysis. GRAS recognition may be based on scientific procedures (safety studies) or through experience based on common use in food prior to 1958. The original determination of GRAS status of Al was based on common use. President Nixon in 1969 directed the FDA to undertake a systematic safety review of all GRAS substances. The Life Sciences Research Office (LSRO) of the Federation of American Societies for Experimental Biology (FASEB) was contracted by the FDA to review and evaluate the available information on each GRAS substance, summarize the available scientific literature, and to recommend what restrictions, if any, on their use would be needed to ensure their safe use in food. A re-review of the GRAS status of Al was conducted by a select committee of the FASEB in 1975, Select Committee on GRAS Substances [SCOGS]. It was based on the previous review, a review of the published literature from 1920 to 1973 related to the safety of Al compounds as food ingredients (Tracor-Jitco 1973), and unpublished studies contracted by producers of SALP which was fed to rats and dogs. The opinion of the select committee was that "There is no evidence in the available literature on ...acidic sodium aluminum phosphate [and other Al forms] ... that demonstrates, or suggests reasonable grounds to suspect, a hazard to the public when they are used at levels that are now current or that might reasonably be expected in the future" (FASEB 1975) (http://www.fda.gov/Food/FoodIngredientsPackaging/GenerallyRecognizedasSafeGRAS/GRASSubstancesSCOGSDatabase/default.htm). Although noting that care should be taken by patients with kidney disease when consuming food containing high levels of Al salts, the authors did not mention dialysis encephalopathy, which has been attributed to Al, or the controversial role of Al in Alzheimer's disease. Description of these clinical problems began about the same time (Crapper, Krishnan, and Dalton 1973; Alfrey, LeGendre, and Kaehny 1976).

In 1982 the FDA issued Toxicological Principles for the "Safety Assessment of Direct Food Additives and Color Additives Used in Food" (a.k.a. the Red Book) to provide testing guidelines to obtain scientific evidence of safety (Humphreys 1992). In 1992 the FDA was in the process of reviewing Al, which was evidently not completed. The FDA Center for Food Safety and Applied Nutrition is responsible for monitoring Al in foods, that which is naturally occurring, and that which is from food additives (Pennington 1992).

3. The current approved uses, purposes for use, and regulation of aluminum-containing food additives

3.1 In the US

Aluminum is used as a direct food additive as a firming agent, carrier, coloring agent, anticaking agent, buffer, neutralizing agent, dough strengthener, emulsifying agent, stabilizer, thickener, leavening agent, curing agent and texturizer. Direct food additives are those that have been intentionally added to food for a functional purpose, up to a few %. These additives are used in milk, processed cheese, yogurt, preserves, jams and jellies, baking soda, sugars, cereals, flours, grains and powdered or crystalline desert products (Pennington 1987). Table 2 lists FDA-approved GRAS food additives containing Al and Table 3 other FDA-approved Al-containing food additives.

The most important uses include the addition of acidic SALP to self-rising flour as an acidifying and leavening agent; basic SALP to processed cheese, cheese food and cheese spread as an emulsifying agent to give them a soft texture and easy melting characteristics (Lione 1983); and sodium aluminosilicate as an anticaking agent. Acidic SALP $NaAl_3H_{14}(PO_4)_8 \cdot 4H_2O$ reacts with $NaHCO_3$ to produce the leavening action by generating CO_2. Basic SALP ($Na_8Al_2(OH_2)((PO)_4)_4$) is one of many "emulsifying salts" added to process cheese, cheese food and cheese spread which react with and change the protein of cheese to produce a smooth, uniform film around each fat droplet to prevent separation and bleeding of fat from the cheese. This produces the desired soft texture of cheese, easy melting characteristics, and desirable slicing properties (Ellinger 1972).

Aluminum is also permitted in lakes (water-soluble artificial colors adsorbed onto alumina), which typically have an Al content of 25% (United Kingdom Ministry of Agriculture Fisheries and Food 1993). Alumina (Al oxide) is the only substratum approved for manufacturing U.S. Food, Drug & Cosmetic approved lakes (Soni et al. 2001). These are listed in Table 3.

Aluminum and some Al compounds are also approved by the FDA as indirect food additives (21CFR) Parts 175, 176, 177, and 178. Indirect food additives are substances used in food-contact articles, and include adhesives and components of coatings (Part 175), paper and paperboard components (Part 176), polymers (Part 177), and adjuvants and production aids (Part 178). In general, these are substances that may come into contact with food and enter food products in small quantities as a result of growing, processing or packaging (http://www.fda.gov/Food/FoodIngredientsPackaging/ucm115333.htm).

- Prior-sanctioned food ingredients. Stabilizers (substances that may migrate from food-packaging material into food) (21 CFR 181.29)
  - Aluminum mono-, di-, and tristearate
- GRAS substances (21 CFR 182)
  - Substances migrating to food from paper and paperboard products (182.90)
    - Alum (double sulfate of aluminum, and ammonium, potassium, or sodium)
    - Aluminum hydroxide
    - Aluminum oleate [1975, 43,1]
    - Aluminum palmitate [1975, 43,1]
    - Sodium aluminate, [1975, 43,1]
    - Sodium phosphoaluminate, [1975, 43,1]
  - Multiple purpose GRAS food substances (Subpart B)
    - Aluminum sulfate (182.1125), [1975, 43,1]
    - Aluminum ammonium sulfate (182.1127), [1975, 43,1], buffer and neutralizing agent
    - Aluminum potassium sulfate (182.1129), [1975, 43,1], buffer and neutralizing agent
    - Aluminum sodium sulfate (182.1131), [1975, 43,1], buffer and neutralizing agent
    - Sodium aluminum phosphate, acidic (182.1781), [1975, 43,1] a leavening agent in cereal foods and related products, self-rising flours & meals
    - Sodium aluminum phosphate, basic (182.1781), [1975, 43,1] an emulsifying agent in pasteurized process cheese, cheese food and cheese spread
  - Anticaking Agents (Subpart C)
    - Aluminum calcium silicate, (182.2122), [1979,1], anticaking agent, < 2% by weight in table salt; (169.179), vanilla powder
    - Sodium aluminosilicate (sodium silicoaluminate) (182.2727), [1979,61,1], anticaking agent, < 2%
    - Sodium calcium aluminosilicate, hydrated (sodium calcium silicoaluminate) (182.2729), [1979,61,1], anticaking agent, < 2%
- Direct food substances affirmed as GRAS (21 CFR 184)
  - Aluminum hydroxide (184.1139) [1975, 43,1]
- Indirect food substances affirmed as GRAS (21 CFR 186)
  - Clay (kaolin), which consists of hydrated aluminum silicate (186.1256)

SCOGS conclusion:

1. There is no evidence in the available information on [substance] that demonstrates, or suggests reasonable grounds to suspect, a hazard to the public when they are used at levels that are now current or might reasonably be expected in the future.

Table 2. Generally recognized as safe (GRAS) food additives containing Al. The 21 CFR for those substances that have a CFR citation is in parentheses. The year of the SGOGS review report, the report number, and its conclusion are in brackets

- Cheeses and related cheese products
  - To bleach milk for cheese, potassium alum, (133.102, 106, 111, 141, 165, 181, 183, and 195), < 0.012%
  - Grated cheeses: Sodium aluminosilicate
  - Pasteurized process cheese: sodium aluminum phosphate, (133.169), < 3% by weight
  - Pasteurized process cheese food, sodium aluminum phosphate, (133.173), < 3% by weight
  - Pasteurized process cheese spread, sodium aluminum phosphate, (133.179), < 3% by weight
- Cereal flours and related products, Subpart B
  - Flour: potassium alum and sodium aluminum sulfate, (137.105)
  - Self-rising flour: sodium aluminum phosphate (137.180)
  - Self-rising white corn meal: sodium aluminum phosphate (137.270)
- Eggs and egg products
  - Dried whole eggs and egg yolks: sodium silicoaluminate, anticaking agent, (160.105), < 2%
  - Food dressings and flavorings
  - Vanilla powder: Aluminum calcium silicate, (169.179), < 2%
- Food additives permitted for direct addition to food for human consumption, Subpart D—Special Dietary and Nutritional Additives
  - As a source of niacin in foods for special dietary use, aluminum nicotinate, dietary supplement, (172.310)
  - Salts of fatty acids: aluminum salts of fatty acids, binder, emulsifier, and anticaking agent (172.863)
  - Food starch-modified: aluminum sulfate, [1979,115,3] , (172.892), < 2.0%
- Defoaming agents: Secondary direct food additives permitted in food for human consumption
  - Aluminum stearate - Processing beet sugar & yeast, antifoaming (or defoaming) agent, (173.340)
- Boiler water additives, in steam contacting food
  - Sodium aluminate, (173.310)
- Aluminum in lakes
  - FD&C Blue #1 aluminum lake
  - FD&C Blue #2 aluminum lake on alumina
  - FD&C Red #40 aluminum lake
  - FD&C Yellow #5 aluminum lake
- Indirect food additives: adjuvants, production aids, and sanitizers, Subpart B - Substances utilized to control the growth of microorganisms, coatings on fresh citrus fruit
  - Aluminum
  - Aluminum acetate
  - Aluminum di(2-ethylhexoate)

Table 3. (continues on next page) Some other FDA-approved sources of Al in food

- Aluminum distearate
- Aluminum, hydroxybis[2,4,8,10-tetrakis(1,1-dimethylethyl)-6-hydroxy12h-ibenzoid[d,g][1,3,2]dioxaphosphocin 6-oxidato]
- Aluminum isodecanoate
- Aluminum linoleate
- Aluminum monostearate
- Aluminum naphthenate
- Aluminum neodecanoate
- Aluminum oxide
- Aluminum potassium silicate
- Aluminum ricinoleate
- Aluminum silicate
- Aluminum sodium sulfate, anhydrous
- Aluminum stearoyl benzoyl hydroxide
- Bis(benzoate-o)(2-propanolato)aluminum
- D&C Red no. 7-aluminum lake
- Dialkyldimethylammonium aluminum silicate
- Gum rosin, aluminum salt
- Gum rosin, disproportionated, aluminum salt
- Rosin, hydrogenated, aluminum salt
- Rosin, partially dimerized, aluminum salt
- Sodium aluminum pyrophosphate
- Soybean oil fatty acids, aluminum salt
- Tall oil fatty acids, aluminum salt
- Tall oil rosin, aluminum salt
- Tall oil rosin, disproportionated, aluminum salt
- Wood rosin, aluminum salt
- Wood rosin, disproportionated, aluminum salt

SCOGS conclusion:

3. While no evidence in the available information on [substance] demonstrates a hazard to the public when it is used at levels that are now current and in the manner now practiced, uncertainties exist requiring that additional studies be conducted.

Table 3. (continued) Some other FDA-approved sources of Al in food

3.2 Countries other than the US

Canada permits the use of Al as a food additive, as described in Part B Foods Division 16 of the Food and Drug Regulations of the Canadian Food and Drugs Act (http://laws.justice.gc.ca/eng/regulations/C.R.C.,_c._870/page-147.html#h-110). These regulations are similar to the FDA's. The fifteen member states of the European Union, as well as Norway and Iceland, allow the use of Al as a food additive. European Parliament and Council Directive 94/36/EC lays down detailed rules on colors and 95/2/EC, as amended by Directives 96/85/EC, 98/72/EC and 2001/5/EC, lays down detailed rules for authorization of all food additives other than colors and sweeteners (http://ec.europa.eu/food/fs/sfp/flav_index_en.html). Approved Al-containing food additives are shown in Table 4. The United Kingdom (http://www.legislation.hmso.gov.uk/si/si1995/Uksi_19953187_en_5.htm#sdiv3) and Australia and New Zealand (http://www.foodstandards.gov.au/mediareleasespublications/publications/shoppersguide/foodadditivesnumeric1680.cfm) also permit Al food additives. Regulations for the former are very similar to the EC's. The approved Al-containing food additives for the latter are shown in Table 5. Japan carries out food safety work under the Food Safety Basic Law (enacted in May 2003) and related laws

94/36/EC (colors):

- Aluminum metal (E 173) is authorized for the external coating of sugar confectionary and for the decoration of cakes and pastries.

95/2/EC:

- E520 Aluminum sulfate, firming agent
- E521 Aluminum sodium sulfate, firming agent
- E522 Aluminum potassium sulfate, acidity regulator
- E523 Aluminum ammonium sulfate, acidity regulator

(Aluminum sulfates (E 520-523) are permitted to be used in egg white up to 30 mg/kg; and candied, crystallized glacé fruit and vegetables, up to 200 mg/kg individually or in combination, expressed as aluminum

- E541 Sodium aluminum phosphate, acidic is permitted to be used in fine bakery wares (scones and sponge wares only) up to 1 g/kg expressed as aluminum.

  [Sodium aluminum phosphate, basic is not authorized in the EU as a food additive]

- E554 Sodium aluminum silicate, anticaking agent
- E555 Potassium aluminum silicate, anticaking agent
- E556 Calcium aluminum silicate, anticaking agent
- E558 Bentonite (hydrated aluminum silicate), carrier in food colors, maximum 5%
- E559 Aluminum silicate (kaolin, anticaking agent)
- E554-E559 are permitted to be used in dried powdered foodstuffs (including sugars); salt and its substitutes; sliced or grated hard, semi-hard and processed cheese and sliced or grated cheese analogues and processed cheese analogues, up to 10 g/kg; chewing gum; rice; sausages as a surface treatment only; seasonings up to 30 g/kg; confectionery excluding chocolate as a surface treatment only; tin-greasing products up to 30 g/kg
- E1452 Starch aluminum octenylsuccinate, permitted encapsulated in vitamin preparations in food supplements in food supplements up to 35 g/kg

94/36/EC on colors for use in foodstuffs:

- Aluminum lakes of the permitted colors
- FD&C Red 40 aluminum lake (E 129)

Table 4. Allowed Al-containing additives for use in foodstuffs by the European Parliament and Council Directives

- Permitted flavoring substances, for the purposes of this Standard
  - Flavoring substances which are listed in at least one of the following publications:
    - Generally Recognized as Safe (GRAS) lists of flavoring substances published by the Flavour and Extract Manufacturers' Association of the United States from 1960 to 2011 (edition 25)
    - Chemically-defined flavoring substances, Council of Europe, November 2000
    - 21 CFR § 172.515
- 559 Aluminum silicate
- 470 Aluminum, calcium, sodium, magnesium, potassium and ammonium salts of fatty acids
- 556 Calcium aluminum silicate
- 554 Sodium aluminosilicate
- 470 Aluminum, calcium, sodium, magnesium, potassium and ammonium salts of fatty acids
- 541 Sodium aluminum phosphate, acidic (baking compound), [Basic SALP is not permitted as a food additive]
- 555 Potassium aluminum silicate, dried milk, milk powder, cream powder
- 555 Potassium aluminum silicate, cheese and cheese products up to 1%
- 173 Aluminum, confectioneries, spirits and liqueurs
- 556 Calcium aluminum silicate, salt

Table 5. Al-containing food additives permitted by the Australia New Zealand Food Standards Code - Standard 1.3.1 . Colours and their aluminium and calcium lakes. A reference to a color listed in Schedules 1, 3 and 4 of this Standard includes a reference to the aluminum and calcium lakes prepared from that color

- Aluminum ammonium sulfate (ammonium alum), raising agent, processing aid
- Aluminum potassium sulfate (potassium alum), raising agent, processing aid
- Food Blue No. 1, aluminum lake, restricted for the purpose of coloring
- Food Blue No. 2, aluminum lake
- Food Green No. 3, aluminum lake
- Food Red No. 2, aluminum lake
- Food Red No. 3, aluminum lake
- Food Red No. 40, aluminum lake
- Food Yellow No. 4, aluminum lake
- Food Yellow No. 5, aluminum lake

Table 6. Al-containing designated food additives deemed not injurious to human health, related to Articles 12 and 21 of the Food Sanitation Law of Japan (http://www.tokio.polemb.net/files/Gospodarka/Handel/food-e.pdf). The conditions of use of all but aluminum potassium sulfate are strictly defined or limited. All lakes are restricted for the purpose of coloring

(including the Food Sanitation Law), under the jurisdiction of the Department of Food Safety under the Pharmaceutical and Food Safety Bureau, in the Ministry of Health, Labour and Welfare. The Standards and Evaluation Division is responsible for the establishment of specifications/standards for food additives. Approved Al-containing food additives are shown in Table 6. The National Standard of the People's Republic of China, Hygiene Standard for Use of Food Additives, GB2760-2007, issued by the Ministry of Health and the Standardization Administration of China, lists the following as allowed food additives: allura, amaranth, brilliant blue, erythrosine, indigo carmine, new red, ponceau 4 r, sunset yellow, and tartrazine aluminum lakes as coloring agents; sodium aluminosilicate as an anticaking agent up to 5 g/kg; aluminum potassium sulfate and aluminum ammonium sulfate as bulking agents and stabilizers with a maximum aluminum residual level of 100 mg/kg dry weight; and starch aluminum octenylsuccinate as a thickener, anticaking agent, and emulsifier (http://www.fas.usda.gov/GainFiles/200803/146294056.pdf). Food additives listed by the Bureau of Food Sanitation, Department of Health, Republic of Taiwan, Scope of Usage and Measurement Standards for Food Additives as Food quality improvement, fermentation and food processing agents include aluminum silicate, aluminum sulfate, bentonite, diatomaceous earth, kaolin and sodium silicoaluminate as leavening agents, as well as ammonium alum, burnt ammonium alum, potassium alum, burnt potassium alum, sodium alum, burnt sodium alum, and acidic SALP (http://www.doh.gov.tw/EN2006/DM/DM2.aspx?now_fod_list_no=6005&class_no=386&level_no=2). The Government of India Prevention of Food Adulteration Act (PFA) of 1954, enforced by the Food Safety and Standards Authority of India, permits acid compounds of Al in baking powder; aluminum silicates and aluminum ammonium, calcium, potassium or sodium myristates, palmitates or stearates as anticaking agents up to 2%; aluminum sulfate (520) and aluminum sodium sulfate (521) as firming agents; aluminum potassium sulfate (522) as an acidity regulator and stabilizer; aluminum ammonium sulfate (523) as a stabilizer and firming agent; acidic and basic sodium aluminum phosphate (541) as an acidity regulator and emulsifier; sodium (up to 0.5% in powdered soft drink concentrate mix/ fruit beverage drinks and up to 10 gm/kg in cocoa powder and lozenges), potassium and calcium aluminosilicate and aluminum silicate (554, 555, 556 and 559) as anticaking agents; aluminum lake of Sunset Yellow FCF in powdered dry beverage mixes; and aluminum lake of Sunset Yellow and aluminum as colors. The Government of South Africa, Department of Health, Foodstuffs, Cosmetics and Disinfectants Act, 1972, Regulations Relating to Food-Grade Salt, permit calcium, magnesium and sodium aluminum and calcium-aluminum silicates as anticaking agents (http://www.doh.gov.za/docs/regulations/2006/reg0114.pdf).

International Numbering System numbers are assigned by the Codex Alimentarius Committee to allow each food additive to be uniquely identified. On packaging in the EC, approved food additives are written with a prefix of 'E'. Some governments, e.g., Australia and New Zealand and India, do not use a prefix letter when listing additives in the ingredients.

4. Al processing and packaging that comes in contact with food

The use of Al cookware began around 1890 (Smith 1928). Al is used in cookware due to its heat conductivity. The surface of Al oxidizes to form a few nm thick layer of Al oxide, which resists corrosion from pH ~ 4.5 to 8.5. Studies on Al mobilization from Al shavings, strips and vessels from 1890 to 1925, reviewed by (Smith 1928), showed Al was solubilized by beverages and foods. Many subsequent studies have shown that Al can be mobilized from Al cookware, particularly by acidic and alkaline foods. These are pH conditions where Al is more soluble than at circumneutral pH (Fimreite, Hansen, and Pettersen 1997; Scancar, Stibilj, and Milacic 2003; Neelam, Bamji, and Kaladhar 2000; Gramiccioni et al. 1996). It has also been demonstrated that NaCl can increase Al mobilization from Al vessels (Datta 1935; Inoue, Ishiwata, and Yoshihara 1988; Takeda, Kawamura, and Yamada 1998; Fukushima and Tanimura 1998).

Most countries of the world, including the EC, do not have specific requirements for light metal alloys in contact with food (Severus 1989). The conclusion of a review by Wuhrer (1939) was that Al is safe and harmless as a material for cooking and household utensils in contact with food. The U.S. FDA came to a similar conclusion (Reilly 1991).

5. The contribution of aluminum food additives, processing and packaging to human aluminum intake

In the UK it was suggested that SALP, sodium Al silicate, and Al lakes contributed ~ 1.4, 0.9 and 3 mg Al to daily intake, respectively (United Kingdom Ministry of Agriculture Fisheries and Food 1993). In the US the two most quantitatively significant food additives containing Al are acidic and basic SALP (Katz et al. 1984; Humphreys and Bolger 1997; Saiyed and Yokel 2005). The use of Al as a food additive can increase the low inherent level of Al in food, particularly in processed cheese and grain products. The foods highest in Al are those with added Al (Delves, Sieniawaska, and Suchak 1993). For example, it was noted that the addition of SALPs to cake mixes could result in 5 to 15 mg/serving and its addition to processed cheese at the permitted concentrations could result in 50 mg Al/slice (Lione 1983). Similarly, the use of sodium Al sulfate in household baking powder could result in 5 mg Al in each serving of a cake made with 1 teaspoon of baking powder (Lione 1983). It has been estimated that this practice can increase food Al content by about five-fold (Humphreys and Bolger 1997).

Women who reported frequent or average use of Al utensils and foil averaged higher duplicate diet Al intakes than those who reported little or no use (Jorhem and Haegglund 1992). They estimated cooking utensils contributed 2 mg Al to the diet (Jorhem and Haegglund 1992). Daily Al intake in China was estimated to be 9 to 12 mg plus ~ 4 mg from Al-ware (Wang, Su, and Wang 1994). It was concluded that food processing and storage are generally not major contributors to Al in food (Pennington 1987; Sherlock 1989). Greger and Sutherland (1997) concluded that "... under typical cooking conditions, it is doubtful that use of Al utensils adds more than 2 mg Al/day to food." However, if daily Al intake is 3.5 to 10 mg (see **Daily consumption of Al in food**, below, and Table 1), this would add or constitute about 20 to 40% of the daily Al intake.

6. Estimates of the extent of aluminum-containing food additive use

Estimated use in the 1970s of Al-containing food additives were ~ 18 million kilograms of SALP; 3.6 million kg of sodium Al sulfate in baking powder; and an unknown amount of Na Al silicate, Al calcium silicate and Na calcium aluminosilicate as anticaking agents (FASEB 1975). Based on 200 million people in the U.S. at that time and 10% of SALP as Al, the consumption of 18 million kg SALP/year resulted in an average daily consumption of ~ 25 mg/person/day. The estimate of daily Al consumption at that time was 20 mg/person (FASEB 1975). The 1977 Survey of Industry on the Use of Food Additives estimated use of ~ 9 million pounds (~ 4 million kilograms) of SALP (NAS 1979). The amount of Al in SALP, sodium aluminosilicate, sodium aluminate, aluminum sulfate, and aluminum ammonium sulfate consumed by the 225 million people in the US produced an estimated daily Al intake at that time of ~ 5 mg/person. Al-containing lakes and other sources would increase this estimate. In 1982 four million pounds of Al were used as food additives in the US, which would average 20 mg/person/day (Committee on Food Additives survey data 1984; Greger 1985). Although these estimates are not very consistent over time, they consistently estimate a large amount of Al intake from FDA-approved food additives in relation to total Al intake in the US. More recent estimates of the use of these approved food additives in the US were not found.

Companies in the global phosphates industry that market SALPs include Innophos, which acquired Rhodia's phosphate business in 2004, maker of acidic SALP (Levair®) and basic SALP (Kasal®); ICL Performance Products LP which acquired Astaris LLC in 2005, maker of Levn-Lite®, Stabil-9® and Pan-O-Lite®; Xuzhou Hengxing Chemical Co., Ltd maker of HENGXING®; Amerisweet Co., Ltd. maker of AmeriPhos®; China Chem Source (HK) Co., Ltd.; Foodchem International Corporation; and Thermphos International. Rhodia completed an expansion in 1998 to its Nashville plant to increase its SALP capacity by 15% (Anon, 1998) and Astaris expanded its capacity in 2001 to boost production of SALP.

7. Daily consumption of aluminum in food

There have been at least 48 reports that measured or estimated daily oral Al consumption from the diet since the FDA's total diet study reported in 1987. They have been conducted using many different methods, including duplicate portion studies of composite diets, total diet studies, calculations based on the foods in a total diet study times their Al content, and market basket surveys. These studies have been conducted in North America (Canada and the US), Europe (England and the U.K., France, Germany, Hungary, Italy, Netherlands, Portugal, Spain and the Canary Islands, Sweden, and Slovenia), Asia (China, India, Japan, and Taiwan), Australia and Brazil. Median daily Al intake for infants, children, teenagers and adults was 0.7, 6, 8.6 and 4.8 mg, respectively. Total Al intake generally relates to total food intake, which partially explains the greater Al intake in adolescents than adults. Another contributor is the selection of foods. Adolescents eat more prepared foods that have Al-containing food additives. For adults in the US and Canada, Europe, China and Japan, median Al intake by adults was 8.5 (n = 3 reports), 3.6 (n = 18), 9.5 (n = 6), and 9 mg (n = 3), respectively. Al food additives are the main source of Al intake in the US (WHO 1997) but of much less importance in other countries, which includes much of Europe, where less Al is added to cereal grain products as raising agents and basic SALP is not permitted to be used in processed cheese (United Kingdom Ministry of Agriculture Fisheries and Food 1993; Müller, Anke, and Illing-Günther 1998; Sherlock 1989; Humphreys and Bolger 1997). This is reflected in the fewer approved Al food additives and lower minimum permissible levels in the EC than the US (Table 4 compared to Tables 2 and 3). For example, the average Swedish daily diet was calculated to contain ~ 0.6 mg Al from unprocessed foods whereas 105 duplicate diets of 15 women living in the Stockholm area were found to provide 1.2 to 99 mg Al/day, averaging 13 mg Al/day. The most important difference between these was a cake made from a mix containing Al phosphate (Jorhem and Haegglund 1992), illustrating the contribution from food additives.

8. The aluminum concentration in commercial products that contain aluminum food additives

Among the three sources of Al in food, naturally occurring, additives and from food preparation and storage, the major food sources of Al in the daily diet of adults ~ 25 years ago were probably grain products with Al additives, processed cheese with Al additives, tea, herbs, spices and salt containing an Al additive (Pennington 1987). It was estimated that grains and grain products, dairy products, desserts, and beverages contributed about 24 to 49%, 17 to 36%, 9 to 26% and 5 to 10%, respectively, of dietary Al in the U.S. diet (Pennington 1987). Similarly, grains, vegetables and tea were estimated to contribute ~ 60 to 70, 25 and 5% of daily Al intake to the Chinese diet (Wang, Su, and Wang 1994; Zhong et al. 1996). The Australian National Nutrition Survey suggested 30, 24, 10 and 8% was from cereals, beverages, vegetables, and meats, respectively (Allen and Cumming 1998). Biego et al. (1998) estimated 36% from milk and dairy products, 29% from fish and crustaceans, 16% from cereals and 8% from vegetables in France. A study from the UK concluded beverages contributed 35%, bread 21%, and cereals 16% of the Al in the diet (Ysart et al. 2000). The Al content of natural cheese and processed cheese containing added Al has been reasonably constant for a few decades. It appears that the use of Al in pancake mixes and prepared pancakes in the US has increased in the past few decades. A report of Al in convenience and fast foods in Spain noted the increased popularity of these foods and considerable Al in them (Lopez et al. 2002).

9. Comparison of aluminum intake to established tolerable intake and minimal risk levels

The Joint Food and Agriculture Organization of the United Nations/World Health Organization Expert Committee on Food Additives (JECFA) withdrew in 2006 its previously established acceptable daily intake (ADI) and provisional tolerable weekly intake (PTWI) of 7 mg/kg body weight to establish a PTWI for Al of 1 mg/kg. It applied to all Al compounds in food, including additives (FAO/WHO 2006; FAO/WHO 2007). The Expert Committee concluded that Al compounds have the potential to affect the reproductive system and developing nervous system at doses lower than those used in establishing the previous PTWI. In setting this limit the Committee assumed a "probable lower bioavailability of the less soluble Al compounds present in food" and noted that the PTWI was likely to be exceeded by some population groups, particularly children, who regularly consume foods containing Al additives. At the 74th meeting of the JECFA (June 14 to 23, 2011, Rome) the previous PTWI of 1 mg/kg body weight was withdrawn and a PTWI of 2 mg/kg body weight was established based on a no-observed-adverse-effect level (NOAEL) of 30 mg/kg body weight per day and application of a safety factor of 100. The PTWI applies to all aluminum compounds in food, including food additives (ftp://ftp.fao.org/ag/agn/jecfa/JECFA_74_Summary_Report_4July2011.pdf).

The Panel on Food Additives, Flavourings, Processing Aids and Food Contact Materials (AFC) of the European Food Safety Authority (EFSA) established a tolerable weekly intake

(TWI) of 1 mg Al/kg/week. They also noted that this TWI is likely to be exceeded in a significant part of the European population by the typical dietary Al intake (Panel on Food Additives 2008). Similarly, the Agency for Toxic Substances and Disease Registry (ATSDR) derived an intermediate-duration oral exposure (15 to 364 days) minimal risk level (MRL) of 1 mg Al/kg/day, based on neurodevelopmental effects in the offspring of mice exposed to aluminum lactate in the diet on gestation day 1 through lactation day 21 followed by pup exposure until postnatal day 35 (Golub and Germann 2001).

At the 42nd session of the Codex Committee on Food Additives (CCFA) of the Joint FAO/WHO Food Standards Programme (March 15-19, 2010, Beijing, China) it was agreed to establish an electronic working group (eWG) to review comments and information submitted and to revise the maximum use levels for the following Al-containing food additives included in the General Standard for Food Additives (GSFA): acidic and basic SALP, aluminum ammonium sulfate, sodium aluminum silicate, calcium aluminum silicate, and aluminum silicate. The eWG recommended that only numerical maximum levels be set for Al-containing food additives and that they be expressed on an Al basis, and that the potential exposure to Al from the food additive intake be assessed and compared to the PTWI and that proposed levels of Al-containing food additive are not acceptable when the contribution of a single portion of the food with its additive reaches the PTWI. Comments to the draft and the compilation of proposals from the eWG and comments from governments and trade organizations provide further insight into the uses of Al-containing food additives (including the results of a Canadian survey of the maximum levels of use of SALP and sodium aluminum silicate reported by various Canadian food industry stakeholders), and concerns about their safety (www.cclac.org) (http://www.docstoc.com/docs/80793084/Comments_Second_Circular_Aluminium) (ftp://ftp.fao.org/Codex/ccfa43/fa43_10e.pdf). The recommendations of the 43rd session of the CCFA (Xiamen, China, March 14 to 18, 2011) were presented to the 34th session of the Codex Alimentarius Commission of the Joint FAO/WHO Food Standards Programme (Geneva, Switzerland, July 4 to 9, 2011) (http://www.cclac.org/documentos/CCFA/2011/1%20Alinorm/REP11_FAe.pdf). The report of the 34th session does not appear to mention these recommendations or whether action was taken on them (ftp://ftp.fao.org/codex/ Reports_2011/REP11_CACe.pdf).

10. The fate of aluminum relevant to its intake by the human in food

10.1 Absorption

Oral bioavailability (fractional absorption, a.k.a. uptake) is the amount absorbed compared to the amount administered. For Al, systemic bioavailability, the fraction that reaches systemic circulation (blood) from which it has access to the target organs of its toxicity, is most relevant.

Oral ^{27}Al bioavailability from water from a municipal water treatment facility was estimated to be 0.36% in a study of 21 humans (Stauber et al. 1999). Two studies that had only two human subjects each estimated oral Al bioavailability to be 0.1 and 0.22% (Hohl et al. 1994; Priest et al. 1998). The bioavailability of hydrophilic substances that are not well absorbed can be determined by comparing the area under the curve (AUC) × time for the test substance given orally and intravenously (Rowland and Tozer 1995). Using a modification of this approach the oral Al bioavailability in the rat averaged 0.28% and 0.29% (Yokel et al.

2001a; Zhou, Harris, and Yokel 2008). These studies indicate oral Al bioavailability from water is ~ 0.1 - 0.3%.

Oral Al bioavailability from food has been estimated to be ~ 0.1 to 0.15% based on average daily urinary Al excretion compared to average daily Al intake from food (Powell and Thompson 1993; Priest 1993; Nieboer et al. 1995; Ganrot 1986; Priest 2004). Using the AUC × time method, oral Al bioavailability in rats that ate ~ 1 gm of biscuit containing [^{26}Al]-labeled acidic SALP averaged ~ 0.12% (Yokel and Florence 2006) and 0.1% to 0.3% from basic SALP incorporated into cheese (Yokel, Hicks, and Florence 2008). Concurrent consumption of citrate, and to a lesser extent other carboxylic acids, can increase oral Al absorption, as can increased solubility of the Al, a more acidic environment, uremia, and perhaps fluoride (Krewski et al. 2007). Absorption of Al from injected Al in vaccines and allergy immunotherapy is probably completely absorbed over time (Flarend et al. 1997).

Water and food consumption provide ~ 0.1 and 3.5 to 10 mg of Al, respectively, to typical daily Al intake by humans (Table 1). The products of the Al contribution from food and water to the diet × the percentage absorbed (for water 0.1 mg Al consumed daily × 0.3% absorption, delivering 0.3 μg; for food 3.5 to 10 mg Al consumed daily × 0.1% absorption, delivering 7.5 μg of Al to systemic circulation) suggest food provides ~ 25-fold more Al to systemic circulation than does drinking water. This suggests food is the largest single source of Al for the typical human. As food additives are the single largest source of Al in food, they are the greatest contributor to the daily Al intake of the human who is not exposed to other major Al sources (oral antacids, immunization and vaccination injections, occupational exposure and total parenteral nutrition).

10.2 Distribution

Normal adult Al content is (in mg/kg wet weight unless stated otherwise): lung: 20, bone: 1 to 3 mg/kg dry weight, liver and spleen: 1, kidney: 0.5, heart: 0.45, muscle: 0.4, brain: 0.35, and blood: 0.002. Approximately 60, 25, 10, 3, 1, 0.3, 0.25 and 0.2% of the Al body burden is in the bone, lung, muscle, liver, brain, heart, kidney and spleen, respectively (Yokel 1997; Priest 2004). Aluminum localizes at the mineralization front and in osteoid of bone. Aluminum in lung may be from environmentally-derived particles, occupational exposure, and distribution from blood. Approximately 80% of an intravenous dose of aluminum citrate was excreted within a week, suggesting the remainder was retained within the body, some of which was excreted over a longer time (Priest et al. 1995). This indicates that Al accumulates in the body during continuous intake, even in subjects with normal renal function. Brain, bone, liver and serum Al concentrations increase with age (Stitch 1957; Zapatero et al. 1995; Markesbery et al. 1984; Roider and Drasch 1999; Shimizu et al. 1994; Hellström et al. 2005).

The terminal half-life ($t_{½}$) of Al has been found to be quite long in the rat and human (reviewed in (Krewski et al. 2007)). The whole-body $t_{½}$ was estimated to be ~50 years in one human who received an intravenous injection of ^{26}Al citrate, based on whole-body ^{26}Al monitoring (Priest 2004). The $t_{½}$ of Al in the rat brain has been estimated to be from ~ 150 days to considerably greater (Yokel et al., 2001b; Yumoto et al., 2003). The lack of a good understanding of allometric scaling of metal elimination rates for rat to human make it difficult to predict the human brain Al $t_{½}$. Given the large percentage of Al in bone, it probably drives the Al concentration and elimination $t_{½}$ throughout the body.

10.3 Biotransformation

Aluminum associates avidly with transferrin in the blood, which binds > 90 of Al in the blood. In the cerebrospinal fluid, and presumably brain extracellular fluid, ~90 of the Al is believed to be associated with citrate (calculated by Dr. Wesley Harris; Yokel and McNamara 2001).

10.4 Excretion

The kidneys account for > 95% of Al excretion, presumably by glomerular filtration of aluminum citrate. Humans who daily consume 3.5 to 10 mg of Al would be expected to excrete 4 to 12 µg, generating a typical urinary Al concentration of 2 to 10 µg/L. Reduced or absent renal function creates the risk of Al accumulation and toxicity. The biliary route accounts for most of the remaining excreted Al.

11. The most susceptible population to aluminum toxicity

Those who have impaired or no renal function are at the greater risk of Al toxicity because of their reduced ability to excrete it. In a study, patients with chronic renal insufficiency who used Al kitchen utensils for > 1 year were divided into two groups, one that continued to do so for 3 months; the other used stainless steel utensils. The latter group showed a significantly greater decrease in serum Al and daily urine Al excretion. These results suggest Al kitchen utensils may be a significant Al source for this population (Lin et al. 1997). When foods provide more Al in the diet than cookware, as is usually the case, one might expect foods that contain considerable Al to have an even greater contribution to the Al body burden of patients with chronic renal insufficiency than produced by Al kitchen utensils.

12. The primary known adverse health effects of aluminum in the human

There is no good evidence that Al is essential or beneficial for the human. Aluminum can produce toxicity to the central nervous, skeletal and hematopoietic systems. It can produce an encephalopathy in renal-impaired humans (dialysis encephalopathy), cognitive deficits in young children, a low-turnover bone disease, a microcytic hypochromic anemia, and has been implicated as an environmental factor that may contribute to some neurodegenerative diseases, including Alzheimer's disease.

The toxicity of Al has been extensively reviewed by the World Health Organization, for the US Department of Health and Human Services, and most extensively by a multi-national group led by Daniel Krewski (WHO 1997; ATSDR 2008; Krewski et al. 2007). When hemodialysis was initially extensively used, some patients developed a progressive encephalopathy that was fatal within 6 months. Dialysis (associated) encephalopathy (aka: dialysis dementia) is characterized by dyspraxia, dysarthria, emotional changes, trembling, ataxia, myoclonus, and fatal convulsions. It is associated with elevated levels of Al in the brain, and serum Al > 80 µg/L (Nieboer et al. 1995). Dialysis encephalopathy was due to Al contamination of the dialysis fluids and administration of Al-based phosphate binders to form an insoluble Al phosphate in the intestine, facilitating phosphate elimination, a goal not well achieved by dialysis. The renal dialysis patient is highly susceptible to Al accumulation and toxicity from Al in dialysis fluids because the Al can diffuse across the dialysis membrane, it rapidly and very strongly binds to transferrin in the blood, and these patients lack the primary route of Al elimination, renal function. Exposure to lower levels of Al than those that caused Al-induced encephalopathy can produce a low turnover bone disease and a microcytic anemia.

12.1 Aluminum-induced bone disease

Aluminum-induced low-turnover bone disease is manifest as osteomalacia and an adynamic bone disease (Bushinsky 1997). Aluminum-induced bone disease in dialysis patients is seen when serum Al is > 30 µg/L and when stains show Al at 30% of the trabecular bone surface (Landeghem et al. 1998; Malluche and Monier-Faugere 1994).

12.2 Aluminum-Induced microcytic anemia

A microcytic, hypochromic anemia is associated with elevated plasma Al in chronic renal failure patients (Jeffery et al. 1996).

12.3 Evidence for and against a role of aluminum in Alzheimer's Disease

Aluminum has been implicated in the etiology of Alzheimer's disease (AD) (Kawahara 2005; Spencer 2000; Gupta et al. 2005; Krewski et al. 2007; Bondy 2010; Tomljenovic 2011). Hallmark neuropathological signs include neurofibrillary tangles (NFTs), senile plaques (SP), and cerebrovascular amyloid. Early onset AD usually has a familial link, due to gene mutations which result in increased secretion of neurotoxic amyloid β protein (Aβ). No specific gene mutations have been associated with late-onset/sporadic forms of AD which account for 85 to 95% of AD cases. The lack of identified hereditary links for the majority of AD cases suggests environmental factors are likely to interact with other factors to cause this disease. Aluminum is one of the suggested environmental contributors. The genesis of the hypothesis that Al plays a role in the etiology of AD was an observation reported in 1965 of neurofibrillary degeneration in rabbit brain after intracerebral Al injection, which resembled, but was not identical to, the NFTs of AD (Klatzo, Wisniewski, and Streicher 1965). Similarly, the neuropathology in dialysis encephalopathy is different from that seen in AD. The observation of elevated Al in post-mortem brain samples (that typically weighed scores of milligrams) of humans with AD reported in 1973 was interpreted as suggesting a role for Al in AD (Crapper, Krishnan, and Dalton 1973). This was followed by many studies, some of which found a few-fold or smaller increase of the Al concentration in AD victim brains than in controls, and some which did not. Studies investigating an elevated level of Al in AD brain using microprobe techniques, such as energy dispersive (electron probe) X-ray microanalysis, secondary ion mass spectrometry, and laser microprobe mass spectroscopy which can quantify Al within a cell, NFT or SP, as well as Al-selective stains, have also produced mixed results (Yokel 2000). If Al is elevated in AD brain, it is not reflected in cerebrospinal fluid Al, which has generally not been found to be elevated (Kapaki et al. 1993). Even if Al is elevated in AD brain, it does not prove cause and effect. The neuronal degeneration of AD may result in accumulation of metals, such as Al.

Another approach to address the potential role of Al in AD is the epidemiological study of the association between the concentration of Al in drinking water and AD incidence, comparing geographic regions where drinking water Al concentrations differ. Again, the results of many such studies are not consistent. The majority reported an increased risk of AD associated with higher drinking water Al concentration. Some of the differences were statistically significant. A major review published in 2007 conducted a risk characterization of the route of Al intake, the exposure level of concern, and Al exposure in the general population and calculated a margin of exposure (ratio of the exposure level of concern to the exposure level) (Krewski et al. 2007). The exposure level of concern was based on an epidemiological study that showed a relative risk of AD of 2.14 associated with a drinking water Al concentration > 100 μg/L (Rondeau et al. 2000).

Given that Al bioavailability from water is not considerably greater than from food, studies assessing a putative link between Al in drinking water and AD might be mis-focused, and studies investigating a possible association between Al in food and cognitive impairment, dementia and AD might be more relevant. However, these are much more difficult to conduct, due to the multiple sources of Al in foods. It has been stated by US FDA employees that no clinical syndrome has been noted for the very low intake of dietary Al (Humphreys and Bolger 1997). It was also suggested that the risk of adverse effects of dietary Al, if there are any, is extremely low (Soni et al. 2001). The only published study addressing the potential association between Al in food and dementia was a preliminary study of 23 newly-diagnosed AD patients and 23 matched non-demented controls to ascertain the relationship between consumption of foods generally high in Al during the previous 5 years and dementia. The results showed increased odds ratios for many food categories. However, the results were only significant for the category containing pancakes, waffles, biscuits, muffins, cornbread and corn tortillas (Rogers and Simon 1999).

13. Determination of aluminum body burden and its treatment

Determination of elevated body burden of Al might be made by quantification of Al in blood. However, this is quite difficult because the blood Al concentration of the normal healthy human is very low, thought to be < 6 μg/ml, and probably ~ 2 μg/ml (Daniela et al. 2002; House 1992; Valkonen and Aitio 1997), and due to its ubiquitous nature, sample contamination is very easily obtained. The desferrioxamine test, a single injection of 5 mg/kg this chelator, has been used to assess Al body burden, as evidenced by an increase in serum Al and urinary Al clearance. The test dose is used because steady-state serum Al concentrations do not correlate well with the Al deposition in bone and soft tissues that results from long-term Al exposure (Yokel 2002; Seo et al. 2007). Up to 5 mg/kg of desferrioxamine once or twice weekly has been shown to be safe and effective for long-term treatment of Al overload, and reduction of Al-induced encephalopathy, bone disease and anemia (Kan et al. 2010).

14. Aluminum toxicity in the premature infant

Another high risk group for Al toxicity is premature infants who are fed intravenously, because they do not tolerate oral feeding. The total parenteral nutrition feeding solutions given intravenously can contain significant Al, which is primarily derived from the calcium gluconate and phosphates used as components of the solution. As this feeding solution is given intravenously, and therefore 100% bioavailable, to premature infants whose kidneys are not fully matured and therefore less able to excrete the Al, they are at risk of sufficient Al accumulation to develop metabolic bone disease, cholestatic hepatitis, and reduction of mental development. To address this concern the U.S. Food and Drug Administration adopted a labeling requirement for Al in large and small volume parenterals used to prepare total parenteral nutrition solutions. This has not solved the problem.

15. Summary

For the typical human, food additives provide the largest source of daily Al intake, exceeding water by several orders of magnitude, the latter having been implicated as a contributing factor to dementia, specifically Alzheimer's disease. The food additive contributing the greatest amount of Al to the diet is SALP, used in its acidic form as a leavening agent in baked goods and in its basic form as an emulsifying agent in cheese. Acidic SALP is an approved food additive in most countries. Basic SALP is approved for use in many fewer countries. There has been debate for a century about the safety of Al as a food additive. This debate continues, and is reflected in recently changed established tolerable intake and minimal risk levels and changes under consideration. It has been noted that there is no good evidence of adverse health effects attributed to Al from food, that this issue has not been adequately directly assessed, and that those who lack good renal function, and therefore unable to efficiently eliminate Al, are at greatest risk of Al-induced toxicity, perhaps from food. As with the controversy concerning a possible role of Al in the etiology of Alzheimer's disease, the controversy about possible adverse effects of Al as a food additive will continue until more definitive demonstration of safety or untoward effect is demonstrated.

16. References

Alfrey, A. C., G. R. LeGendre, and W. D. Kaehny. 1976. The dialysis encephalopathy syndrome. Possible aluminum intoxication. *NEJM* 294 (4):184-188.

Allen, J. L., and F. J. Cumming. 1998. Aluminum in the food and water supply: An Australian perspective. Melbourne: Urban Water Research Association of Australia, a division of Water Services Association of Australia.

Anon. 1998. Rhodia adds SALP. *Chemical Week*, Sep. 23, 1998, 73.

Anon. 1913. Culinary and chemical experiments with aluminium cooking vessels. *The Lancet* i:843.

ATSDR. 2008. Toxicological profile for aluminum. US Department of Health and Human Services, Public Health Service, Agency of Toxic Substances and Disease Registry.

Biego, G. H., M. Joyeux, P. Hartemann, and G. Debry. 1998. Daily intake of essential minerals and metallic micropollutants from foods in France. *Sci Total Environ* 217 (1-2):27-36.

Bohrer, D., S. M. Oliveira, S. C. Garcia, P. C. Nascimento, and L. M. Carvalho. 2010. Aluminum loading in preterm neonates revisited. *J Pediatr Gastroenterol Nutr* 51 (2):237-241.

Bondy, S. C. 2010. The neurotoxicity of environmental aluminum is still an issue. *Neurotoxicology* 31 (5):575-581.

Bushinsky, D. A. 1997. Bone disease in moderate renal failure: cause, nature and prevention. *Ann Rev Med* 48:167-176.

Committee on Food Additives survey data, National Research Council (U.S.). 1984. *Poundage update of food chemicals, 1982*. 2 vols., PB 84-16214, Washington, D.C.: National Academy Press.

Crapper, D. R., S. S. Krishnan, and A. J. Dalton. 1973. Brain aluminum distribution in Alzheimer's disease and experimental neurofibrillary degeneration. *Science* 180 (85):511-513.

Daniela, B. B., D. F. Antonella, L. Antonela, F. Maurizio, and P. Augusta. 2002. Aluminum contamination in home parenteral nutrition patients. *JPEN J Parenter Enteral Nutr* 26 (Suppl):S30-S31.

Datta, N. C. 1935. Metallic contaminations of foods. II. Effect of cooking and storage on foodstuffs in aluminum vessels. *Proc - Indian Acad Sci, Section A* 2B:322-332.

Delves, H. T., C. E. Sieniawaska, and B. Suchak. 1993. Total and bioavailable aluminium in foods and beverages. *Anal Proc* 30 (9):358-360.

Döllken, von. 1898. Ueber die wirkung des aluminiums mit besonderer beriicksichtigung der durch das aluminium verursachten lasionen im centralnervensystem. *Arcr Exp Path Pharmacol* 40:98-120.

Ellinger, R.H. 1972. *Phosphates as food ingredients*. Edited by R. C. Weast. Cleveland, OH.: CRC Press.

FAO/WHO. 2006. Summary and conclusions of the sixty-seventh meeting of the Joint FAO/WHO Expert Committee on Food Additives (JECFA)

FAO/WHO, Joint FAO/WHO Expert Committee on Food Additives. 2007. Evaluation of certain food additives and contaminants *World Health Organization technical report series* 940:92pp.

FASEB, Life Sciences Research Office, Federation of American Societies for Experimental Biology. 1975. Evaluation of the health aspects of aluminum compounds as food ingredients, Contract No. FDA 223-75-2004, U.S. FDA Report FDA/BF-77/24, NTIS-PB 262 655,

Fimreite, N., O. O. Hansen, and H. C. Pettersen. 1997. Aluminum concentrations in selected foods prepared in aluminum cookware, and its implications for human health. *Bull Environ Contam Toxicol* 58 (1):1-7.

Flarend, R., T. Bin, D. Elmore, and S. L. Hem. 2001. A preliminary study of the dermal absorption of aluminium from antiperspirants using aluminum-26. *Food Chem Toxicol* 39:163-168.

Flarend, R. E., S. L. Hem, J. L. White, D. Elmore, M. A. Suckow, A. C. Rudy, and E. A. Dandashli. 1997. In vivo absorption of aluminium-containing vaccine adjuvants using ^{26}Al. *Vaccine* 15 (12-13):1314-1318.

Fukushima, Masako, and Akio Tanimura. 1998. Aluminium absorption by food from packaging materials. *Nippon Kasei Gakkaishi (J. Home. Econ. Jpn.)* 49 (12):1313-1317.

Ganrot, P. O. 1986. Metabolism and possible health effects of aluminum. *Environmental Health Perspectives* 65:363-441.

Gies, W.J. 1911. Some objections to the use of alum baking-powder. *JAMA* 57:816-821.

Gitelman, H. J., F. R. Alderman, M. Kurs-Lasky, and H. E. Rockette. 1995. Serum and urinary aluminium levels of workers in the aluminium industry. *Ann Occup Hyg* 39 (2):181-191.

Golub, M. S., and S. L. Germann. 2001. Long-term consequences of developmental exposure to aluminum in a suboptimal diet for growth and behavior of Swiss Webster mice. *Neurotoxicology and Teratology* 23 (4):365-72.

Gramiccioni, L., G. Ingrao, M. R. Milana, P. Santaroni, and G. Tomassi. 1996. Aluminium levels in Italian diets and in selected foods from aluminium utensils. *Food Add Contam* 13 (7):767-774.

Greger, J. L. 1985. Aluminum content of the American diet. *Food Technol* 39 (May):73-80.

Greger, J. L., and J. E. Sutherland. 1997. Aluminum exposure and metabolism. *Crit Rev Clin Lab Sci* 34 (5):439-474.

Gupta, Veer Bala, S. Anitha, M. L. Hegde, L. Zecca, R. M. Garruto, R. Ravid, S. K. Shankar, R.; Stein, P. Shanmugavelu, and K. S. Jagannatha. Rao. 2005. Aluminium in Alzheimer's disease: Are we still at a crossroad? *Cell Mol Life Sci* 62 (2):143-158.

Hellström, H-O., B. Mjöberg, H. Mallmin, K. Michaëlsson. 2005. The aluminum content of bone increases with age, but is not higher in hip fracture cases with and without dementia compared to controls. *Osteoporosis Int* 16 (12):1982-1988.

Hohl, Ch., P. Gerisch, G. Korschinek, E. Nolte, and T. H. Ittel. 1994. Medical application of ^{26}Al. *Nucl Instrument Meth Physics Res* B 92:478-482.

Horemans, B., A. Worobiec, A. Buczynska, K. Van Meel, and R. Van Grieken. 2008. Airborne particulate matter and BTEX in office environments. *J Environ Monit* 10 (7):867-876.

House, R. A. 1992. Factors affecting plasma aluminum concentrations in nonexposed workers. *J Occup Med* 34 (10):1013-1017.

Humphreys, S.H. 1992. The GRAS review process and aluminum salts. Paper read at Proceedings of the Second International Conference on Aluminum and Health, Feb 2-6, 1992, at Tampa, FL.

Humphreys, S., and P. M. Bolger. 1997. A public health analysis of dietary aluminium. In *Aluminium toxicity in infants' health and disease,* edited by P. F. Zatta and A. C. Alfrey. Singapore, Singapore: World Scientific.

Inoue, T., H. Ishiwata, and K. Yoshihara. 1988. Aluminum levels in food-simulating solvents and various foods cooked in aluminum pans. *J Agric Food Chem* 36:599-601.

Jeffery, E. H., K. Abreo, E. Burgess, J. Cannata, and J. L. Greger. 1996. Systemic aluminum toxicity: effects on bone, hematopoietic tissue, and kidney. *J Toxicol Environ Health* 48 (6):649-665.

Jones, K. C., and B. G. Bennett. 1986. Exposure of man to environmental aluminium--an exposure commitment assessment. *Sci Total Environment* 52 (1-2):65-82.

Jorhem, L., and G. Haegglund. 1992. Aluminium in foodstuffs and diets in Sweden. *Zeitschrift für Lebensmittel-Untersuchung und-Forschung* 194 (1):38-42.

Kan, W. C., C. C. Chien, C. C. Wu, S. B. Su, J. C. Hwang, and H. Y. Wang. 2010. Comparison of low-dose deferoxamine versus standard-dose deferoxamine for treatment of aluminium overload among haemodialysis patients. *Nephrol Dial Transplant* 25 (5):1604-1608.

Kapaki, E. N., C. P. Zournas, I. T. Segdistsa, D. S. Xenos, and C. T. Papageorgiou. 1993. Cerebrospinal fluid aluminum levels in Alzheimer's disease. *Biolog Psychiatry* 33 (8-9):679-681.

Katz, A. C., D. W. Frank, M. W. Sauerhoff, G. M. Zwicker, and R. I. Freudenthal. 1984. A 6-month dietary toxicity study of acidic sodium aluminium phosphate in beagle dogs. *Food Chem Toxicol* 22 (1):7-9.

Kawahara, M. 2005. Effects of aluminum on the nervous system and its possible link with neurodegenerative diseases. *J Alzheimers Dis* 8 (2):171-182.

Klatzo, I., H. Wisniewski, and E. Streicher. 1965. Experimental production of neurofibrillary degeneration. 1. Light microscopic observation. *J Pathology* 24:187-199.

Kovalchik, M. T., W. D. Kaehny, A. P. Hegg, J. T. Jackson, and A. C. Alfrey. 1978. Aluminum kinetics during hemodialysis. *J Lab Clin Med* 92:712-720.

Krewski, D., R. A. Yokel, E. Nieboer, D. Borchelt, J. Cohen, J. Harry, S. Kacew, J. Lindsay, A. M. Mahfouz, and V. Rondeau. 2007. Human health risk assessment for aluminium, aluminium oxide, and aluminium hydroxide. *J Toxicol Environ Health, Part B: Crit Rev* 10, Suppl 1:1-269.

Landeghem, G. F., P. C. D'Haese, L. V. Lamberts, L. Djukanovic, S. Pejanovic, W. G. Goodman, and M. E. De Broe. 1998. Low serum aluminum values in dialysis patients with increased bone aluminum levels. *Clin Nephrol* 50 (2):69-76.

Lin, J. L., Y. J. Yang, S. S. Yang, and M. L. Leu. 1997. Aluminum utensils contribute to aluminum accumulation in patients with renal disease. *Am J Kidney Dis* 30 (5):653-658.

Lione, A. 1983. The prophylactic reduction of aluminium intake. *Food Chem Toxicol* 21 (1):103-109.

Lopez, F. E., C. Cabrera, M. L. Lorenzo, and M. C. Lopez. 2002. Aluminum levels in convenience and fast foods: in vitro study of the absorbable fraction. *Sci Total Environment* 300 (1-3):69-79.

Malluche, H. H., and M. C. Monier-Faugere. 1994. The role of bone biopsy in the management of patients with renal osteodystrophy. *J Am Soc Nephrol* 4 (9):1631-42.

Markesbery, W. R., W. D. Ehmann, M. Alauddin, and T. I. M. Hossain. 1984. Brain trace element concentrations in aging. *Neurobiol Aging* 5 (1):19-28.

Müller, M., M. Anke, and H. Illing-Günther. 1998. Aluminium in foodstuffs. *Food Chem* 61 (4):419-428.

NAS, Committee on the GRAS List Survey-Phase III - Estimates of daily intake. 1979. The 1977 survey of industry on the use of food additives. Washington, D.C.: National Academy of Sciences

Neelam, M.S. Bamji, and M. Kaladhar. 2000. Risk of aluminium burden in the Indian population: contribution from aluminium cookware. *Food Chem* 70:57-61.

Nieboer, E., B. L. Gibson, A. D. Oxman, and J. R. Kramer. 1995. Health effects of aluminum: a critical review with emphasis on aluminum in drinking water. *Environ Rev* 3 (1):29-81.

Panel on Food Additives, Flavourings, Processing Aids and Food Contact Materials, European Food Safety Authority. 2008. Safety of aluminium from dietary intake[1] - Scientific Opinion of the Panel on Food Additives, Flavourings, Processing Aids and Food Contact Materials (AFC). In *Question number: EFSA-Q-2006-168, EFSA-Q-2008-254*, edited by E. F. S. Authority: EFSA Journal, doi:10.2903/j.efsa.2008.754.

Pennington, J.A.T. 1987. Aluminium content of foods and diets. *Food Addit Contam* 5 (2):161-232.

Pennington, J. A. T. 1992. Dietary exposure to aluminum. Paper read at Proceedings of the Second International Conference on Aluminum and Health, Feb 2-6, 1992, at Tampa, Fl.

Pierre, F., F. Baruthio, F. Diebold, and P. Biette. 1995. Effect of different exposure compounds on urinary kinetics of aluminium and fluoride in industrially exposed workers. *Occupat Environ Med* 52 (6):396-403.

Poole, R. L., L. Schiff, S. R. Hintz, A. Wong, N. Mackenzie, and J. A. Kerner, Jr. 2010. Aluminum content of parenteral nutrition in neonates: measured versus calculated levels. *J Pediatr Gastroenterol Nutr* 50 (2):208-211.

Powell, J. J., and R. P. Thompson. 1993. The chemistry of aluminium in the gastrointestinal lumen and its uptake and absorption. *Proceed Nutri Soc* 52 (1):241-253.

Priest, N. D. 1993. The bioavailability and metabolism of aluminium compounds in man. *Proceed Nutr Soc* 52 (1):231-240.

Priest, N.D. 2004. The biological behaviour and bioavailability of aluminium in man, with special reference to studies employing aluminium-26 as a tracer: review and study update. *J Environ Monit* 6 (5):375-403.

Priest, N. D., D. Newton, J. P. Day, R. J. Talbot, and A. J. Warner. 1995. Human metabolism of aluminium-26 and gallium-67 injected as citrates. *Hum Exp Toxicol* 14 (3):287-293.

Priest, N. D., R. J. Talbot, D. Newton, J. P. Day, S. J. King, and L. K. Fifield. 1998. Uptake by man of aluminium in a public water supply. *Hum Exp Toxicol* 17 (6):296-301.

Reilly, C. 1991. *Metal contamination of food*. Vol. Second Edition. Essex, England: Elsevier Science Publishers Ltd.

Riihimäki, V., H. Hänninen, R. Akila, T. Kovala, E. Kuosma, H. Paakkulainen, S. Valkonen, and B. Engström. 2000. Body burden of aluminum in relation to central nervous system function among metal inert-gas welders. *Scand J Work, Environ Health* 26 (2):118-130.

Rogers, M.A.M., and D.G. Simon. 1999. A preliminary study of dietary aluminium intake and risk of Alzheimer's disease. *Age Ageing* 28:205-209.

Roider, G., and G. Drasch. 1999. Concentration of aluminum in human tissues - investigations on an occupationally non-exposed population in Southern Bavaria (Germany). *Trace Elem Electrolytes* 16 (2):77-86.

Rondeau, V., D. Commenges, H. Jacqmin-Gadda, and J. F. Dartigues. 2000. Relation between aluminum concentrations in drinking water and Alzheimer's disease: an 8-year follow-up study. *A J Epidemiol* 152 (1):59-66.

Rowland, M. , and T. N. Tozer. 1995. *Clinical Pharmacokinetics. Concepts and Applications*. third ed. Media, PA: Williams & Wilkins.

Saiyed, S. M., and R. A. Yokel. 2005. Aluminium content of some foods and food products in the USA, with aluminium food additives. *Food Addit Contam* 22 (3):234-244.

Scancar, J, V Stibilj, and R Milacic. 2003. Determination of aluminium in Slovenian foodstuffs and its leachability from aluminium-cookware. *Food Chem* 85 (1): 151-157.

Schaeffer, G., G. Pontes, E. Le Breton, Ch. Oberling, and L. Thivolle. 1928. The dangers of certain mineral baking powders based on alum, when used for human nutrition. *J Hygiene* 28:92-99.

Seo, Y. S., H. W. Gil, J. O. Yang, E. Y. Lee, and S.-Y. Hong. 2007. The clinical study on aluminum levels in patients undergoing hemodialysis. *Korean J Nephrol* 26 (4):435-439.

Severus, H. 1989. The use of aluminium - especially as packaging material - in the food chemistry. In *Aluminium in food and the environment; the proceedings of a symposium organised by the environment and Food Chemistry groups of the Industrial Division of the Royal Society of Chemistry, London, 17th May 1988 - (Special publication, no. 73)*, pp. 88-101, edited by R. C. Massey and D. Taylor. Cambridge: The Royal Society of Chemistry, Thomas Graham House.

Sherlock, J.C. 1989. Aluminium in foods and the diet. In *Aluminium in food and the environment; the proceedings of a symposium organised by the environment and Food Chemistry groups of the Industrial Division of the Royal Society of Chemistry, London, 17th May 1988 - (Special publication, no. 73)*, pp. 68-75, edited by R. C. Massey and D. Taylor. Cambridge: The Royal Society of Chemistry, Thomas Graham House.

Shimizu, H, T Mori, M Koyama, M Sekiya, and H Ooami. 1994. [A correlative study of the aluminum content and aging changes of the brain in non-demented elderly subjects]. *Jap J Geriatrics* 31:950-960.

Sjögren, B., C.-G. Elinder, A. Iregren, D. R. C. McLachlan, and V. Riihimäki. 1997. Occupational aluminum exposure and its health effects. pp. 165-183, In *Research Issues in Aluminum Toxicity*, edited by R. A. Yokel and M. S. Golub. Washington, D.C.: Taylor & Francis.

Smith, E. E. 1928. *Aluminum compounds in food, including a digest of the report of the Referee board of scientific experts on the influence of aluminum compounds on the nutrition and health of man*. New York: P.B. Hoeber, Inc., 378 pp.

Soni, M. G., S. M. White, W. G. Flamm, and G. A. Burdock. 2001. Safety evaluation of dietary aluminum. *Reg Toxicol Pharmacol* 33 (1):66-79.

Speerhas, R. A., and D. L. Seidner. 2007. Measured versus estimated aluminum content of parenteral nutrient solutions. *Am J Health-System Pharmacy* 64 (4):740-746.

Spencer, P.S. 2000. Aluminum and its compounds. In *Experimental and clinical neurotoxicology*, edited by P. S. Spencer and H. H. Schaumburg. New York: Oxford University Press.

Stauber, J. L., T. M. Florence, C. M. Davies, M. S. Adams, and S. J. Buchanan. 1999. Bioavailability of Al in alum-treated drinking water. *Journal AWWA (American Water Works Association)* 91:84-93.

Stitch, S.R. 1957. Trace elements in human tissues. 1. A semi-quantitative spectrographic survey. *Biochem J* 67:97-109.

Takeda, Y., Y Kawamura, and T Yamada. 1998. [Dissolution of aluminium from aluminium foil into foods and effect of food components on the dissolution]. *Shokuhin Eiseigaku Zasshi (J Food Hygienic Society Japan)* 39 (4):266-271.

Tomljenovic, L. 2011. Aluminum and Alzheimer's disease: after a century of controversy, is there a plausible link? *J Alzheimers Dis* 23 (4):567-98.

Tracor-Jitco. 1973. Scientific literature reviews on generally recognized as safe (GRAS) food ingredients - aluminum compounds. Rockville, MD: Tracor-Jitco.

United Kingdom Ministry of Agriculture Fisheries and Food, (MAFF) 1993. Aluminium in food. The thirty ninth report of the Steering Group on Chemical Aspects of Food Surveillance. Food Surveillance Paper No. 39. London: HMSO (Her Magesty's Stationery Office).

Valkonen, S., and A Aitio. 1997. Analysis of aluminium in serum and urine for the biomonitoring of occupational exposure. *Sci Total Environ* 199:103-110.

Wang, L, D Z Su, and Y F Wang. 1994. Studies on the aluminium content in Chinese foods and the maximum permitted levels of aluminum in wheat flour products. *Biomed Environ Sci: BES* 7 (1):91-99.

WHO, International Programme on chemical safety. 1997. *Aluminium*. Vol. 194, *Environmental Health Criteria*. Geneva: World Health Organization.

Wiley, H. W. 1928. The baking powder controversy. *Science* 68:159-162.

Wiley, H. W. 1929. *History of a crime against the food law: The amazing story of the national food and drugs law intended to protect the health of the people, perverted to protect adulteration of foods and drugs*. Washington, 413 pp, pages 400-402 on objection to use of alum in foods.

Wuhrer, J. 1939. Consideration of aluminum from the sanitary, especially from the food-hygienic point of view. *Korrosion und Metallschutz* 15:15-24.

Yokel, R. A. 1997. The metabolism and toxicokinetics of aluminum relevant to neurotoxicity. In *Mineral and Metal Neurotoxicology*, pp.81-89, edited by M. Yasui, M. Strong, K. Ota and M. A. Verity. Boca Raton: CRC Press.

Yokel, R. A. 2000. The toxicology of aluminum in the brain: A review. *NeuroToxicology* 21 (5):813-828.

Yokel, R.A. 2002. Aluminum chelation principles and recent advances. *Coord Chem Rev* 228:97-113.

Yokel, R.A., and R. L. Florence. 2006. Aluminum bioavailability from the approved food additive leavening agent acidic sodium aluminum phosphate, incorporated into a baked good, is lower than from water. *Toxicol* 227:86-93.

Yokel, R.A., and M.S. Golub, eds. 1997. *Research issues in aluminum toxicity*. Washington, D.C.: Taylor & Francis.

Yokel, R.A., C.L. Hicks, and R. L. Florence. 2008. Aluminum bioavailability from basic sodium aluminum phosphate, an approved food additive emulsifying agent, incorporated in cheese. *Food Chem Toxicol* 46:2261-2266.

Yokel, R.A., and P.J. McNamara. 2001. Aluminum toxicokinetics: An updated mini-review. *Pharmacol Toxicol* 88:159-167.

Yokel, R.A., S.S. Rhineheimer, R.D. Brauer, P. Sharma, D. Elmore, and P.J. McNamara. 2001a. Aluminum bioavailability from drinking water is very low and is not appreciably influenced by stomach contents or water hardness. *Toxicology* 161:93-101.

Yokel, R.A., S.S. Rhineheimer, P. Sharma, D. Elmore, and P.J. McNamara. 2001b. Entry, half-life and desferrioxamine-accelerated clearance of brain aluminum after a single ^{26}Al exposure. Toxicol Sci 64:77-82.

Ysart, G., P. Miller, M. Croasdale, H. Crews, P. Robb, M. Baxter, C. de L'Argy, and N. Harrison. 2000. 1997 UK Total Diet Study--dietary exposures to aluminium, arsenic, cadmium, chromium, copper, lead, mercury, nickel, selenium, tin and zinc. *Food Addit Contamin* 17 (9):775-786.

Yumoto, S., H. Nagai, K. Kobayashi, A. Tamate, S. Kakimi, and H. Matsuzaki. 2003. ^{26}Al incorporation into the brain of suckling rats through maternal milk. J Inorg Biochem 97 (1):155-160.

Zapatero, M. D., A. Garcia de Jalon, F. Pascual, M. L. Calvo, J. Escanero, and A. Marro. 1995. Serum aluminum levels in Alzheimer's disease and other senile dementias. *Biol Trace Element Res* 47 (1-3):235-240.

Zhong, C., Y. Wang, H. Xie, Y. Zhao, F. Cai, and Z. Zhang. 1996. [A study on aluminum intake of inhabitants in Nanjing]. *Nanjng Yixueyuan Xuebao* 16 (1):50-53.

Zhou, Y., W. R. Harris, and R.A. Yokel. 2008. The influence of citrate, maltolate and fluoride on the gastrointestinal absorption of aluminum at a drinking water-relevant concentration: A ^{26}Al and ^{14}C study *J Inorganic Biochem* 102:798-808.

13

The Use of Blood and Derived Products as Food Additives

Jack Appiah Ofori and Yun-Hwa Peggy Hsieh
Department of Nutrition, Food and Exercise Sciences,
420 Sandels Building Florida State University, Tallahassee, Florida
USA

1. Introduction

Blood is a rich source of iron and proteins of high nutritional and functional quality. Because of the high protein content of blood, generally about 18% [1], it is sometimes referred to as "liquid protein" [2]. It has been estimated that the approximately 1,500,000 tons of porcine blood produced yearly in China has a protein content equivalent to that of 2,000,000 tons of meat or 2,500,000 tons of eggs [3]. Thus, a valuable protein source is lost if animal blood is discarded as waste and this is compounded by the resulting serious environmental pollution problems. Many countries require that animal blood be disposed off in an environmentally friendly manner, which is a capital intensive process. Accordingly, to eliminate a sizeable pollution hazard and prevent the loss of a valuable protein source, efforts have been made to ensure the utilization of animal blood on a massive scale. A further incentive is the increased profits to be made through adding value to the blood.

The environmental, nutritional and economic benefits derived from the maximal utilization of animal blood, coupled with recent advances in blood collection and processing techniques, have led to a myriad of blood protein ingredients becoming available for use in foods and dietary supplements to serve specific needs (Table 1). By 2001, it was estimated that the food industry was utilizing 30% of the blood produced in slaughterhouses [4]. Current utilization likely surpasses this figure, though there is no data to confirm this. Plasma, the liquid portion of blood remaining after the blood cells [white blood cells (WBCs), red blood cells (RBCs), and platelets] have been removed, is most widely used in the food industry because it is neutral in taste and devoid of the dark color associated with the red blood cells (and hence whole blood) [5]. The traditional use of whole blood and its separated red blood cells as ingredients in food products is generally restricted to such products as blood sausages where the black color is both expected and acceptable [2]. The cellular portion of blood (and for that matter whole blood) has not enjoyed widespread usage in the food industry as a result of its heme component, which imparts an undesirable color, odor and metallic taste to the final product [6,7]. Even in trace quantities hemoglobin imparts a dark-brown color to foods [8]. Their use is also restricted because the cellular fraction is thought to have a higher microbial load, although this has been disproved [9]. The amount of cellular fraction-derived products used in food has traditionally been limited to 0.5 to 2% of the product in order not to jeopardize the sensory qualities of the final

Product	Company	Source of blood	Description	Usage
Fibrimex®	Sonac BV, Netherlands	Porcine or bovine	Thrombin and fibrinogen protein isolate	Cold set binder for meat products
Plasma Powder FG	Sonac BV, Netherlands	Porcine or bovine	Plasma with increased fibrinogen concentration	Cold set binder for meat products
Harimix (C, P or P+)	Sonac BV, Netherlands	Porcine or bovine	Stabilized hemoglobin	Coloring for meat products
Hemoglobin	Sonac BV, Netherlands	Porcine or bovine	Frozen or powder hemoglobin	Natural coloring for meat products
PP	Sonac BV, Netherlands	Porcine or bovine	Frozen or powder plasma	Heat set binder for meat products
Prolican 70	Lican Functional Protein Source, Chile	Bovine	Spray-dried bovine plasma concentrate	Emulsifier, gelling and binding agent in meat-based products, fish-based products, pasta and bakery products
Prietin	Lican Functional Protein Source, Chile	Porcine	Spray-dried porcine whole blood	Emulsifier, gelling, binding and coloring agent in blood sausages, cured meats and pates
Myored	Lican Functional Protein Source, Chile	Porcine or bovine	Natural colorant obtained from the red pigments of blood	Enhance meat color and increase contrast between fat and meat
ImmunoLin®	Proliant, USA	Bovine	Bovine serum concentrate	Immune system supplement
B7301	Proliant, USA	Bovine	Spray-dried bovine red blood cells	Enhance color and iron supplementation for meat products
AproRed	Proliant, USA	Porcine	Stabilized hemoglobin	Coloring for meat products
Aprofer 1000®	APC Europe, Spain	Porcine or bovine	Heme iron polypeptide	Iron supplementation
Proferrin®	Colorado Biolabs Inc., USA	Bovine	Heme iron polypeptide	Iron supplementation
Vepro 95 HV	Veos NV, Belgium	Bovine	Globin (hemoglobin with the heme group removed)	Emulsifier in meat products
Plasma	Veos NV, Belgium	Bovine or porcine	Liquid, powder, frozen or flaked plasma	Gelling and binding agent in meat products

Table 1. Examples of blood-derived protein ingredients used as food additives and dietary supplements

product [10]. This notwithstanding, recent efforts have focused on improving the utility of the cellular fraction in food proteins by removing the heme component using various techniques [6, 11-16] to produce an off-white product known in the food industry as globin or decolorized blood. Many of these treatments, however, are either not cost-effective, impart a salty or bitter taste to the globin recovered [17], or are only partially successful in alleviating the dark color associated with the heme component [8]. Thus, further processing methods continue to be devised to produce globin that lacks the bitter taste and has a whiter appearance. One such method involves treating RBCs recovered from blood with papain to release the heme, followed by further treatment with a bleaching agent, sodium hypochlorite. The isolated globin is almost tasteless with a white color [17].

Despite the technical successes achieved, animal blood proteins remain underutilized, largely due to consumer concerns. This paper reviews some of the specific uses of blood proteins in both the meat and non-meat industries and discusses the consumer issues hindering its maximal utilization.

2. Meat industry

The meat industry uses the bulk of the blood proteins employed as ingredients in the food industry, mainly as a binder but also as natural color enhancers, emulsifiers, fat replacers and meat curing agents. These will be discussed in turn below. However, some limited usages, for example as biodegradable natural casings for sausage products, will not be considered as such blood-based films have a relatively high solubility which may limit their application [18].

2.1 Binder

Binders have traditionally been used in meat products to counter the textural and sensorial changes brought about by processing. In addition to absorbing the moisture that is released from meat during thermal processing, they are used to bind water and fat to stabilize meat emulsions in ground meat products [19]. The development of comminuted or emulsion type meat systems that incorporate binders allows more efficient utilization of tough meat from spent animals. The partial replacement of meat with binders reduces product costs while improving, or at least maintaining, the nutritional and organoleptic qualities of the final product [20]. Binders have a macromolecular structure that has the capacity to form matrices that retain aroma and nutrients and also entrap large amounts of water in such a manner that exudation is prevented [21]. They are also used in the restructuring of meat, which is geared towards the transformation of lower value cuts and quality trimmings into consumer-ready products of higher value that resemble intact muscles such as steaks, chops or roasts. Meat restructuring is also driven by increasing consumer demand for exact portions in commercial meat and fish products.

Blood plasma and isolated blood proteins are both used as binders in the meat industry. Plasma functions as a binder in meat systems due to its ability to form gels upon heating, while the performances of plasma proteins are comparable, if not superior, to other binders. In terms of binding properties, blood plasma may be an alternative to the egg albumen traditionally used in the food industry for binding purposes [22], and remains the standard by which to judge other binders. Interestingly, plasma powders have been reported to be less effective than egg white powder, though still better than other binders (meat powders, gelatin, wheat gluten, isolated soy protein, and a combination of sodium alginate and calcium carbonate [23]). However, this study pointed out that the superiority of egg white powder as a binder over plasma powder was based solely on muscle to muscle binding; accurate muscle food product binding should consider not only muscle-to-muscle binding but also muscle-to-fat and fat-to-fat binding and plasma powders exhibited higher muscle to fat and fat to fat binding compared to egg white powder [23]. At acid pH, plasma produces soft and exuding gels and so is not effective as a gelling agent or capable of water retention in products with a low pH such as fermented products, as these functional properties are strongly affected by changes in the protein structure brought about by the acidic environment [24]. Subjecting porcine plasma to simultaneous treatment with microbial transglutaminase and high hydrostatic pressure has been shown to improve the gelling properties of porcine plasma under acidic conditions [25].

Besides their excellent functionality as binders, the utilization of blood proteins in the preparation of restructured meats offers a number of health and industry benefits. In the past, preparation of restructured muscle products required tumbling meat pieces with salt to extract soluble protein for meat binding. This tumbling action, however, damages the texture of the product and the use of salt is also a concern for consumers who are aware of the link between the sodium content of food and the increased risk of cardiovascular and bone disease [26, 27]. Conventionally, the process also involved the use of heat to bind the myofibrillar proteins extracted from the meat by the combined salt and tumbling action, and as such was referred to as hot-set binding [28]. The problem with the hot-set method is that the product must be sold either pre-cooked or frozen because product binding is very low in the raw state. A cold set binding agent that relies on the physiological clot forming action of the plasma proteins fibrinogen and thrombin is available commercially as Fibrimex®, and can be obtained from both bovine and porcine sources. Fibrimex® is produced through a patented process by the Dutch company Sonac BV and its binding action is based on the transformation of fibrinogen into fibrin by the action of thrombin. The fibrin produced then interacts with collagen, enabling the binding of meat pieces in reconstituted meat [29]. Fibrimex® produces restructured meat that can be sold in the chilled or raw state, with an eating quality comparable to that of cuts from intact muscles. The cold-set method utilized reduces oxidative rancidity [30] and discoloration [31], both of which are common with the hot-set method. A further advantage is that it develops sufficient bonds to eliminate the cavities [29] invariably created in the meat pieces during deboning. Products made with this cold-set binder are more versatile, as they can be treated and cooked in ways similar to fresh meat. Sonac BV also produces a cold set binder known as Plasma Powder FG, a plasma product with an increased fibrinogen concentration.

Two other commonly used cold set binders in the industry are Activa™ TG-RM (a transglutaminase obtained from microbial sources) and alginate (a polysaccharide that works in combination with a divalent cation). Compared to Activa™ TG-RM, Fibrimex® has been reported to produce less binding in both raw and cooked restructured pork [32]. Several studies have reported that restructured beef steaks formulated with Fibrimex® produced a weaker bind compared with alginate. However, once cooked, Fibrimex® steaks were found to have a binding strength equal or stronger than that of steaks restructured with alginate [33, 34]. Interestingly, other researchers have reported the binding strength of both raw and cooked steaks formulated with Fibrimex® to be stronger than those reformulated with alginate [35]. In this study, whereas some of the sensory panelists scored alginate formulated meat samples as bland or lacking uniformity in texture, no such unfavorable comments were made regarding the Activa™ and Fibrimex™ samples. Nevertheless, all three binders were rated as producing satisfactory binding in re-formed steaks and having overall acceptability [35]. These perceived differences may arise as a result of differences in the experimental procedure, for example the type of muscle used and the amount and proportion of binder used. More research is needed to identify the optimum amount of plasma proteins to utilize in particular meat products to achieve the desired results.

2.2 Natural color enhancer

Color is known to be one of the main factors used by consumers when evaluating the quality and freshness of meat products. The trend towards natural colorants due to adverse reactions to some artificial food dyes (and their impurities) [36, 37] has resulted in the substitution of artificial colorants with natural ones where technologically feasible. Although there are many safe natural sources of colorants, their suitability for use in foodstuffs depends largely on the availability of the raw material. In this sense, hemoglobin, which is obtained from slaughtered animals, represents a good source of natural red colorant given the large quantities of blood generated daily. It is important to note that the use of hemoglobin as a colorant has the added benefit of combating iron deficiency, which will be discussed in detail in a later section of this paper. However, its red color is unstable and largely dependent on the oxidation state of the heme iron. Oxyhemoglobin, the dioxygen ferrous form is bright red in color but when the heme iron is oxidized to Fe^{3+} the resulting hemoprotein, known as methemoglobin (metHb), has a characteristic brown color that is undesirable. Fortunately, the concentration of metHb in mammalian red blood cells is typically only 1 to 3% of the total pigment, mainly as a result of the presence of metHb reductase [38]. However, the processing techniques utilized to preserve the shelf life and ensure the safety of the product, greatly increase the amount of metHb due to hemoglobin autooxidation [39]. Although spray-drying is a good way to preserve the red blood cell fraction, even better than freeze drying [39], and is a necessary step to ensure the microbial quality of blood-derived ingredients, oxidation of the hemoglobin may continue during the storage of the spray-dried powder. Because of the unstable nature of hemoglobin, its use as a food ingredient is only possible once it has been stabilized. Hence, hemoglobin powders that are used as food colors are treated with protecting agents that form complexes with hemoglobin, prevent oxygen accessibility to heme iron and ultimately reduce hemoglobin auto-oxidation. The following few paragraphs will look at some of the hemoglobin-stabilizing techniques employed to make hemoglobin more useful as a natural colorant in meat products.

Chelating agents such as nicotinic acid and nicotinamide that have the ability to form complexes with the heme moiety have been reported to be effective in preventing hemoglobin auto-oxidation during drying and subsequent storage [40]. Nicotinic acid at a concentration of 2% w/v has been found to considerably retard the deterioration of hemoglobin concentrate during both spray-drying and subsequent storage. A positive effect was also detected with 2.5% w/v of nicotinamide, though the results were not as impressive as for nicotinic acid. However, in the presence of starch gels nicotinamide proved better at stabilizing the red color of hemoglobin in comparison with nicotinic acid [39]. The use of these chelating agents is not without problems. Nicotinamide, for example, has been found to be a harmful vasodilator [41]. Conversion of hemoglobin into the more stable and organoleptically acceptable carboxyhemoglobin by saturation with carbon monoxide (CO) has been proposed as an alternative method to stabilize hemoglobin. The underlying principle of the technique is based on the strong affinity between CO and hemoglobin [42] and the low dissociation of the complex [43] that provides a strong and enduring linkage of the CO to hemoglobin, which can thus tolerate cooking and/or dehydration. In addition, CO-treated blood has a redder color and hence greater potential for use in meat product formulations [44]. Studies have shown that CO-treated porcine spray-dried blood stored in low oxygen transmission rate (OTR) bags retained an acceptable reddish color after 12 weeks of storage compared to untreated dry blood, which turned brown immediately upon drying. Comparatively, CO-treated porcine spray-dried blood in high oxygen transmission rate (OTR) packaging presented color indices similar to those of untreated blood after 12 weeks of storage. The authors explained that the observed difference could be as a result of the oxygen permeability of the high OTR bags and light oxidation, which caused the otherwise stable carboxyhemoglobin to turn brown [45]. Liquid porcine blood treated with CO at different pH levels (7.4, 6.7, and 6.0) and kept under refrigeration for 4 days maintained a more stable and attractive red color than fresh blood. The study was restricted to 4 days of refrigerated storage as this is the typical period for which the bacterial count remains acceptably low and hence the blood maintains a safe microbiological profile, allowing its safe utilization as a food ingredient [44]. There are currently several such hemoglobin-containing products available for use as natural color enhancers in meat products. However, effective stabilizing methods for hemoglobin remain an issue. It is possible that shifting the focus to the development of better packaging technology may avoid the need to use stabilizing agents, resulting in the production of hemoglobin-based colorants that are both safe and more natural.

2.3 Emulsifier

Emulsifiers are utilized in emulsified meat products such as frankfurters, pate, luncheon meat and mortadella to bind meat proteins, fat and water in a stable emulsion. Emulsifiers also help redistribute the fat finely throughout the product so the right texture is obtained. Emulsifiers often substitute for fat in association with other ingredients in low-fat meat products to give the final product sensory qualities comparable to those of their full-fat versions. The food industry has access to a wide range of emulsifiers suitable for use in foods and, in particular, meat products though these tend to be proteins. Proteins are the most common emulsifying agents in the food industry because they occur naturally and are generally both non-toxic and widely available [46]. Proteins that are used as emulsifiers are obtained from a variety of sources, both animal and plant. Processors, however, generally prefer animal proteins because they are better emulsifiers than plant proteins. Amongst the emulsifiers obtained from animal sources, casein and its salt derivatives are the most popular and are widely used in the food industry for this purpose. Less expensive protein sources with comparable emulsifying capacities are being sought to replace casein, which is expensive because of its high processing costs [47]. This is particularly important in developing countries, where milk products as casein must be imported.

Blood proteins have been found to have emulsifying properties that are comparable, and even in some situations, superior to those of casein and can therefore adequately replace casein as the emulsifying agent in meat products. Silva and others [47] compared the emulsifying properties of sodium caseinate (SC) and globin isolated either by the acidified acetone (AG) method or the carboxymethyl cellulose (CG) method in the absence and presence of salt. In the absence of salt AG and CG showed a higher emulsifying capacity (EC), which measures the ability of the emulsifier to migrate to the oil-water interface, than SC. When salt was present, however, SC showed a higher EC than globins. The emulsifying activity index (EAI), which measures the ability of the emulsifier to stay in the oil-water interface immediately after the emulsion has formed, was greatest for AG with salt present. In the absence of salt, both SC and AG exhibited EAIs that were greater than that of CG. At no and very low salt concentrations there was no difference in emulsion stability (ES) (a measure of the ability of the emulsifier to stay in the oil-water interface after storage for a while or heating of the emulsion) among all three proteins. ES was, however, greater for CA for a salt concentration of 0.25mol/L. Since the emulsifying property is a cooperative function of EC, EAI, and ES, it seems that casein as an emulsifier has an edge over globin in high salt products. This would, however, not be an advantage in the food industry, which is tending to minimize the use of salt in its products because of the problems associated with high salt consumption mentioned above.

Globin also compares favorably to other proteins used as emulsifiers in meat products and food products in general. Crenwelge and others [48] compared the emulsifying capacities of globin with those of cotton seed, soy and milk proteins and reported that globin had the best emulsifying capacity. The emulsifying activity of globin was greater (EAI of 28) than that of either hemoglobin or ovalbumin, with EAI values of 14 and 8, respectively. Other authors have also reported globin to have emulsifying activity superior to that of hemoglobin and ovalbumin, although BSA, which had an EAI value of 31, was higher. A combination of peptic hydrolysis and the addition of carboxymethyl cellulose (CMC) to globin successfully raised the emulsifying activity of globin to above that of bovine serum albumin (BSA) [49]. All the blood proteins (globin, BSA, hemoglobin) have been shown to be better emulsifiers than egg (ovalbumin), which is generally considered to have good emulsifying properties. However, plasma is reported to have better emulsifying capacity than globin when incorporated into sausages and it can be incorporated into meat products at higher inclusion levels than globin without jeopardizing the visual appeal of the product because the decolorized product retains a reddish hue that is transferred to the final product [50].

In summary, blood proteins have excellent emulsifying capacity and can adequately replace casein and egg in emulsified meat systems. Blood proteins such as hemoglobin, when used as emulsifiers, have the added advantage of providing a good source of heme iron. Compared to casein and egg, both of which are potent allergens and are thus among the "big eight" allergens covered under the Food Allergen Labeling and Consumer Protection Act (FALCPA) law, blood proteins do not present a problem of inducing allergic reactions. In situations involving highly salted products, however, plasma or globin that has undergone further processing such as a combination of peptic hydrolysis and CMC addition to enhance its emulsifying capacity are likely to be the best option.

2.4 Fat replacer

Fat plays an essential role in the diet as a source of vitamins, essential fatty acids and energy [51]. Besides its nutritional role, fat has a major effect on the tenderness, mouth feel, binding properties, juiciness and overall appearance and palatability of processed meats [52]. On the other hand, empirical data has shown a correlation between dietary fat and cardio-vascular disease (CVD) and some types of cancers (colon, breast and prostate). Thus, for today's health conscious consumers, reducing dietary fat is a major consideration. The meat industry has been seriously affected by adverse publicity due to the high fat content and unhealthy fatty acid composition of their products, particularly those that include emulsified meat. For example, the minimum fat content in finely ground meat emulsion type products such as frankfurters, dry fermented sausages, and spreadable dry sausages are about 10%, 20-30%, and 20%, respectively [53] and this includes considerable amounts of saturated fatty acids. In response to consumer demand, the food industry has developed an assortment of low-fat meat products but given the important physico-chemical and sensory role that is played by fat, replacing fat in food products is a daunting task: if not properly done, the sensory appeal of the product will be severely jeopardized. Thus, low-fat meat varieties that are not acceptable in terms of taste and appearance will simply not sell, regardless of any health benefits [54].

Studies have shown blood proteins to have potential as fat replacers in meat products, contributing soluble proteins while at the same time reducing costs. They also reduce the caloric content of food for every gram of fat replaced, since proteins contain fewer calories than fat. In addition blood proteins are a natural product and hence in line with present day consumer demands for natural foodstuffs. In a study by Viana and others [55], replacing the fat with globin, plasma, and a combination of plasma and globin resulted in ham pates with 25%, 35% and 35% less fat, respectively. Among the treatments, the reduction in fat observed was not statistically different. The use of these fat replacers also resulted in an increase in moisture and protein content with no observed change in aroma, taste and consistency. The color of pate was, however, reduced in all the treatments containing blood proteins in comparison with control samples with no added blood protein. Reduction in color on addition of fat replacers is not peculiar to blood proteins as this occurrence has also been reported with the incorporation of other fat replacers in meat products by several authors, as cited by [56]. In short, fat influences the color of meat products and reducing its level will invariably affect the color attributes of the product.

The use of blood proteins as fat replacers compared favorably with other commercially available fat replacers, as shown by the results of several studies. Confrades and others [57] compared the effect of plasma and soy fiber as fat replacers in bologna sausage, concluding that plasma protein had a greater influence on the binding and textural properties of bologna than soy fiber and was hence a better choice. Desmond and others [58] evaluated the sensory, chemical and physical properties of low-fat beef burgers formulated with nineteen different commercially available fat replacers including Plasma Powder U70 and Protoplus® U70, both of which are blood-derived products. They reported that beef burgers formulated with these blood proteins had poor overall quality and flavor attributes. It is worth noting, however, that burgers containing these blood proteins had individual sensory traits that were superior to burgers containing other fat replacers. For example, the moistness and juiciness of burgers formulated with Protoplus® U70 was superior to 15 of the 17 non-blood derived fat replacers tested. Burgers containing Protoplus® U70 were rated as having the best overall texture, together with three other non-blood fat replacers. Inclusion of these blood proteins also produced physical characteristics that were superior to most of the other fat replacers used. For example, beef burger containing these blood proteins had % cook yields that were greater than 13 of the 17 other fat replacers tested. Similarly, the WHC of beef burgers containing Plasma Powder U70 and Protoplus70 were superior to 15 and 12 other fat replacers, respectively.

Blood proteins therefore provide a cheaper alternative protein for use as fat replacers in the production of low-fat meat products. However, this may not satisfy some consumers due to the non-meaty flavor resulting from their inclusion. Guzman and others [59] have attributed the lower meaty scores when blood proteins are incorporated to the specific flavor characteristics contributed by each type of blood protein. Since a single ingredient is unlikely to produce the desired effect as a fat-replacer in the production of low-fat meat products [57], combining plasma proteins with other fat replacers with superior meaty flavor characteristics promises to be a worthwhile approach.

2.5 Curing agent

Nitrite plays a multifunctional role in the curing process of meat. It is responsible for the red color [60, 61], and characteristics and desirable flavor of cured meats [62, 63]. It may also influence the texture of the product through the promotion of protein cross-linking [64]. Furthermore, nitrite acts as an antimicrobial agent [65, 66]. Although nitrite continues to be the most widely used of all food additives, its use has come under scrutiny in recent years. This is because nitrite has a tendency to react with amines, amides, amino acids and related compounds present in the meat to produce carcinogenic N-nitroso compounds. In addition, ingested nitrite (in the form of residual nitrite) may react with various substances in the gastrointestinal tract (GIT) to produce these N-nitroso compounds [67, 68]. Various studies have confirmed the presence of volatile N-nitrosamines in cured meat compounds, though there are differences in both the type and the quantities reported by different authors [69]. To address these issues, efforts are being made to reduce the amount of nitrite used in the curing system or to develop alternative methods of curing that avoid the use of nitrite altogether. The latter approach may be the best option; the most attractive and reliable method for preventing the formation of N-nitroso compounds is the total elimination of nitrite from the curing process [70]. However, finding a viable alternative for nitrite in the meat curing process is problematic as it is extremely difficult to find a single compound that can fully reproduce the multifunctional role of nitrite described above. To date, no single agent has been identified that is capable of replacing nitrite. The best option may be the use of a composite mixture containing no nitrite but instead comprising a colorant, an antioxidant, and an antimicrobial agent, among others, to produce nitrite-free cured meats [71].

One of the colorants suggested for use in a composite nitrite-free cocktail for meat curing is cooked cured meat pigment (CCMP), which when added to meat prior to cooking duplicates the color typical of nitrite-cured meats. CCMP, a mono and/or di-nitric oxide hemochrome, is synthesized directly or indirectly through a hemin intermediate. The pigment is prepared by reacting bovine or porcine red blood cells with a nitrosating agent and at least one reducing agent at a suitable elevated temperature. The resultant product is either spray-dried or freeze-dried to obtain a powder. The pigment may also be treated with a starch and a binding agent (e.g. gum) to stabilize the pigment [72]. Addition of CCMP to comminuted pork meat at levels of 3 to 30mg/kg produced a pink color after heat-treatment that was visually similar to that of nitrite-treated pork [73]. Although the best performance of CCMP is in meat systems containing low to intermediate concentrations of myoglobin, its performance in dark meats has been found to be satisfactory [71]. Results support CCMP as a healthy substitute for nitrite salt, as CCMP has been found not to be mutagenic compared to nitrite. The use of CCMP in meat emulsions therefore results in a drastic reduction in the residual nitrite that contributes to the formation of carcinogenic N-nitroso compounds in the GIT [69].

Though CCMP was initially developed for use in cured meat products to duplicate the characteristic red color associated with the use of nitrite, it has also been found to have antioxidant properties similar to those of nitrite [74]. At the time of writing, CCMP is not yet available commercially, possibly due to its unstable nature although this is being improved with the use of stabilizers. Further studies to include the addition of antimicrobials in the formulation of CCMP to duplicate all three major functions of nitrite in cured meats may make possible the commercialization of this product in the near future.

3. Non-meat usage

Although the meat industry remains the predominant user of blood proteins, other sectors of the food industry also utilize them in the production of such items as baked goods, dietary supplements and functional foods, usually as egg replacers, protein supplements, iron supplements or a source of bioactive compounds. Some of these uses will be reviewed in this section.

3.1 Egg replacer

In baked goods such as cakes, egg proteins play an important role in defining such final product characteristics as cake volume and texture due to the unique foaming, emulsifying, and heat coagulation properties of eggs [75]. Among the various ingredients utilized in baked goods, eggs are the most costly [76] and they are also a significant source of cholesterol. The use of alternative ingredients to partially or totally replace eggs in baked goods is therefore of interest to processors not only to reduce cost but to provide lower cholesterol baked goods varieties. Blood proteins have been found to serve as useful egg replacers, as reported by several authors. Spray-dried plasma protein concentrates marketed for diverse use in food processing, including as egg replacers, are sold at a cost of only about one-third that of spray-dried egg whites [76].

Lee and others [76] evaluated the sensory and physical properties of white layer cakes made by substituting egg (as dried egg white) either partially (25%, 50% and 70%) or totally (100%) with spray-dried plasma. The authors reported that cake volume was not affected by egg replacement, although cake symmetry decreased at 75% and 100% substitution. In terms of shrinkage, cakes formulated using partial replacement had lower shinkage values than control cakes made entirely with dried egg white; those formulated with total replacement of egg with plasma produced shrinkage that did not differ significantly from the control cakes. This suggests that so far as critical quality indices such as volume and shrinkage are involved, replacing egg with plasma produced superior products, particularly at partial replacement. It is worth noting that the authors reported that replacing egg with plasma at all levels resulted in products with darker and tanner crust and crumb, but these differences notwithstanding, trained panels liked the control cakes and those prepared with partial or total replacement of egg with plasma equally. The same study [76] utilized a blend of 90%

enzymatically hydrolyzed plasma and 10% beef stock as a replacement for eggs in the formulation of devil's food cake, a whole-egg product. The results are not relevant here, however, due to the inclusion of beef stock which makes it difficult to deduce which product effects can be ascribed to the inclusion of plasma only. In a similar study, Johnson and others [77] investigated the feasibility of replacing egg white and whole egg with freeze-dried plasma in the manufacture of white and yellow layer cakes, respectively. In the latter case, 1.3% lecithin was added to compensate for the natural phospholipid content of whole egg. The control cakes consisted of 14% egg white solids and 3% added salt, whereas the experimental cakes contained 18.6% plasma solids with no added salt. The formulation was selected to ensure equivalent protein and salt concentrations for the plasma and egg-white control cakes. The authors reported that cakes made with plasma were at least equal if not greater in volume and also shrank less than the egg-white control cakes. In the case of the yellow layer cakes, the volumes of those made with plasma solids was between 7 and 15% less than those made with whole egg at equivalent protein levels. They concluded that blood plasma was effective in replacing all the egg white or whole egg in cakes and hence shows potential as a low cost egg substitute for use by the baking industry.

Decolorized bovine blood has also been considered as a replacement for egg white in white layer cakes. Although cakes made with decolorized blood had volumes comparable to those made with egg white, cakes made with decolorized blood had an objectionable flavor [78]. The poor sensory quality of cakes formulated with decolorized blood may be due to residual heme. Raeker and Johnson [79] also compared the baking properties of egg white and plasma and their component proteins, reporting that cakes made with egg whites had slightly greater volumes than their counterparts made with freeze or spray-dried plasma, though the differences were not significant. Among the blood proteins, only γ-globulin had baking properties superior to those of plasma. They also noted that separating fibrinogen (which by itself produced the smallest cake volume among the blood plasma proteins) from plasma increased cake volumes. However, given that the primary purpose of using blood proteins as a substitute for eggs in baked goods is to reduce costs, the additional expense incurred by fractionating plasma into its component proteins appears counterproductive and there appears to be no incentive for using fractionated plasma proteins for this purpose.

In summary, plasma proteins can be very effective low cost alternatives for replacing eggs in baked goods. The use of blood proteins derived from the cellular fraction, as decolorized bovine blood leads to the production of baked goods with objectionable odor, however, and further research is needed to counter the flavor defect associated with its use. Also, since the main objective of replacing eggs with plasma is to reduce cost, the use of expensive fractionated blood proteins for this purpose cannot be justified. The removal of fibrinogen from plasma, leading to an increase in cake volumes, may be a practicable means of improving the efficiency of plasma for this purpose, as separation of this component from plasma by centrifugation would be relatively inexpensive.

3.2 Protein supplement

Protein malnutrition is a major problem in developing countries during the transitional phase of weaning in infants, retarding their physical and mental development. Introducing weaning foods containing the right quality and quantity of protein during this transitional period has been recommended by international bodies such as the WHO and FAO as a necessary preventive measure [80]. In many developing countries, diets that are used to supplement breast feeding during the transition from exclusive breast feeding to a mixed diet (between the ages of 6 to 8 months) and thereafter as a major breakfast meal (between age 1 to 6 years) are usually produced from maize flour [81, 82]. Kwashiorkor, a severe protein deficiency disease that is common in some parts of the world has been linked to the quality and quantity of protein in maize when it is the sole source of protein for infants [83]. Unfortunately, animal proteins such as milk and meat that contain high quality proteins in good quantities tend to be expensive and not readily available in developing countries. Consequently, concerted efforts by researchers, government and international bodies to eliminate protein malnutrition have generally been directed towards the use of plant proteins (e.g. from legumes and soy) as substitutes for animal proteins to supplement grains and cereals in the formulation of weaning diets. However, these plant proteins have lower protein levels, lack essential amino acids, and/or contain antinutritive factors [84-86]. Processing plant proteins such as soy to eliminate their antinutritive factors greatly increases the cost and hence defeats the purpose of their use as a low-cost high quality protein to substitute for animal proteins. The extreme processes involved also reduce the protein quality of the final product due to denaturation [87]. Studies with newly weaned pigs have established that incorporating blood proteins into their feeds accelerates daily weight gains [88-89]. Similar results have been observed in newly weaned mice [90], indicating that the benefits of including blood proteins in the diet is not species-specific. Thus, blood proteins, which are abundant, cheap, readily available, devoid of antinutrients, have a good amino acid profile, and a proven track record in animal nutrition, have been suggested for use as protein supplements in infant formula to tackle protein malnutrition. Several studies that have investigated the feasibility of using blood proteins as protein supplements to combat protein malnutrition in developing countries are highlighted below.

Begin and others [91] compared the effect of supplementing a maize diet with bovine serum concentrate (BSC) alone or in combination with multiple micronutrients (MMN), or with whey protein concentrate alone or with MMN (as the control) on the physical growth and micronutrient status of 132 Guatemalan infants between the ages of 6 to 8 months who completed the 8-month long study. Over the period of the study, the subjects gained about 1.8kg in weight and about 8cm in length, which corresponds to approximately 63% and 75% of the expected weight and length gain, respectively, for their age counterparts in North America. The increase in growth by treatment group was not, however, significant. So far as micronutrient status was concerned, the authors reported that infants who received a combination of WPC and MMN showed higher final serum ferritin concentrations than those receiving a combination of BSC and MMN. All other markers of micronutrient status were the same. Their conclusion was that supplementation with BSC or with BSC and MMN was not effective in preventing growth stunting of young children in a peri-urban setting in Guatemala, although they did not specify the basis for this conclusion. It seems likely that their conclusions were based on a comparison with North American children, which is not a good yardstick as growth charts in these regions differ from those used by the WHO. They did, however, note that the children who received the supplements containing bovine serum concentrate consumed about 12% less of the amount administered in comparison to those who received diets containing WPC. This suggests that the flavor of BSC was less popular with the children. It is possible that if more of the maize diet supplemented with BSC had been consumed by the children, the growth rate recorded for this group may have surpassed that of the control group consuming whey supplements. Nonetheless, the results of this study demonstrate that blood proteins can adequately replace other animal proteins such as whey.

The adequacy of blood proteins as protein supplements has also been demonstrated by Oshodi and others [92] who investigated in vitro protein digestibility (using a multienzyme digesting system comprising trypisn, chymotrypsin, and peptidase), amino acid profile and available iron of an infant weaning food prepared from maize flour and bovine blood. Their results indicated an improved digestibility when blood protein concentrate was added to the maize. The amino acid profile of the formula was also satisfactory in meeting the essential amino acid requirement for infants as specified by the FAO/WHO joint committee [80]. The iron content of the formula was double that of maize alone. In this study [92], the bovine blood powder concentrate used was so bland in taste that its addition to the maize flour did not alter the overall flavor. This contrasts with the study discussed earlier where the blends containing BSC appeared to be less liked than those containing WPC, which is understandable as the subjects were still suckling and therefore not accustomed to unusual tastes (in this case BSC; WPC is similar to milk). In a related study by Lembcke and others [93], older children (between the ages of 9 and 25 months) admitted for rehabilitation for severe cases of protein deficiency (manifested as marasmus and kwashiorkor) were fed diets in which 25% or 50% of the dietary protein normally provided by milk was substituted with an equivalent protein from spray-dried bovine serum. Here, the formulas containing the spray-dried bovine serum were well accepted by the children Also, weight gain on the test diets was no different from those on the control diet. It is important to note, however, that in this study [93] the milk fat in the spray-dried bovine serum diet was partially replaced by vegetable oil. The addition of fat, which plays a critical role in the organoleptic quality of foods, could also be responsible for the greater acceptability of the product in this case.

The studies outlined above support the use of bovine blood proteins as efficient protein supplements in cereal and grain based weaning and infant diets as a measure to tackle protein malnutrition problems in developing countries. However, if the target consumers are infants that are starting to consume solid foods, the addition of vegetable oil or other flavors may be necessary to improve the palatability and consequent acceptability of formulas. Where anemic children are involved, the use of bovine blood instead of plasma or serum offers a useful way to combat this condition owing to the heme content of blood.

3.3 Iron supplement

Iron deficiency is the most common nutritional deficiency in the world affecting about 20% of the world's population, with women and children at greater risk [94]. It is induced in part by plant-based diets that typically contain non-heme iron, a form of iron that is poorly absorbed [95, 96]. In infants, iron deficiency can delay normal infant motor and mental function [97. 98]. Iron deficiency manifesting as anemia in pregnant women increases the risk of small or preterm babies [99], which are more likely to suffer from health problems or die in the first year of life. It can also cause fatigue in adults, impairing their ability to do physical work [100]. Several strategies are commonly directed at combating iron deficiency, including supplementation with capsules and tablets and fortification of processed foods. However, in developing countries, food-based strategies remain the most sustainable approach for addressing iron and other micronutrient deficiencies [101]. Fortifying staple foods with heme-iron, which is better absorbed than non-heme iron because its absorption is essentially unaffected by other dietary factors [102], has therefore been suggested as a measure to overcome the problem of iron deficiency. Heme iron is known to result in less gastrointestinal discomfort and oxidative stress than non-heme iron [103, 104]. Dietary sources of heme iron are only of animal origin such as meat and liver, which is often a problem in developing countries where animal products are expensive and not readily available. As bovine blood has the largest amount of heme iron than any animal source [105] and is readily available, it has therefore been suggested for the fortification of staple foods. The success of blood-fortified foods in addressing iron deficiency has been reported by several researchers.

Kikafunda and Sserumaga [105] investigated the physical, chemical, microbiological and shelf life characteristics of bovine blood powder and the sensory attributes of a bean sauce fortified with it and found that the fortified bean sauce was less liked than the non-fortified counterpart in terms of all the sensory attributes (color, flavor and overall acceptability) considered. This is understandable, as the presence of the heme moiety generally imparts an objectionable color and flavor to the final product. It is worth noting, however, that although the fortified bean sauce was less liked it was not rejected outright. As children are among the most vulnerable groups affected, the introduction of heme iron fortified foods into school lunch programs becomes a viable way to combat this problem. The suitability of such an approach has been demonstrated by Walter and others [106], who investigated the effect of bovine-hemoglobin-fortified cookies on the iron status of school children in a nationwide school lunch program in Chile. Average serum ferritin values in both boys and girls were greater in the heme-fortified cookie. In addition, the fortified cookie group had a remarkably low prevalence of anemia. These benefits were achieved at virtually no cost as the researchers estimated that the additional expense of fortifying the cookies increased the cost by only $0.53 per child per year. As far as taste was concerned, the fortified cookies were virtually indistinguishable from the non-fortified cookies. It is possible that the different ingredients utilized in the formulation of the hemoglobin-fortified cookies may have masked the characteristic flavor of hemoglobin, rendering it imperceptible in this case. Some skeptics have opined that it is impractical to use hemoglobin as an iron supplement because it is low in iron (0.35%) [107], poorly soluble at low gastric pH [107], and its absorption is usually less than iron absorption from muscle [108, 109]. Heme iron polypeptide (HIP), a soluble heme moiety with an attached polypeptide obtained from the enzymatic digestion of bovine or porcine hemoglobin has been developed to overcome these concerns. The hydrolysis action enhances iron absorption by preventing the formation of large insoluble heme polymers [107]. Examples of commercially available HIP products are Proferrin® (Colorado Biolabs Inc., Frederick, CO) and Aprofer 1000® (APC Europe, S.A., Barcelona, Spain). The effectiveness of HIP as an iron supplement has been demonstrated by several researchers.

Quintero-Gutierrez and others [110] evaluated the bioavailability of a heme iron concentrate product incorporated into a chocolate flavor biscuit filling in sandwich-type biscuits using the pig as a human model. The heme iron was prepared by isolating hemoglobin from porcine blood followed by enzyme hydrolysis and subsequent separation of globin from the heme group using ultrafiltration. Twenty pigs with slightly induced iron deficiency at weaning were divided into the study group (fed a low iron diet [452.7mg/kg] and sandwich biscuits with the heme iron fortified chocolate flavor filling) and the control group (fed normal food containing 537.1mg/kg of ferrous sulphate). The bioavailability of heme iron in the study group was 23% greater than that of the control group despite the fact that the control group consumed greater amounts of iron. The study group also exhibited better weight increase and reduced mortality compared to the control group. The product had a good sensory and functional appeal as it had a chocolate color and smell, creamy appearance, and appropriate spreadability. The same research group went on to evaluate the acceptability and bioavailability of the heme iron enriched chocolate flavor biscuit filling in sandwich-type biscuits in adolescent girls living in a rural area of Mexico [111]. The girls were divided into three groups: the placebo control (PC) group (consisting of girls with the highest hemoglobin levels at baseline, who were given non-fortified biscuits), and two study groups one of which was given an iron sulfate fortified biscuit and the other a heme iron fortified biscuit. The iron content for both types of fortified biscuits was the same. The two study groups included both anemic and non-anemic girls. The researchers observed that iron bioavailability in the heme fortified biscuit group was 23.7% higher than that in the iron sulfate fortified biscuit group, indicating that the heme fortification was well absorbed and tolerated. At the end of the study, serum ferritin concentrations, a measure of the amount of iron stored in the body, were reduced in the control group and the iron sulfate fortified biscuit group, but not in the heme iron fortified biscuit group. Other studies have confirmed that supplementing with HIP produces a better iron status in terms of better iron absorption and higher storage (serum ferritin levels) compared to iron salts [112, 113].

The findings of these studies agree that hemoglobin provides a low-cost heme-iron source that can conveniently be used to fortify commonly consumed staples in order to address problems associated with iron deficiency. However, though enzymatic hydrolysis increases the iron absorption, the process will add to the cost and may not be suitable in all developing countries. Other ways to increase the iron absorption of hemoglobin without adding to the cost therefore requires further study. Also, in certain situations the addition of flavor components may be desirable to make the final product more acceptable.

3.4 Bioactive compounds

Functional foods have begun to attract a lot of attention because of the growing belief that foods should possess health-promoting qualities. Recent physiological and biochemical research has shown that the protein in food not only furnishes amino acids but also provides bioactive peptides after digestion or food processing [114]. Consequently, bioactive peptides produced from both animal and plant sources are now being widely investigated and have been reported to have antibacterial [115, 116], opioid [117], antitumor [118], antioxidant [119, 120], and angiotensin I-converting enzyme (ACE) inhibitory (anti-hypertensive) [121, 122] activities. Bioactive peptides derived from food proteins are considered milder and safer than synthetic drugs and are easily absorbed [123]. Non-conventional food sources are also being investigated; animal blood, which is both abundant and readily available but greatly underutilized as a protein source, is therefore being actively studied as a potential source of bioactive peptides.

Present day dietary habits have led to an increase in the prevalence of cardiovascular diseases, as evidenced by the fact that hypertension-related diseases make up half of the most common causes of death [124]. The renin-angiotensin-aldosterone system controls blood pressure and ACE plays a critical role by converting the non-active angiotensin I in blood into its active form, angiotensin II through hydrolysis. This, in turn, causes blood vessels to contract and blood pressure to rise. In addition, ACE deactivates bradykinin, a nonapeptide that causes blood vessels to extend, by removing amino acids from its C-terminus, contributing to the rise in blood pressure. Thus, repressing ACE activity with ACE inhibitors lowers blood pressure [125]. However, ACE inhibitor drugs such as captopril, enalapril, alacepril and lisinopril that are utilized as anti-hypertensive drugs have a number of side effects such as cough, taste disturbance and skin rashes [126]. Accordingly, natural food-based anti-hypertensive alternatives may offer a valuable alternative.

Wei and Chiang [127] investigated the possibility of hydrolyzing porcine blood proteins using an enzyme mixture in a membrane reactor for the production of bioactive peptides. Red blood cells, plasma and defibrinated blood isolated from porcine blood were used as the substrates for hydrolysis. Of the three fractions, the red blood cells had the highest ACE inhibitor and antioxidant activities. The unpleasant dark red color of blood was also lost during the process, resulting in the production of a pleasant golden yellow colored product and rendering it more useful in the production of anti-hypertensive functional foods. Yu and others [123] also described the isolation and characterization of ACE inhibitor peptides from porcine hemoglobin. Three enzymes (papain, trypsin and pepsin) were used in the digestion of the globin isolated from hemoglobin and the most active hydrolysates from the peptic digestion identified. The peptides exhibiting ACE inhibitor activity were identified as LGFPTTKTYFPHF and VVYPWT, with the former been identified as a novel peptide. The two ACE-inhibitor peptides both competitively inhibited ACE and maintained the inhibitory activity even after incubation with GIT proteases. There is now a protein supplement product, Rifle High Powered Protein Powder, Chocolate (Theta Brothers Sports Nutrition Inc., Brick, NJ) on the market for body builders that contains a patented bioactive peptide from hemoglobin hydrolysate designated as VVYP. Although the manufacturers do not mention the role of this peptide in the formula, its inclusion is probably designed to offset any rise in blood pressure brought about by intense body building exercises. The ACE inhibitor activity of blood seems not to be species specific, as in addition to its demonstrated effect in porcine and bovine blood it has also been demonstrated in chicken blood hydrolysate [128], where of the three enzymes used (Alcalase, Prozyme 6, and Protease N), the alcalase digested chicken blood produced the highest ACE-inhibition activities.

According to the CDC (Centers for Disease Control), three pathogens (*Salmonella, Listeria,* and *Toxoplasma*), are responsible for 1,500 deaths each year in the US, representing more than 75% of those caused by known pathogens. Given the current context of food safety, protection utilizing natural products as preservatives could be beneficial for safe storage and distribution of meat products [129], which are often responsible for outbreaks of food-borne diseases. This is particularly important as some of the proposed means of addressing food-borne diseases such as irradiation are unacceptable for many consumers, who are increasingly demanding natural products. Blood is known to contain important elements (antibodies and leukocytes) that fight against infection. Turning to blood proteins as sources of anti-bacterial peptides is, therefore, justified and a number of researchers have investigated the anti-bacterial properties of peptides isolated from animal blood. Nedjar-Arroume and others [129] identified four antibacterial peptides in bovine hemoglobin. The total hemoglobin isolate obtained by peptic hydrolysis was found to inhibit the microorganisms *Micrococcus luteus, Listeria innocua, Escherichia coli,* and *Salmonella enteritidis.* Antibacterial activity towards the latter two organisms is of particular importance, as both organisms are frequently implicated in food-borne illnesses. Froidevaux and others [130] have also demonstrated the anti-bacterial activity of pepsin-digested bovine hemoglobin. The antibacterial peptide in this case was the 1-23 fragment (VLSAADKGNVKAAWGKVGGHAAE) of the alpha chain of bovine hemoglobin and it exhibited antibacterial activity towards *Micrococcus luteus*, which is a bacterial strain commonly used as a sensitive strain for the detection of antibiotic substances. They did not, however, test the efficacy of the isolated peptide against common food borne pathogens, which is important because antibacterial activity against *Micrococcus luteus* does not necessary imply antibacterial activity towards other organisms, as was pointed out by Nedjar-Arroume and others [29].

Lipid peroxidation is a serious concern for food manufacturers because it results in the production of undesirable off-flavors and potentially toxic reaction products. Oxidation of membranes and lipoproteins in the human circulatory system has been identified as the culprit in the pathogenesis of vascular diseases such as atherosclerosis and hypertension [131]. Many synthetic antioxidants, including butylated hydroxytoluene (BHT), butylated hydroxyanisole (BHA) and propyl gallate, are commonly utilized in food formulations to inhibit lipid peroxidation. However, the use of these synthetic antioxidants is strictly regulated because they represent a potential health hazard. The antioxidant activity of blood protein peptides has therefore been studied and demonstrated to offer an effective natural alternative to synthetic antioxidants. Xu and others [132] investigated the antioxidant activity of porcine plasma hydrolysate obtained by pepsin and papain digestion in a peroxidation system of aqueous linoleic acid. Both the pepsin and papain digested hydrolysates exhibited significant activities against linoleic acid oxidation and good DPPH free radical scavenging ability. However, the chelating power of the pepsin-digested hydrolysate was greater than that of the papain-digested hydrolysate. Their recommendation was therefore that pepsin digested porcine plasma hydrolysate has the potential to serve as a potent natural antioxidant in foodstuffs.

Albumin, a major protein in blood [133] is considered a major circulating antioxidant in the blood and Bishop and others [134] have reported that albumin protects cultured cells from oxygen radical damage. An in vitro study has also shown that albumin protects human low-density lipoproteins from oxidation [135]. Wang and others [136] therefore sought to examine the antioxidant ability of crude plasma and globulin hydrolysates and also that of the peptide fractions. Of these, the peptide fractions were found to have better lipid peroxidation inhibitory activity than vitamin E, a known antioxidant, while the antioxidant properties of both the crude hydrolysates and small molecular weight peptide fractions of albumin were superior to those of globulin. The authors noted that the hydrolyzation process was both easy and economical, and hence amenable to large-scale production.

Blood clearly has the potential to serve as a useful source of valuable peptides with anti-hypertensive, antioxidant, and anti-bacterial properties, among others. Several bioactive peptides that offer other health benefits such as analgesic and antinociception (reduction in pain sensitivity) activities have also been isolated from blood proteins; the review article by Gomes and others [137] provides more detailed information on some of these peptides that have been isolated from blood. However, there is still work to be done in the area of in vivo studies, as most of these demonstrated activities of blood peptides have been in vitro.

4. Consumer concerns

Despite the immense efforts devoted to ensuring that blood proteins are maximally utilized in the food sector, consumer concerns related to religious prohibitions, the belief that these blood proteins are unsafe, the possibility of their fraudulent usage, and ethical issues prevent the realization of this objective. Interestingly, the available scientific data suggests that some of these concerns are actually groundless, as will be explained below. However, the anxiety expressed by consumers is real and warrants a concerted effort by government, processors and researchers to allay these fears and hence ensure the maximal utilization of blood proteins.

4.1 Fraudulent usage

The largest user of blood proteins in food is the meat industry. The economic advantage to be derived from their fraudulent use makes this a significant concern. Partial replacement of lean meat with blood plasma content offers a major economic incentive to the manufacturer, as the addition of 2% of blood plasma to a meat product can boost yield by 4 to 5% and substitute for up to 10% of the lean meat content [138]. It is therefore tempting for rogue manufacturers and food service providers to fraudulently utilize these blood proteins for economic gain. A typical case is the use of Fibrimex® as a meat binder, discussed earlier in Section 2.1. In May 2010, the EU voted to ban the use of Fibrimex® as the EU believes the product has no proven benefit and its usage carries an unacceptably high risk of misleading consumers. The concern was that Fibrimex®-reconstituted meat products would find their way into meat dishes served in restaurants, given the higher prices that can be obtained for pieces of meat sold as a single meat product. Legislators considered that consumers should be able to trust that the meat they are buying is a real steak and not simply pieces of meat glued together. The ban, however, never took effect as in accordance with Commission directive 2010/67/EU, Fibrimex® is permitted for use as a food additive for reconstituting food. In the US, where Fibrimex® is permitted at usage levels up to 10% [139] doubts about its usage such as those expressed in the EU have not been an issue. Interestingly, at no point did other products such as Activa™ TG-RM and alginate used for the same purpose as Fibrimex® come under scrutiny in the EU regarding their potential for fraudulent use. The same is true for Plasma FG, which is also derived from blood and produced by the same company for the same purpose. Such inconsistencies in regulations lead consumers to doubt the quality of these blood proteins. Efforts to ensure that their usage is not abused for economic gain at the restaurant level are, however, necessary as the problem has not yet been properly addressed. The development of effective methods for monitoring their presence in food products, along with frequent restaurant auditing by the appropriate authorities, is necessary to address this concern.

4.2 Religion and ethical reasons

Certain individuals, for example Jews and Muslims, do not consume blood because of the religious dictates enshrined in the Kosher and Halal dietary laws, respectively. Whereas some of these dietary laws concerning certain foods are not explicit and are therefore subject to individual interpretation, both codes unequivocally forbid the consumption of blood. In most parts of the world, particularly in the developing countries, animal slaughter is almost exclusively performed by Muslims, who condemn the use of blood as a food item. This is probably one of the main reasons why blood consumption by humans globally is very low [105].

Others such as vegans avoid consuming products of animal origin for ethical reasons. It is therefore imperative that appropriate labeling laws compelling manufacturers to declare the presence of these blood proteins in simple and understandable layman's terms (as is the case with allergen labeling), are enforced to help the above-mentioned group of individuals to make the right food choices. In this respect, whilst many of the labeling laws are explicit and adequate, others may have to be revised. For example, beef blood is an acceptable ingredient for beef patties provided the product name is qualified as "Beef and Blood Patties" or "Beef Patties with Blood" [139]. Clear and explicit labeling is extremely helpful in protecting the interests of individuals who must avoid consuming food containing blood. This is not always the case, however; the permitted usage of Fibrimex® (as thrombin and fibrinogen) is contingent on the conspicuous display of the terms "Beef Fibrinogen and Thrombin", "Beef Fibrin" or "Fibrin" as product name qualifiers on the label [139]. Such technical labeling provisions are unhelpful and may have to be revised.

The situation is even worse when these blood proteins are used as ingredients in dietary supplements. Several of these dietary supplements have brand names on the label that include no indication that they are blood-derived. For example, Proferrin® lists no explicit information on the label that indicates HIP is derived from bovine hemoglobin (and hence from bovine blood). The only information given as to the source of the heme is that "Heme is a natural form of iron derived from animal sources and is intended for individuals seeking to maintain normal iron levels." A similar issue pertains to some of the dietary supplements on the market that contain Immunolin®; while some products such as Immune Advantage (Now Foods, Bloomingdale, IL) clearly state on the label that Immunolin® is derived from bovine serum, other products such as Daily Immune Defense (Distributed by Doctor's Best, Inc., San Clemente, CA) and ImmunAssure (Schiff Nutrition Group, Inc., Salt Lake City, UT) give no indication that Immunolin® is blood-derived. Interestingly, as the labeling regulations stand these manufacturers are not violating the law, as the labeling laws require them only to list the names of ingredients. Declaring the source of the ingredients is voluntary [140].

Certainly, the lax labeling regulations regarding the use of blood proteins as food ingredients, particularly in dietary supplements, fails to adequately protect individuals seeking to avoid consuming these products for religious and ethical reasons. Better labeling is vital to ensure that the interests of such people are protected and to boost consumer confidence in such products, thus promotinge efforts to ensure the full-scale utilization of blood proteins.

4.3 Health reasons

The belief that blood provides a haven for pathogens and toxic metabolites is another reason why some people avoid consuming blood. This concern, though genuine, is not exclusive to blood products; in general, foods of animal origin are easily contaminated with spoilage microorganisms and possibly pathogens through improper processing and handing [141]. Blood taken from a healthy animal is essentially sterile, so any contamination must be due to the bleeding technique and the drainage system employed during collection [142]. Both manufacturers and processors have instituted measures to guarantee the safety of these blood proteins. In the US, federal regulation 9 CFR 310.20 monitors blood that enters the food chain, ensuring that it originates from official establishments whose livestock and carcasses have passed inspection. The use of closed-draining systems in the collection of blood ensures the blood is siphoned directly from the animal into collecting tanks without exposure to the atmosphere. As a further safety measure, the collected blood is associated with the individual animals or batches of animals from which the blood was sourced until inspected to prevent uninspected blood from entering the food chain [143]. Collected and inspected blood is transported in refrigerated and dedicated isothermal stainless steel trucks to the processing plant to prevent any possible re-contamination. On arrival at the manufacturing plant, quality assurance (QA) and quality control (QC) procedures are performed on the received blood prior to spray-drying at high temperatures to inactivate any pathogens that may be present. Using Fibrimex® as an example, it is the opinion of the European Commission Scientific Panel on Food Additives, Flavourings, Processing Aids and Materials in Contact with Food that its use as a food additive is not of concern from the safety point of view [144].

There are situations where liquid blood (or plasma) is utilized as an ingredient in foods, although such usage is only allowed in formulations that will undergo heat-treatment. However, considering the highly perishable nature of blood, where blood is used in the liquid form extra precautions through a mandatory HACCP system may be necessary. Ramos-Clamont and others noted that the use of a HACCP system in a pilot plant ensured that isolated blood fractions had excellent microbiological quality [145]. Thus, there is still scope for more improvements to be implemented to guarantee the quality of these blood ingredients. For example, it has been argued that the use of blood proteins as binders in meat products has a relatively high risk of bacterial contamination because pathogenic and spoilage microorganisms on the surfaces of the meat and trim pieces become internalized. This is compounded by the fact that these binders work on the cold-set method. Although such health concerns arise because of the process and not the ingredients per se, they nonetheless require attention. Incorporation of bacteria growth inhibitors such as nisin (a broad-spectrum bacteriocin) in the formulation of restructured meats, which has been found to be effective in reducing undesirable bacteria in such products [146], may be one way to address this problem.

5. Conclusions

As this review has shown, blood proteins provide an economic and readily available alternative source of proteins and irons for use in foods and dietary supplements to address a wide range of functional and nutritional needs. Its additional benefits as a source of bioactive peptides with anti-hypertensive, anti-bacterial, analgesic, and antinociception properties have the potential to provide safer and cheaper alternatives to conventional drugs, which tend to be expensive and have unwelcome side effects. Efforts to ensure large-scale utilization of blood proteins as food additives should be encouraged because of the economic, nutritional, health and environmental benefits conveyed. This is particularly true in developing countries, where traditional sources of proteins and irons are either expensive or are unavailable. However, even as the industry strives to achieve greater utilization of blood as a food additive, consumer concerns related to religious and ethical issues must be addressed through the concerted effort of regulators, producers and researchers to develop accurate and concise labeling based on adequate labeling regulations and the development of effective blood detection methods to enforce regulations. On the issue of the safety of these products, making HACCP a mandatory requirement for plants that produce blood proteins, providing third world countries with the necessary expertise and equipment for their production, and educating the public about the safety of these blood proteins, will encourage consumer confidence in these products and, consequently, their greater utilization.

6. References

[1] F. W. Putnam, *The plasma proteins: structure, function, and genetic control*. New York: Academic press 1975.
[2] H. W. Ockerman and C. L. Hansen, "Animal by-product processing and utilization," ed Lancaster Technomic Publishing company Inc, 2000, pp. 325-353.
[3] J. Z. Wang, *et al.*, "Changes of chemical and nutrient composition of porcine blood during fermentation by Aspergillus oryzae," *World Journal of Microbiology and Biotechnology*, vol. 23, pp. 1393–1399, 2007.
[4] R. Gatnau, *et al.*, "Plasma protein antimicrobial substitution at negligible risk," in *Feed manufacturing in the Mediterranean region. Improving safety: From feed to food*. vol. 54, B. J., Ed., ed Zaragoza: CIHEAM-IAMZ, 2001, pp. 141-150.
[5] PPIMLA. (2001). *Restructured meat using bovine plasma products*. Available: http://www.meatupdate.csiro.au/infosheets/Restructured%20Meat%20using%20Bovine%20Plasma%20Products.pdf
[6] R. T. Duarte, *et al.*, "Bovine blood components: fractionation, composition, and nutritive value," *Journal of Agricultural and Food Chemistry*, vol. 47, pp. 231-236, 1999.
[7] X. Q. Liu, *et al.*, "Physicochemical properties of aggregates of globin hydrolysates," *Journal of Agricultural and Food Chemistry*, vol. 44, pp. 2957-2961, 1996.
[8] J.-H. Yang and C.-W. Lin, "Functional properties of porcine blood globin decolourized by different methods," *International Journal of Food Science and Technology*, vol. 33, pp. 419-427, 1998.
[9] B. Nowak and T. von Mueffling, "Porcine blood cell concentrates for food products: hygiene, composition, and preservation," *Journal of Food Protection*, vol. 69, pp. 2183-2192, 2006.
[10] E. Slinde and M. Martens, "Changes in Sensory Properties of Sausages When Small Amounts of Blood Replace Meat," *Journal of the Science of Food and Agriculture*, vol. 33, pp. 760-762, 1982.
[11] K. Autio, *et al.*, "The effect of processing method on the functional behavior of globin protein," *Journal of Food Science* vol. 49, pp. 369-370, 1984.
[12] S. Hayakawa, *et al.*, "Effect of Heat Treatment on Preparation of Colorless Globin from Bovine Hemoglobin Using Soluble Carboxymethyl Cellulose " *Journal of Food Science*, vol. 51, pp. 786-796, 1986.
[13] B. Houlier, "A process of discolouration of slaughter house blood: some technical and economical results," *Proceedings of 32nd European meeting of meat research workers, Ghent, Belgium*, pp. 91-94, 1986.
[14] Y. Sato, *et al.*, "Preparation of blood globin through carboxymethyl cellulose chromatography " *Journal of Food Technology*, vol. 16, pp. 81-91, 1981.
[15] P. T. Tybor, *et al.*, "Functional Properties of Proteins Isolated from Bovine Blood by a Continuous Pilot Process," *Journal of Food Science*, vol. 40, pp. 155-159, 1975.
[16] J.-H. Yang and C.-W. Lin, "Effects of Various Viscosity Enhancers and pH on Separating Haem from Porcine Red Blood Cells," *Journal of the Science of Food and Agriculture*, vol. 70, pp. 364-368, 1996.

[17] C. Gómez-Juárez, *et al.*, "Protein recovery from slaughterhouse wastes," *Bioresource Technology,* vol. 70, pp. 129-133, 1999.
[18] P. Nuthong, *et al.*, "Effect of phenolic compounds on the properties of porcine plasma protein-based film," *Food Hydrocolloids,* vol. 23, pp. 736–741, 2009.
[19] I. P. Devadason, *et al.*, "Effect of different binders on the physico-chemical, textural, histological, and sensory qualities of retort pouched buffalo meat nuggets," *Journal of Food Science,* vol. 75, pp. S31-S35, 2010.
[20] V. K. Modi, *et al.*, "Quality of buffalo meat burger containing legume flours as binders," *Meat Science,* vol. 66 pp. 143-149, 2004.
[21] M. J. Chen and C. W. Lin, "Factors affecting the water-holding capacity of fibrinogen/plasma protein gels optimized by response surface methodology," *Journal of Food Science,* vol. 67, pp. 2579-2582, 2002.
[22] D. W. Hickson, *et al.*, "A comparison of heat-induced gel strengths of bovine plasma and egg albumen proteins," *Journal of Animal Science,* vol. 51, pp. 69-73, 1980.
[23] G. H. Lu and T. C. Chen, "Application of egg white and plasma powders as muscle food binding agents," *Journal of Food Engineering* vol. 42, pp. 147-151, 1999.
[24] D. Parés, *et al.*, "Functional properties of heat induced gels from liquid and spray dried porcine blood plasma as influenced by pH," *Journal of Food Science,* vol. 63, pp. 958-961, 1998.
[25] N. Fort, *et al.*, "Cold storage of porcine plasma treated with microbial transglutaminase under high pressure. Effects on its heat-induced gel properties," *Food Chemistry,* vol. 115, pp. 602-608, 2009.
[26] G. A. MacGregor and H. E. de wardener, "Salt, blood pressure and health," *International Journal Epidemiology,* vol. 31, pp. 320-327, 2002.
[27] B. Teucher, *et al.*, "Sodium and bone health: impact of moderately high and low salt intakes on calcium metabolism in postmenopausal women," *Journal of Bone and Mineral Research,* vol. 23, pp. 1477-85, 2008.
[28] G. R. Schmidt and G. R. Trout, "Chemistry of meat binding," in *Meat Science and Technology International Symposium Proceedings*, Lincoln, NE, 1982, p. 265.
[29] G. Wijngaards and E. J. C. Paardekooper, "Preparation of a composite meat product by means of an enzymatically formed protein gel," in *Trends in Modern Meat Technology 2 Proceedings of the international symposium,* Den Dolder, Netherlands, 1987, pp. 125-129.
[30] S. Raharjo, *et al.*, "Influence of Meat Restructuring Systems on Lipid Oxidation in Beef," *Lebensmittel-Wissenschaft & Technologie,* vol. 22, pp. 199-203, 1989.
[31] W. J. Means and G. R. Schmidt, "Algin Calcium Gel as a Raw and Cooked Binder in Structured Beef Steaks," *Journal of Food Science,* vol. 51, pp. 60-65, 1986.
[32] N. C. Flores, *et al.*, "Instrumental and consumer evaluation of pork restructured with activaTM or with fibrimexTM formulated with and without phosphate," *Food Science and Technology,* vol. 40, pp. 179-185, 2007.
[33] J. A. Boles and P. J. Shand, "Effects of raw binder system, meat cut and prior freezing on restructured beef," *Meat Science,* vol. 53, pp. 233-239, 1999.
[34] C. H. Shao, *et al.*, "Functional, sensory and microbiological properties of restructured beef and emu steaks," *Journal of Food Science,* vol. 64, pp. 1052-1054, 1999.
[35] A. M. Lennon, *et al.*, "Performance of cold-set binding agents in re-formed beef steaks," *Meat Science,* vol. 85, pp. 620-624, 2010.
[36] H. Ben Mansour, *et al.*, "Evaluation of genotoxicity and pro-oxidant effect of the azo dyes: acids yellow 17, violet 7 and orange 52, and of their degradation products by

Pseudomonas putida mt-2," *Food and Chemical Toxicology,* vol. 45, pp. 1670-1677, 2007.
[37] M. C. Gennaro, *et al.*, "High-performance liquid chromatography of food colours and its relevance in forensic chemistry," *Journal of Chromatography A,* vol. 674, pp. 281-299, 1994.
[38] E. R. Jaffé, *Metabolic processes involved in the formation and reduction of methaemoglobin in human erythrocytes in The red blood cells.* New York: Academic Press, 1964.
[39] E. Saguer, *et al.*, "Colour stabilization of spray-dried porcine red blood cells using nicotinic acid and nicotinamide," *Food Science and Technology International,* vol. 9, pp. 301-307, 2003.
[40] P. Salvador, *et al.*, "Color stabilization of porcine hemoglobin during spray-drying and powder storage by combining chelating and reducing agents," *Meat Sci,* vol. 83, pp. 328-333, 2009.
[41] E. Press and L. Yaeger, "Food poisoning due to sodium nicotinate-report of an outbreak and a review of the literature," *American Journal of Public Health,* vol. 52, pp. 1720–1728, 1962.
[42] R. A. Mancini and M. C. Hunt, "Current research in meat color," *Meat Science,* vol. 71, pp. 100–121, 2005.
[43] O. Sørheim, *et al.*, "Technological, hygienic and toxicological aspects of carbon monoxide used in modified-atmosphere packaging of meat," *Trends in Food Science and Technology,* vol. 8, pp. 307–312., 1997.
[44] P. R. Fontes, *et al.*, "Color evaluation of carbon monoxide treated porcine blood," *Meat Science,* vol. 68, pp. 507–513., 2004.
[45] P. R. Fontes, *et al.*, "Composition and color stability of carbon monoxide treated dried porcine blood," *Meat Science,* vol. 85, pp. 472-480, 2010.
[46] P. Wilde, *et al.*, "Proteins and emulsifiers at liquid interfaces," *Advances in Colloid and Interface Science,* vol. 108-109, pp. 63-71, 2004.
[47] J. G. Silva, *et al.*, "Comparative study of the functional properties of bovine globin isolates and sodium caseinate," *Food Research International* vol. 36, pp. 73-80, 2003.
[48] D. D. Crenwelge, *et al.*, "A comparison of the emulsification capacities of some protein concentrates," *Journal of Food Science,* vol. 39, pp. 175-177, 1974.
[49] R. Nakamura, *et al.*, "Emulsifying Properties of Bovine Blood Globin: A Comparison with Some Proteins and Their Improvement," *Food Science,* vol. 49, pp. 102-104, 1984.
[50] H. A. Caldironi and H. W. Ockerman, "Incorporation of blood proteins into sausage," *Journal of Food Science,* vol. 47, pp. 405-408, 1982.
[51] D. J. Mela, "The basis of dietary fat preference," *Trends in Food Science and Technology,* vol. 1, pp. 55-78, 1990.
[52] A. Drewnowski, "Sensory properties of fats and fat replacements," *Nutrition Reviews,* vol. 50, pp. 17-20, 1992.
[53] F. J. Colmenero, "Technologies for developing low-fat meat products," *Trends in Food Science & Technology,* vol. 7, pp. 41-48, 1996.
[54] O. Tokusoglu and M. K. Unal, "Fat Replacers in Meat Products," *Pakistan Journal of Nutrition,* vol. 2, pp. 196-203, 2003.
[55] F. R. Viana, *et al.*, "Quality of ham pate containing bovine globin and plasma as fat replacers," *Meat Science,* vol. 70, pp. 153-160, 2005.
[56] E. Hughes, *et al.*, "Effects of fat level, tapioca starch and whey protein on frankfurters formulated with 1-25% and 12% fat," *Meat Science,* vol. 48, pp. 169-180, 1998.

[57] S. Cofrades, *et al.*, "Plasma Protein and Soy Fiber Content Effect on Bologna Sausage Properties as Influenced by Fat Level " *Journal of Food Science,* vol. 65, pp. 281-287, 2000.
[58] E. M. Desmond, *et al.*, "Comparative studies of nonmeat adjuncts used in the manufacture of low-fat beef burgers," *Journal of Muscle Foods,* vol. 9, pp. 221-241, 1998.
[59] J. C. Guzman, *et al.*, "Texture, Color and Sensory Characteristics of Ground Beef Patties Containing Bovine Blood Proteins," *Journal of Food Science,* vol. 60, pp. 657-660, 1995.
[60] J. B. Fox, Jr., "The chemistry of meat pigments," *Journal of Agricultural and Food Chemistry,* vol. 14, pp. 207-210, 1966.
[61] D. B. MacDougall, *et al.*, "Contribution of nitrite and nitrate to the colour and flavour of cured meats," *Journal of the Science of Food and Agriculture,* vol. 26, pp. 1743–1754, 1975.
[62] J. O. Igene, *et al.*, "Mechanisms by which nitrite inhibits the development of warmed-over flavour (WOF) in cured meat," *Food Chemistry,* vol. 18, pp. 1-18, 1985.
[63] L. A. Freybler, *et al.*, "Nitrite stabilization of lipids in cured pork," *Meat Science,* vol. 33, pp. 85-96, 1993.
[64] J. B. Fox, Jr., "Role of cure accelerators," in *Proceedings of the Meat Industry Research Conference*, Chicago, IL, 1974, pp. 17-21.
[65] J. N. Sofos, *et al.*, "Sodium nitrite and sorbic acid effects on Clostridium botulinum spore germination and total microbial growth in chicken frankfurter emulsions during temperature abuse," *Applied and Environmental Microbiology,* vol. 37, pp. 1103-1109, 1979.
[66] R. C. Benedict, "Biochemical basis for nitrite-inhibition of Clostridium botulinum in cured meat," *Journal of Food Protection,* vol. 43, pp. 877-891, 1980.
[67] N. P. Sen, *et al.*, "Volatile nitrosamines in various cured meat products: Effect of cooking and recent trends," *Journal of Agricultural and Food Chemistry,* vol. 27, pp. 1354-1357, 1979.
[68] N. P. Sen, *et al.*, "Volatile and nonvolatile nitrosamines in fish and the effect of deliberate nitrosation under simulated gastric conditions," *Journal of Agricultural and Food Chemistry,* vol. 33, pp. 264-268, 1985.
[69] M. Stevanovic, *et al.*, "Genotoxicity Testing of Cooked Cured Meat Pigment (CCMP) and Meat Emulsion Coagulates Prepared with CCMP," *Journal of Food Protection,* vol. 63, pp. 945–952, 2000.
[70] F. Shahidi and R. B. Pegg., "Nitrite-free meat curing systems: update and review," *Food Chemistry,* vol. 43, pp. 185–191, 1992.
[71] R. B. Pegg and F. Shahidi, "Unraveling the Chemical Identity of Meat Pigments," *Critical Reviews in Food Science and Nutrition,* vol. 37, pp. 561-589, 1997.
[72] F. Shahidi and R. B. Pegg, "Powdered Cooked Cured- Meat Pigment Which is a Non-Nitrite Meat Preservative," U. S. Patent 5,230,915 Patent, 1991.
[73] F. Shahidi and R. B. Pegg, "Colour characteristics of cooked cured-meat pigment and its application to meat," *Food Chemistry,* vol. 38, pp. 61-68, 1990.
[74] M. Wettasinghe and F. Shahidi, "Antioxidant activity of preformed cooked cured-meat pigment in a β-carotene/linoleate model system," *Food Chemistry,* vol. 58, pp. 203-207, 1997.
[75] A. Ashwini, *et al.*, "Effect of hydrocolloids and emulsifiers on the rheological, microstructural and quality characteristics of eggless cake," *Food Hydrocolloids,* vol. 23, pp. 700–707, 2009.

[76] C. C. Lee, *et al.*, "Sensory and Physical Properties of Cakes with Bovine Plasma Products Substituted for Egg," *Cereal Chemistry,* vol. 70, pp. 18-21, 1993.
[77] L. A. Johnson, *et al.*, "Bovine Plasma as a Replacement for Egg in Cakes," *Cereal Chemistry,* vol. 56, pp. 339-342, 1979.
[78] R. M. Myhara and G. Kruger, "The performance of decolorized bovine plasma protein as a replacement for egg white in high ratio white cakes," *Food Quality and Preference,* vol. 9, pp. 135-138, 1998.
[79] M. O. Raeker and L. A. Johnson, "Cake-Baking (High-Ratio White Layer) Properties of Egg-White, Bovine Blood-Plasma, and Their Protein-Fractions," *Cereal Chemistry,* vol. 72, pp. 299-303, 1995.
[80] FAO/WHO, "Report of a Joint FAO/WHO Expert Consultation on Energy and Protein Requirement.," Geneva, Switzerland. 1985.
[81] K. A. Dualeh and F. T. Henry, "Breast milk-the life saver: observation from recent studies," *Food Nutrition Bulletin,* vol. 11, pp. 43-47, 1989.
[82] P. Mensah, *et al.*, "Fermented cereal gruel: towards a solution of the weaning dilemma," *Food Nutrition Bulletin,* vol. 13, pp. 50-55, 1991.
[83] I. A. Akinrele and C. C. A. Edward, "An assessment of the nutritive value of maize soya mixture. "Soya-ogi" as a weaning food in Nigeria," *British Journal of Nutrition,* vol. 26, pp. 177-185, 1971.
[84] A. H. El Tinay, *et al.*, "Proximate Composition and Mineral and Phytate Contents of Legumes Grown in Sudan," *Journal of Food Composition and Analysis,* vol. 2, pp. 69-78, 1989.
[85] A. D. Ologhobo and B. L. Fetuga, "Distribution of phosphorus and phytate in some Nigerian varieties of legumes and some effects of processing," *Journal of Food Science,* vol. 49, pp. 199-201, 1984.
[86] S. Fallon and M. Enig, "Soy products: Tragedy and Hype," *NEXUS Magazine* vol. 7, pp. 1-19, 2000.
[87] G. M. Wallace, *et al.*, "Studies on the processing and properties of soymilk: II.—Effect of processing conditions on the trypsin inhibitor activity and the digestibility in vitro of proteins in various soymilk preparations," *Journal of Science and Food Agriculture,* vol. 22, pp. 526–531, 1971.
[88] J. L. Pierce, *et al.*, "Effects of spray-dried animal plasma and immunoglobulins on performance of early weaned pigs," *Journal of Animal Science,* vol. 83, pp. 2876-2885, 2005.
[89] E. M. Weaver, *et al.*, "The effect of spray-dried animal plasma fractions on performance of newly weaned pigs," *Journal of Animal Science,* vol. 73, p. 81, 1995.
[90] J. E. Thomson, *et al.*, "Effect of spray-dried porcine plasma protein on feed intake, growth rate, and efficiency of gain in mice," *Journal of Animal Science,* vol. 72, pp. 2690-2695, 1994.
[91] F. Begin, *et al.*, "Effects of bovine serum concentrate, with or without supplemental micronutrients, on the growth, morbidity, and micronutrient status of young children in a low-income, peri-urban Guatemalan community," *European Journal of Clinical Nutrition,* vol. 62, pp. 39-50, 2008.
[92] A. A. Oshodi, *et al.*, "In vitro protein digestibility, amino acid profile and available iron of infant weaning food prepared from maize flour and bovine blood," *Food Research International,* vol. 30, pp. 193-197, 1997.

[93] J. L. Lembcke, *et al.*, "Acceptability, safety, and digestibility of spray-dried bovine serum added to diets of recovering malnourished children," *Journal of Pediatric Gastroenterology and Nutrition*, vol. 25, pp. 381-384, 1997.

[94] N. Martınez-Navarretea, *et al.*, "Iron deficiency and iron fortified foods—a review," *Food Research International*, vol. 35, pp. 225–231, 2002.

[95] S. A. Chiplonkar, *et al.*, "Fortification of vegetarian diets for increasing bioavailable iron density using green leafy vegetables," *Food Research International*, vol. 32, pp. 169-174, 1999.

[96] Y.-C. Huang, "Nutrient intakes and iron status of vegetarians," *Nutrition*, vol. 16, pp. 147-148, 2000.

[97] J. K. Friel, *et al.*, "A double-masked, randomized control trial of iron supplementation in early infancy in healthy term breast-fed infants," *Journal of Pediatrics*, vol. 143, pp. 582–586, 2003.

[98] B. Lozoff, *et al.*, "Behavioral and developmental effects of preventing iron-deficiency anemia in healthy full-term infants," *Pediatrics*, vol. 112, pp. 846–54, 2003.

[99] T. O. Scholl, *et al.*, "Anemia vs iron deficiency: increased risk of preterm delivery in a prospective study.," *American Journal of Clinical Nutrition*, vol. 55, pp. 985-988, 1992.

[100] J. D. Haas and T. Brownlie, IV, "Iron deficiency and reduced work capacity: a critical review of the research to determine a causal relationship," *Journal of Nutrition*, vol. 131, pp. 676S-690S, 2001.

[101] FAO and ILSI, "Preventing micronutrient malnutrition a guide to food-based approaches - Why policy makers should give priority to food-based strategies ", Washington, DC1997.

[102] C. Uzel and M. E. Conrad, "Absorption of heme-iron," *Seminars in Hematology*, vol. 35, pp. 27-34, 1998.

[103] E. Frykman, *et al.*, "Side-Effects of Iron Supplements in Blood-Donors - Superior Tolerance of Heme Iron," *Journal of Laboratory and Clinical Medicine*, vol. 123, pp. 561-564, 1994.

[104] E. K. Lund, *et al.*, "Oral ferrous sulfate supplements increase the free radical-generating capacity of feces from healthy volunteers," *American Journal of Clinical Nutrition*, vol. 69, pp. 250-255, 1999.

[105] J. K. Kikafunda and P. Sserumaga, "Production and use of a shelf-stable bovine blood powder for food fortication as a food-based strategy to combat iron deficiency anaemia in Sub-Saharan Africa," *African journal of food agriculture nutrition and development* vol. 5, pp. 1-18, 2005.

[106] T. Walter, *et al.*, "Effect of Bovine-Hemoglobin Fortified Cookies on Iron Status of Schoolchildren - a Nationwide Program in Chile," *American Journal of Clinical Nutrition*, vol. 57, pp. 190-194, 1993.

[107] N. Vaghefi, *et al.*, "Influence of the extent of hemoglobin hydrolysis on the digestive absorption of heme iron. An in vitro study," *Journal of Agricultural and Food Chemistry*, vol. 50, pp. 4969-4973, 2002.

[108] L. Hallberg, *et al.*, "Inhibition of haem-iron absorption in man by calcium," *British Journal of Nutrition*, vol. 69, pp. 533-540, 1992.

[109] I. Pallares, *et al.*, "Supplementation of a cereal-milk formula with haem iron palliates the adverse effects of iron deficiency on calcium and magnesium metabolism in rats " *Annals of Nutrition and Metabolism*, vol. 40, pp. 81-90, 1996.

[110] A. G. Quintero-Gutierrez, *et al.*, "Bioavailability of heme iron in biscuit filling using piglets as an animal model for humans," *International Journal of Biological Science,* vol. 4, pp. 58-62, 2008.

[111] G. Gonzalez-Rosendo, *et al.*, "Bioavailability of a heme-iron concentrate product added to chocolate biscuit filling in adolescent girls living in a rural area of Mexico," *Journal of Food Science,* vol. 75, pp. H73-H78, 2010.

[112] A. R. Nissenson, *et al.*, "Clinical evaluation of heme iron polypeptide: sustaining a response to rHuEPO in hemodialysis patients," *American Journal of Kidney Diseases,* vol. 42, pp. 325-330, 2003.

[113] P. A. Seligman, *et al.*, "Clinical studies of hip: An oral heme-iron product," *Nutrition Research,* vol. 20, pp. 1279-1286, 2000.

[114] D. A. Clare and H. E. Swaisgood, "Bioactive milk peptides: a prospectus," *Journal of Dairy Science,* vol. 83, pp. 1187-1195, 2000.

[115] D. G. Lee, *et al.*, "Fungicidal effect of antimicrobial peptide, PMAP-23, isolated from porcine myeloid against Candida albicans," *Biochemical and Biophysical Research Communications,* vol. 282, pp. 570-574, 2001.

[116] P. L. Yu, *et al.*, "Purification and characterization of the antimicrobial peptide, ostricacin," *Biotechnology Letters,* vol. 23, pp. 207–210, 2001.

[117] E. A. Perpetuo, *et al.*, "Biochemical and pharmacological aspects of two bradykinin-potentiating peptides obtained from tryptic hydrolysis of casein," *Journal of Protein Chemistry,* vol. 22, pp. 601-606, 2003.

[118] A. Qureshi, *et al.*, "Microsclerodermins F–I, antitumor and antifungal cyclic peptides from the lithistid sponge Microscleroderma sp.," *Tetrahedron,* vol. 56, pp. 3679–3685, 2000.

[119] S. K. Kim, *et al.*, "Isolation and characterization of antioxidative peptides from gelatin hydrolysate of Alaska pollack skin," *Journal of Agricultural and Food Chemistry,* vol. 49, pp. 1984-199, 2001.

[120] J. R. Liu, *et al.*, "Antimutagenic and antioxidant properties of milk-kefir and soymilk-kefir," *Journal of Agricultural and Food Chemistry,* vol. 53, pp. 2467-2474, 2005.

[121] T. Matsui, *et al.*, "Gastrointestinal enzyme production of bioactive peptides from royal jelly protein and their antihypertensive ability in SHR," *Journal of Nutritional Biochemistry,* vol. 13, pp. 80-86, 2002.

[122] K. Arihara, *et al.*, "Peptide inhibitors for angiotensin I-converting enzyme from enzymatic hydrolysates of porcine skeletal muscle proteins," *Meat Science,* vol. 57, pp. 319–324, 2001.

[123] Y. Yu, *et al.*, "Isolation and characterization of angiotensin I-converting enzyme inhibitory peptides derived from porcine hemoglobin," *Peptides,* vol. 27, pp. 2950-2956, 2006.

[124] K. Miura, *et al.*, "Relationship of blood pressure to 25-year mortality due to coronary heart disease, cardiovascular diseases, and all causes in young adult men: the Chicago Heart Association Detection Project in Industry," *Archives of Internal Medicine,* vol. 161, pp. 1501-1508, 2001.

[125] D. Coates, "he angiotensin converting enzyme (ACE)," *International Journal of Biochemistry and Cell Biology,* vol. 35, pp. 769-773, 2003.

[126] A. B. Atkinson and J. I. S. Robertson, "Captopril in the treatment of clinical hypertension and cardiac failure," *Lancet,* vol. 314, pp. 836-839, 1979.

[127] J.-T. Wei and B.-H. Chiang, "Bioactive peptide production by hydrolysis of porcine blood proteins in a continuous enzymatic membrane reactor," *Journal of the Science of Food and Agriculture,* vol. 89, pp. 372-378, 2009.
[128] S.-C. Huang and P.-J. Liu, "Inhibition of Angiotensin I - Converting Enzymes by Enzymatic Hydrolysates from Chicken Blood," *Journal of Food and Drug Analysis,* vol. 18, pp. 458-463, 2010.
[129] N. Nedjar-Arroume, *et al.*, "Isolation and characterization of four antibacterial peptides from bovine hemoglobin," *Peptides,* vol. 27, pp. 2082-2089, 2006.
[130] R. Froidevaux, *et al.*, "Antibacterial activity of a pepsin-derived bovine hemoglobin fragment.," *FEBS Letters,* vol. 491, pp. 159-63, 2001.
[131] B. Halliwell, "Free radicals, antioxidants, and human disease: Curiosity, cause, or consequence," *Lancet,* vol. 344, pp. 721-724, 1994.
[132] X. Xu, *et al.*, "Antioxidant activity of hydrolysates derived from porcine plasma," *Journal of the Science of Food and Agriculture,* vol. 89, pp. 1897-1903, 2009.
[133] B. Halliwell, " Albumin - an important extra cellular antioxidant?," *Biochemical Pharmacology,* vol. 37, pp. 569-571, 1988.
[134] C. T. Bishop, *et al.*, "Free radical damage to culture porcine aortic endothelial cells and lung fibroblasts: Modulation by culture conditions," *In Vitro Cellular and Developmental Biology,* vol. 21, pp. 229-236, 1985.
[135] E. Bourdon, *et al.*, "Glucose and free radicals impair the antioxidant properties of serum albumin," *FASEB Journal,* vol. 13, pp. 233-244, 1999.
[136] J.-Z. Wang, *et al.*, "Antioxidant activity of hydrolysates and peptide fractions of porcine plasma albumin and globulin," *Journal of Food Biochemistry,* vol. 32, pp. 693-707, 2008.
[137] I. Gomes, *et al.*, "Hemoglobin-derived peptides as novel type of bioactive signaling molecules," *AAPS Journal,* vol. 12, pp. 658-669, 2010.
[138] K. D. Hargin, "Authenticity Issues in Meat and Meat Products," *Meat Science,* vol. 43, pp. S277-S289, 1996.
[139] USDA/FSIS. (2005). *Food Standards and Labeling Policy Book.* Available: http://www.fsis.usda.gov/OPPDE/larc/Policies/Labeling_ Policy_Book_082005.pdf
[140] FDA, "Food Labeling Regulation, Amendments; Food Regulation Uniform Compliance Date; and New Dietary Ingredient Premarket Notification; Final Rules," *Federal Register,* vol. 62, pp. 49825-49858, 1997.
[141] M. Al-Bachir and A. Mehio, "Irradiated luncheon meat: microbiological, chemical and sensory characteristics during storage," *Food Chemistry,* vol. 75, pp. 169-175, 2001.
[142] M. N. Riaz, "Fundamentals of halal foods and certification," *Prepared foods,* vol. 179, pp. 71-76, 2010.
[143] CSIRO. (2003). *Blood products in Meat Technology Update Newsletter.* Available: http://www.meatupdate.csiro.au/data/MEAT_TECHNOLOGY_UPDATE_03-5.pdf
[144] R. Anton, *et al.*, "Opinion of the Scientific Panel on food additives, flavourings, processing aids and materials in contact with food (AFC) related to use of an enzyme preparation based on thrombin:fibrinogen derived from cattle and/or pigs as a food additive for reconstituting food " *The EFSA Journal,* vol. 3, pp. 1-8, 2005.
[145] G. Ramos-Clamont, *et al.*, "Functional properties of protein fractions isolated from porcine blood," *Food and Chemical Toxicology,* vol. 68, pp. 1196-1200, 2003.
[146] C. N. Cutter and G. R. Siragusa, "Incorporation of nisin into a meat binding system to inhibit bacteria on beef surfaces," *Letters in Applied Microbiology,* vol. 27, pp. 19-23, 1998.